Basic Concepts in Information Theory and Coding
The Adventures of Secret Agent 00111

Applications of Communications Theory
Series Editor: R. W. Lucky, *Bellcore*

Recent volumes in the series:

BASIC CONCEPTS IN INFORMATION THEORY AND CODING:
The Adventures of Secret Agent 00111
Solomon W. Golomb, Robert E. Peile, and Robert A. Scholtz

COMPUTER COMMUNICATIONS AND NETWORKS
John R. Freer

COMPUTER NETWORK ARCHITECTURES AND PROTOCOLS
Second Edition • Edited by Carl A. Sunshine

DATA COMMUNICATIONS PRINCIPLES
Richard D. Gitlin, Jeremiah F. Hayes, and Stephen B. Weinstein

DATA TRANSPORTATION AND PROTECTION
John E. Hershey and R. K. Rao Yarlagadda

DIGITAL PHASE MODULATION
John B. Anderson, Tor Aulin, and Carl-Erik Sundberg

DIGITAL PICTURES: Representation and Compression
Arun N. Netravali and Barry G. Haskell

FUNDAMENTALS OF DIGITAL SWITCHING
Second Edition • Edited by John C. McDonald

AN INTRODUCTION TO BROADBAND NETWORKS: LANs, MANs, ATM,
B-ISDN, and Optical Networks for Integrated Multimedia Telecommunications
Anthony S. Acampora

AN INTRODUCTION TO PHOTONIC SWITCHING FABRICS
H. Scott Hinton

OPTICAL CHANNELS: Fibers, Clouds, Water, and the Atmosphere
Sherman Karp, Robert M. Gagliardi, Steven E. Moran, and Larry B. Stotts

PRACTICAL COMPUTER DATA COMMUNICATIONS
William J. Barksdale

SIMULATION OF COMMUNICATIONS SYSTEMS
Michel C. Jeruchim, Philip Balaban, and K. Sam Shanmugan

A Continuation Order Plan is available for this series. A continuation order will bring delivery of each new volume immediately upon publication. Volumes are billed only upon actual shipment. For further information please contact the publisher.

Basic Concepts in Information Theory and Coding
The Adventures of Secret Agent 00111

Solomon W. Golomb
Departments of Electrical Engineering and Mathematics
University of Southern California
Los Angeles, California

Robert E. Peile
Racal Research, Limited
Reading, Berkshire, United Kingdom

Robert A. Scholtz
Department of Electrical Engineering
University of Southern California
Los Angeles, California

PLENUM PRESS • NEW YORK AND LONDON

```
Library of Congress Cataloging-in-Publication Data

Golomb, Solomon W. (Solomon Wolf)
   Basic concepts in information theory and coding : the adventures
of secret agent 00111 / Solomon W. Golomb, Robert E. Peile, Robert
A. Scholtz.
      p.   cm. -- (Applications of communications theory)
   Includes bibliographical references and index.
   ISBN 0-306-44544-1
   1. Coding theory.  2. Information theory.   I. Peile, Robert E.
II. Scholtz, Robert A.   III. Title.   IV. Series.
QA268.G575  1994
003'.54--dc20                                              93-48869
                                                               CIP
```

ISBN 0-306-44544-1

©1994 Plenum Press, New York
A Division of Plenum Publishing Corporation
233 Spring Street, New York, N.Y. 10013

All rights reserved

No part of this book may be reproduced, stored in a retrieval system, or transmitted in any form or by any means, electronic, mechanical, photocopying, microfilming, recording, or otherwise, without written permission from the Publisher

Printed in the United States of America

Preface

Basic Concepts in Information Theory and Coding is an outgrowth of a one-semester introductory course that has been taught at the University of Southern California since the mid-1960s. Lecture notes from that course have evolved in response to student reaction, new technological and theoretical developments, and the insights of faculty members who have taught the course (including the three of us). In presenting this material, we have made it accessible to a broad audience by limiting prerequisites to basic calculus and the elementary concepts of discrete probability theory. To keep the material suitable for a one-semester course, we have limited its scope to discrete information theory and a general discussion of coding theory without detailed treatment of algorithms for encoding and decoding for various specific code classes.

Readers will find that this book offers an unusually thorough treatment of noiseless self-synchronizing codes, as well as the advantage of problem sections that have been honed by reactions and interactions of several generations of bright students, while Agent 00111 provides a context for the discussion of abstract concepts.

Information theory and coding has progressed from its earliest beginnings in Shannon's epochal paper, "A Mathematical Theory of Communication," and the initial work on error-correcting codes by Hamming and Golay (all in the late 1940s) to revolutionize all aspects of information handling, storage, and communication and transform how *information* is viewed in fields as diverse as biology, linguistics, and philosophy. It is our hope that this book will introduce the fascinating subject of information theory to many new readers.

>Solomon W. Golomb
>*Los Angeles*
>
>Robert E. Peile
>*Reading, England*
>
>Robert A. Scholtz
>*Los Angeles*

Contents

CHAPTER 1. Introduction

1.1.	Agent 00111	1
1.2.	Uncertainty and Measurement	2
	1.2.1. Agent 00111 Ponders Pricing	2
	1.2.2. Axioms for Uncertainty and the Entropy Function	4
1.3.	Information Gain	13
	1.3.1. Agent 00111's Sliding Scale of Success Charges	13
	1.3.2. Mutual Information and Equivocation	16
1.4.	Handling Large Amounts of Information	21
	1.4.1. The Problem for Agent 00111	21
	1.4.2. A General Model for an Information Source	22
1.5.	Tutorial on Homogeneous Markov Sources	27
1.6.	The Number of Typical Sequences	45
	1.6.1. Agent 00111 Uncovers a Puzzle	45
	1.6.2. List Length and Entropy	46
1.7.	The Utility of Information Source Models	53
	1.7.1. Agent 00111 and Language Generation	53
	1.7.2. Language Models and Generation	53
1.8.	Notes	59
	References	61

CHAPTER 2. Coding for Discrete Noiseless Channels

2.1.	The Problem	63
	2.1.1. Agent 00111's Problem	63
	2.1.2. Problem Statement	64

2.2.	An Algorithm for Determining Unique Decodability in the U_F and U_S Cases	67
2.3.	A Simple Coding Theorem for Fixed-Rate Sources	74
2.4.	The Significance of Information Theory	80
2.5.	Tree Codes	83
2.6.	A Coding Theorem for Controllable Rate Sources	91
2.7.	Huffman's Coding Procedure	95
2.8.	Efficiently Encoding Markov Sources	101
2.9.	Variable Symbol Duration Channels	105
2.10.	Lempel–Ziv Coding Procedure	116
	2.10.1. Agent 00111's Problem	116
	2.10.2. The Algorithm	116
	2.10.3. The Lempel–Ziv Algorithm and Entropy	123
	2.10.4. The Lempel–Ziv Approach and Sequence Complexity Results	127
2.11.	Notes	128
	References	129

CHAPTER 3. Synchronizable Codes

3.1.	An Untimely Adventure	131
3.2.	Identifying U_I Dictionaries	132
3.3.	The Hierarchy of Synchronizable Codes	140
3.4.	A Bound on U_I Dictionary Size	143
3.5.	Fixed-Word-Length U_I Dictionaries	147
	3.5.1. Maximal Comma-Free Codes	148
	3.5.2. Prefixed Comma-Free Codes	151
	3.5.3. Path-Invariant Comma-Free Codes	153
	3.5.4. Lexicographic U_I Codes	155
3.6.	Comparing Fixed-Word-Length Synchronizable Codes	160
3.7.	Variable-Word-Length Synchronizable Codes	164
3.8.	Necessary Conditions for the Existence of U_I Dictionaries	170
3.9.	Cyclic Equivalence Class Occupancy and the Sufficiency of Iteratively Constructed Codes	174
3.10.	Constructing Maximal Comma-Free Codes of Odd Word Length	182
3.11.	Automating Binary Bounded Synchronization Delay Codes	189
	3.11.1. Cyclic Equivalence Class Representations	189
	3.11.2. Encoding and Decoding	196
3.12.	Notes	199
	Appendix: The Möbius Inversion Formula	200
	References	204

Contents

CHAPTER 4. Infinite Discrete Sources

4.1.	Agent 00111 Meets the Countably Infinite	207
4.2.	The Leningrad Paradox	208
4.3.	Mean vs. Entropy in Infinite Discrete Distributions	211
4.4.	Run-Length Encodings	219
4.5.	Decoding Run-Length Codes	222
4.6.	Capacity-Attaining Codes	223
4.7.	The Distribution Waiting Times and Performance of Elias–Shannon Coding	226
4.8.	Optimal, Asymptotically Optimal, and Universal Codes	230
4.9.	The Information-Generating Function of a Probability Distribution	237
	4.9.1. Uniqueness of the Inverse	238
	4.9.2. Composition of Generating Functions	240
4.10.	Notes	241
	References	241

CHAPTER 5. Error Correction I: Distance Concepts and Bounds

5.1.	The Heavy-Handed Cancellation Problem	243
5.2.	Discrete Noisy Channels	244
5.3.	Decoding Algorithms	248
5.4.	A Hamming Distance Design Theorem	256
5.5.	Hamming Bound	260
5.6.	Plotkin's Bound	265
5.7.	Elias Bound	269
5.8.	Gilbert Bound	273
5.9.	Perfect Codes	275
5.10.	Equidistant Codes	280
5.11.	Hamming Distance Enumeration	284
5.12.	Pless Power Moment Identities and the Welch, McEliece, Rodemich, and Rumsey (WMR) Bound	293
5.13.	Finite State Channels	295
5.14.	Pure Error Detection	302
5.15.	Notes	306
	References	307

CHAPTER 6. Error Correction II: The Information-Theoretic Viewpoint

6.1.	Disruption in the Channel	309
6.2.	Data-Processing Theorem and Estimation Problems	310

6.3.	An Upper Bound on Information Rate for Block-Coding Schemes	315
6.4.	The Chernoff Bound	323
6.5.	Linked Sequences	331
6.6.	Coding Theorem for Noisy Channels	336
6.7.	The Situation for Reliable Communication	346
6.8.	Convex Functions and Mutual Information Maximization	348
6.9.	Memoryless Channel Capacity Computations	358
6.10.	Notes	367
	References	368

CHAPTER 7. Practical Aspects of Coding

7.1.	Agent 00111 Is Not Concerned	369
7.2.	Types of Practical Codes	369
	7.2.1. Convolutional Codes	370
	7.2.2. Block Codes: A General Overview	374
	7.2.3. Reed–Solomon Codes	377
	7.2.4. Interleaving	382
	7.2.5. Concatenated Codes	383
7.3.	Coding and Modulation	388
7.4.	Hybrid Forward Error Correction (FEC) and Retransmission (ARQ) Schemes	389
	7.4.1. General Comments	390
	7.4.2. Some Basic ARQ Strategies	390
	7.4.3. Type-1 Hybrid FEC/ARQ	391
	7.4.4. Type-2 Hybrid ARQ/FEC	393
	7.4.5. Chase Code Combining	395
	7.4.6. Coding and Future Networks	395
7.A.	Appendix: Lattices and Rings	397
	References	416

Author Index 419

Subject Index 423

Basic Concepts in Information Theory and Coding
The Adventures of Secret Agent 00111

1

Introduction

1.1. Agent 00111

Secret Agent 00111 was in a most uncharacteristic mood; he was thinking about his career and remembering details that he was trained to forget. With some particularly sordid exceptions, it was not a story of universal appeal. However, as he neared the end of his service, he had been approached with several financial offers for the secrets of his legendary success. The process was all wrong, he thought glumly. The true secret of his success was more second nature than a mathematical equation, and it was probably not so salable as his backers believed. Oh well, he could always lie. . . . However, since he had been asked, he pondered what precisely he had bought and brought to the field of espionage.

The answer was a fresh look at the information business. Agent 00111 had indeed had a long and tortuous history of trafficking in information. At various times during his nefarious past, he had done almost everything conceivable to and with information. He had bought, sold, and priced it. He had gathered it, sifted it, evaluated it, analyzed it, compressed it, coded it for transmission, decoded it on reception, distorted it, and destroyed it. Once, he recalled with a shudder, he had even understood it, although this was not necessary for his operation. Agent 00111 was superior to his rivals in one essential regard: He understood how to evaluate and price his services in a manner acceptable both to himself and his many admiring clients.

Certainly no one ever accused Agent 00111 of being a philosopher or an abstract theoretician. However, in his pricing policies and execution of his work, he had invented, discovered, or plagiarized virtually every important fact about information theory and coding. For him, it was strictly a matter of practical necessity: If you are going to be in the business of buying and selling information, you need an accurate way of measuring it. And if you have to communicate information in precarious surroundings over unreliable channels, you certainly need ingenious ways of encoding and decoding it and measuring the amount of information lost and retained.

In this book, we will retrace some of the exciting and bizarre adventures in the life of Agent 00111 that led to conclusions underpinning communication technology as we know it today.

1.2. Uncertainty and Its Measurement

1.2.1. Agent 00111 Ponders Pricing

The beginning of a project for Agent 00111 normally involved lengthy negotiation over pricing and definition of the project's aims. Customers came to Agent 00111 when they were interested in an area but were uncertain about the true state of affairs or wanted to reduce their uncertainty about the area. Although this sounds elementary enough, it was the basis of Agent 00111's superiority. His competitors either did not take pricing seriously, or they failed to set prices that reflected their customers's concerns. Before Agent 00111 accepted a mission, he and the customer had to agree on two issues: first, on how many "units of uncertainty" had to be resolved; second, on a price for each unit of uncertainty that Agent 00111 succeeded in removing. Agent 00111 and his more astute clients were fully aware that most projects never eliminated all uncertainty; however, an agent of Agent 00111's quality hardly ever failed to remove some of the uncertainty, so that most projects resulted in partial clarification.

By establishing a measurement of uncertainty, Agent 00111 avoided an acrimonious and prolonged period of postmission haggling about payment. Of course, reaching agreement on a per unit price was critical, since sometimes a customer did not set a high enough price per unit of uncertainty to justify the danger that Agent 00111 perceived in removing the uncertainty. Sometimes the customer was afraid that if Agent 00111 was detected trying to remove uncertainty, the effect could be to change the whole scenario.

The key to success was measuring information and establishing how many units of information were on the bargaining table, and this was where Agent 00111 surpassed his competitors. For example, suppose that Agency X and Agent 00111 are negotiating. First they have to establish their present level of knowledge, i.e., the state of affairs prior to Agent 00111's endeavors. Agency X might ask Agent 00111 to determine which one of m disjoint events had actually occurred behind the Black Curtain. Let the set $\mathbf{M} = \{M_1, M_2, \ldots, M_m\}$ represent the possible events. Pricing depends on what the agency

Introduction

believes the relative probabilities to be prior to further investigation. The set of events is assumed to cover all possible situations, so that

$$\sum_{i=1}^{m} Pr(M_i) = 1$$

(if necessary, the last event M_m can be designated as "other"). The task in measuring uncertainty consists of moving from the *a priori* probabilities to a single real number representing the units of uncertainty. Once this number, $H[Pr(M_1), Pr(M_2), \ldots, Pr(M_m)]$, say, had been established, Agent 00111 could compute his charges so as not to exceed

$$\text{Price} = k_s H[Pr(M_1), Pr(M_2), \ldots, Pr(M_m)]$$

This price represents the maximum amount, given complete success, that the customer will be charged. The significance constant k_s represents the price factor, in dollars per unit of uncertainty removed, in the information supply process.

The question of what $H(\ )$ actually equals remains, and indeed, Agent 00111 had no concrete proof of the answer—only some intuitions and animal cunning served him in this regard. In fact, many years ago Agent 00111 had stumbled on an *ad hoc* formula that had served him well throughout his career. He computed $H[Pr(M_1), Pr(M_2), \ldots, Pr(M_m)]$ as

$$H[Pr(M_1), Pr(M_2), \ldots, Pr(M_m)] = -\sum_{i=1}^{m} Pr(M_i) \log[(Pr(M_i)] \quad (1)$$

Having little trust in formulas, Agent 00111 still relied heavily on his intuition. However, he had observed that the formula in Equation 1 passed every test it had faced. Moreover, his clients had on occasion taken his formula away and asked various members of their technical staff if the pricing was reasonable. Agent 00111 was not privy to their reports, but apart from an occasional bluff, the customers had never refused to accept his pricing.

So what were the intuitions that led Agent 00111 to the formula in Equation 1? In any particular case, Agent 00111 was aware that probabilities were important to the amount of information, and his most difficult jobs involved clients who had absolutely no idea about the *a priori* probabilities. In such a case, Agent 00111 set $Pr(M_i) = 1/m$ for all i, which represented maximum uncertainty and the highest price. Furthermore, in this equally likely case, as m increased Agent 00111 had more alternatives to eliminate and the amount of uncertainty and his price rose accordingly. If events were not considered to be equally likely, Agent 00111 and his clients had a hunch (right or wrong) as to the outcome, and the *a priori* uncertainty was reduced.

Another of Agent 00111's intuitions was that the uncertainty eliminated in two independent projects equals the sum of the uncertainties eliminated in each project. A related intuition was that the amount of uncertainty in a project was not a function of how the project was handled and any attempt to charge the customer for the *method* used to eliminate uncertainty risked being undercut by a competitor's superior strategy. Agent 00111 thought with relish of how an application of this independence principle had brought him a lot of valuable business over the years. In particular, he had once been asked to find the most dissatisfied scientist in country D. From this unworkably vague request, he had persuaded his client that it was sufficient to find the most underfunded senior scientist in the most underfunded research laboratory in country D. Since there were a hundred such laboratories, each with a thousand scientists and no *a priori* idea of the funding, less experienced competitors thought they were being asked for a case-by-case elimination of 100,000 scientists and overbid the job. Agent 00111 based pricing differently: First, he applied the independence principle to eliminate uncertainty about the relative funding at the laboratories and then to eliminate uncertainty about funding within that establishment. In other words,

$$H(100{,}000) = H(100) + H(1000)$$

where $H(m)$ is the amount of uncertainty in m equally likely events.

Chuckling at the memory, Agent 00111 idly wondered about his formula for $H(\ )$. Why had it proved to be so useful? He had always half-expected it to fail in some cases.

1.2.2. Axioms for Uncertainty and the Entropy Function

In this section, we derive the entropy function $H(\ )$ from an axiomatic standpoint. These axioms consist of the three axioms discussed in Section 1.2.1 from Agent 00111's viewpoint. Furthermore, there is an axiom required to make the definitions extend continuously from the rational to real number values of the probabilities. We begin with the same notation.

Let the possible events be disjoint and elements of a set of size m, i.e.,

$$\mathbf{M} = \{M_1, M_2, \ldots, M_m\}$$

Let the *a priori* probability of M_i be $Pr(M_i)$, $1 \leq i \leq m$, where the events are exhaustive.

Introduction

$$\sum_{i=1}^{m} Pr(M_i) = 1$$

The task in measuring uncertainty involves moving from the *a priori* probabilities to a single real number giving the units of uncertainty. This number, $H[Pr(M_1), Pr(M_2), \ldots, Pr(M_m)]$, is known as either the *uncertainty function* or the *entropy* function; the terms are synonymous.

Actually it is not difficult to quantify the uncertainty function once a few simple axioms are stated concerning its behavior. Consider the following two axioms.

Axiom 1. If the events are all equally likely, the uncertainty function increases as the number of events in the set increases. Mathematically speaking, $H(1/m, 1/m, \ldots, 1/m)$ is monotone increasing with m.

Axiom 2. If $\{M_1, M_2, \ldots, M_m\}$ and $\{N_1, N_2, \ldots, N_n\}$ are statistically independent sets of events each composed of equally likely disjoint events, then the uncertainty of the sets of events $\{M_\mu \cap N_\nu; \mu = 1, \ldots, m; \nu = 1, \ldots, n\}$ is the sum of the uncertainties of the separate independent sets.

Mathematically, we write

$$H\left(\frac{1}{mn}, \frac{1}{mn}, \ldots, \frac{1}{mn}\right) = H\left(\frac{1}{n}, \frac{1}{n}, \ldots, \frac{1}{n}\right) + H\left(\frac{1}{m}, \frac{1}{m}, \ldots, \frac{1}{m}\right) \quad (2)$$

If we simplify notation by defining

$$h(m) = H\left(\frac{1}{m}, \frac{1}{m}, \ldots, \frac{1}{m}\right) \quad (3)$$

our two axioms are equivalent to

$$h(m) \text{ is monotone increasing in } m \quad (4)$$

$$h(mn) = h(m) + h(n) \quad (5)$$

Equation 5 implies several things about the structure of the uncertainty function. First, notice that $m = n = 1$ gives

$$h(1) = h(1) + h(1) \quad (6)$$

which implies that
$$h(1) = 0 \tag{7}$$

Hence there is no uncertainty for an event set composed of exactly one certain (sure) event. It is also apparent from Equations 4 and 5 that

$$h(m) = \lambda \log_c m \tag{8}$$

is a satisfactory solution for $h(m)$, where λ is a positive constant when the base c of the logarithm is greater than 1. We will now prove that Equation 8 indicates the only possible functional form for $h(m)$.

Let a, b, and c be positive integers greater than 1. Then there exists a unique integer d such that

$$c^d \leq a^b < c^{d+1} \tag{9}$$

On taking logarithms,

$$d \log c \leq b \log a < (d+1) \log c \tag{10}$$

Since $c > 1$ and $b > 1$,

$$\frac{d}{b} \leq \frac{\log a}{\log c} < \frac{d+1}{b} \tag{11}$$

Applying Equation 4 to Equation 9 gives

$$h(c^d) \leq h(a^b) < h(c^{d+1}) \tag{12}$$

Using Equation 5 yields

$$dh(c) \leq bh(a) < (d+1)h(c) \tag{13}$$

Again, using Equation 4 and noting that $h(1)$ is zero, we have

$$\frac{d}{b} \leq \frac{h(a)}{h(c)} < \frac{d+1}{b} \tag{14}$$

Introduction

Equations 11 and 14 indicate that

$$\left| \frac{\log a}{\log c} - \frac{h(a)}{h(c)} \right| < \frac{1}{b} \tag{15}$$

Since b is an arbitrary positive integer,

$$\frac{h(a)}{h(c)} = \frac{\log a}{\log c} \tag{16}$$

Hence,

$$\frac{h(a)}{\log a} = \frac{h(c)}{\log c} \tag{17}$$

Since a and c are arbitrary

$$\frac{h(a)}{\log a} = \lambda = \frac{h(c)}{\log c} \tag{18}$$

where λ is a constant. Hence Equation 8 is the unique solution for h.

To handle uncertainty computations when events are not equally likely, we need Axiom 3.

Axiom 3. The total uncertainty eliminated by indicating which event actually occurred does not depend on the method of indication.

For example,

let j_i, $i = 0, 1, \ldots, n$ be integers where $0 = j_0 \leq j_1 < j_2 < \cdots < j_n = m$

Divide the set of disjoint events $\{M_1, \ldots, M_m\}$ into n sets of events

$$S_1 = \{M_1, M_2, \ldots, M_{j_1}\}$$
$$S_2 = \{M_{j_1+1}, M_{j_1+2}, \ldots, M_{j_2}\}$$
$$\vdots$$
$$S_n = \{M_{j_{n-1}+1}, M_{j_{n-1}+2}, \ldots, M_m\}$$

If the information is specified by first indicating the subset S_k containing the true event and then indicating the element within the subset, then the uncertainty relation takes the form

$$H[Pr(M_1), Pr(M_2), \ldots, Pr(M_m)] = H[Pr(S_1), Pr(S_2), \ldots, Pr(S_n)]$$
$$+ \sum_{i=1}^{n} Pr(S_i) H[Pr(M_{j_{i-1}+1} | S_i), Pr(M_{j_{i-1}+2} | S_i), \ldots, Pr(M_{j_i} | S_i)] \quad (19)$$

Hence, we postulate that the uncertainty of a collection of events equals the uncertainty of which subset contains the true event plus the average uncertainty remaining after being told which subset contains the true event.

Axiom 3 often is called the *grouping axiom.* It can be shown that Axiom 2 is actually a special case of Axiom 3.

The grouping axiom allows us to derive an expression for uncertainty when all the event probabilities are rational. Suppose that we now evaluate Equation 19, assuming that

$$Pr(M_i) = \frac{1}{m}, \quad i = 1, \ldots, m$$

Let n_k be the number of events in the set S_k. Then

$$n_k = j_k - j_{k-1} \quad \text{for } k = 1, \ldots, n$$
$$Pr(S_k) = \frac{n_k}{m} \quad \text{for } k = 1, \ldots, n \quad (20)$$
$$Pr(M_i | S_k) = \frac{1}{n_k} \quad \text{for } j_{k-1} < i \leq j_k$$

Equation 19 thus reduces to

$$h(m) = H[Pr(S_1), Pr(S_2), \ldots, Pr(S_n)] + \sum_{i=1}^{n} Pr(S_i) h(n_i) \quad (21)$$

Solving for H in Equation 21 and using the value for h given in Equation 8 yields

Introduction

$$H[Pr(S_1), \ldots, Pr(S_n)] = -\sum_{i=1}^{n} Pr(S_i)[h(n_i) - h(m)]$$

$$= -\sum_{i=1}^{n} Pr(S_i)\left[\lambda \log \frac{n_i}{m}\right]$$

$$= -\lambda\left[\sum_{i=1}^{n} Pr(S_i) \log Pr(S_i)\right] \quad (22)$$

Since our development of Equation 22 constrains $Pr(S_i)$ to be rational for all i, a fourth axiom is required.

Axiom 4. $H(p_1, p_2, \ldots, p_n)$ is a continuous function of its arguments; that is, the uncertainty measure is insensitive to minute changes in the probabilities.

Then Equation 22 applies equally well to irrational probabilities, and it is the appropriate measure of uncertainty for all probability distributions over finite sets of possible outcomes.

It is apparent from Equation 22 that uncertainty about a set of events, as presented here, does not depend on the significance or type of event. Uncertainty depends only on the set of event probabilities, and not on the events to which the set of event probabilities is assigned. Hence if $\{p_1, \ldots, p_m\}$ is a set of probabilities summing to 1, the uncertainty of which of the corresponding n events is true is given in shortened notation by

$$H(p_1, \ldots, p_m) = -\sum_{i=1}^{m} p_i \log p_i \quad (23)$$

Notice that the scale factor λ in Equation 22 is assumed to have been absorbed in the base of the logarithm in Equation 23, since

$$\log_b x = (\log_b a)(\log_a x) = \lambda \log_a x \quad (24)$$

Thus choosing a base for the logarithm in an uncertainty computation is equivalent to choosing units for uncertainty. If the base of the logarithms is 2, the uncertainty is said to be measured in bits. If natural logarithms are used, then the uncertainty is measured in nats.

$$1 \text{ nat} = \log_2 e \text{ bits} \approx 1.443 \text{ bits} \quad (25)$$

The entropy function for two events with probabilities p and $1 - p$ is shown in Figure 1.1.

$$H(p, 1 - p) = -p \log p - (1 - p) \log (1 - p) \quad (26)$$

When the occurrence of either event becomes certainty, the entropy function is zero, since

$$\lim_{p \to 0} p \log p = 0$$
$$\lim_{p \to 1} p \log p = 0 \quad (27)$$

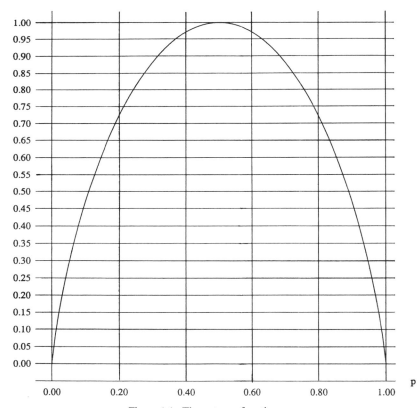

Figure 1.1. The entropy function.

Introduction

The entropy function assumes a maximum value when the two events are equiprobable. This relationship may be demonstrated for larger numbers of events by using the following inequalities for natural logarithms:

$$\ln x \leq x - 1 \tag{28}$$

$$\ln \frac{1}{x} \geq 1 - x \tag{29}$$

Inequalities 28 and 29 become equalities if and only if $x = 1$, where $\ln x$ indicates $\log_e(x)$, with $e = 2.71828\cdots$. Then

$$H\left(\frac{1}{m}, \frac{1}{m}, \ldots, \frac{1}{m}\right) - H(p_1, p_2, \ldots, p_m) = \log_b m + \sum_{i=1}^{m} p_i \log_b p_i$$

$$= \log_b e \sum_{i=1}^{m} p_i \ln m p_i \geq \log_b e \sum_{i=1}^{m} p_i \left(1 - \frac{1}{m p_i}\right) = 0 \tag{30}$$

Hence equiprobable events provide maximum uncertainty; that is, if a situation has m possible outcomes, the uncertainty is greatest when all m outcomes are believed to be equally likely.

Exercises

1. Compute the entropy in bits for the sets of events having the following probabilities:

    (a) 1/2, 1/2

    (b) 1/3, 1/3, 1/3

    (c) 1/2, 1/4, 1/4

    (d) 1/3, 1/4, 1/4, 1/12, 1/12

    Order these sets of probabilities according to their entropy. Could you have performed this ordering without knowing the results of the numerical computations? Explain.

2. Agent 00111 was offered \$100/bit to determine which one of 10,000 scientists alone knew the secret formula for Grapefruit Ignatius (a space age refreshment). His buyer had no *a priori* information about the probability of any particular scientist knowing the answer. Agent 00111 reckoned that

the actual cost of obtaining the information was $3000. Would this be a profitable venture for Agent 00111?

3. Show that the magnitude of the slope of the entropy function $H(p, 1-p)$ becomes infinite as p approaches 0 or 1. Does a similar statement for the entropy function of m events hold? Explain.

4. What are the uncertainties of the following sets of events?

    (a) {(You are dealt a bridge hand containing four aces.), (You are dealt a bridge hand that does not contain all four aces.)}

    (b) {(You are dealt a bridge hand containing an ace, king, queen, and jack in the same suit.), (You are dealt a bridge hand that does not contain an ace, king, queen, and jack in any one suit.)}

    (c) {all bridge hands}

5. The parlor game Twenty Questions allows a contestant to ask 20 questions about, for example, another contestant's occupation. Each question must be answerable by a yes or no. Let us suppose that there are 2^{15} possible occupations.

    (a) What is the most uncertainty that a contestant can have about another person's occupation?

    (b) What is the maximum *change* in uncertainty that learning a yes or no answer to a question can effect when averaged over the possible answers?

    (c) State an algorithm for playing the game that will always determine the correct occupation. To how many questions does the algorithm require an answer?

6. Show that the entropy function for m events is nonnegative. Hint: Use Equation 29.

7. Show that the grouping axiom (Axiom 3) implies that the entropy of two independent sets of events is the sum of the entropies of each set of events computed separately.

8. Let p_i, $i = 1, \ldots, n$ and q_i, $i = 1, \ldots, n$ be two sets of nonnegative numbers, each summing to unity.

$$\sum_{i=1}^{n} p_i = \sum_{i=1}^{n} q_i = 1$$

Introduction

Using logarithmic inequalities, verify that the *divergence inequality* given by

$$\sum_{i=1}^{n} p_i \log \frac{p_i}{q_i} \geq 0$$

is true. What is the condition for equality?

9. One of the problems in using the axiomatic approach is determining the set of axioms on which to base the measure. Consider the following:

   (a) $H(\mathbf{P}) = H(p_1, \ldots, p_n)$ is a symmetric function of its variables for all $n = 1, 2, 3, \cdots$ [i.e., the arguments $p_1, \ldots, p_n$ can be permuted without changing the value of $H(p_1, \ldots, p_n)$].

   (b) $H(p, 1-p)$ is a continuous function of p for $0 \leq p \leq 1$.

   (c) $H(1/2, 1/2) = 1$

   (d) $H(\mathbf{P} \times \mathbf{Q}) = H(\mathbf{P}) + H(\mathbf{Q})$, where $\mathbf{P}$ and $\mathbf{Q}$ are independent alphabets of events.

   All of these axioms are satisfied by the entropy function (Equation 23).

   (i) Show that the function

   $$H(p_1, \ldots, p_n) = \frac{1}{(1-\alpha)} \cdot \log_2 \left(\sum_{k=1}^{n} p_k^\alpha \right)$$

   also satisfies the above axioms for $\alpha > 0$, $\alpha \neq 1$.

   (ii) To what limit (if any) does the preceding function converge as $\alpha \to 1$?

   (iii) Can you find any other functions satisfying the preceding set of axioms?

1.3. Information Gain

1.3.1. Agent 00111's Sliding Scale of Success Charges

As we explained in Section 1.2.1, Agent 00111 stumbled across the entropy or uncertainty function as one part of a formula for estimating the price of a completely successful mission. If this amount exceeded his potential employer's budget or failed to meet his minimal charge, further negotiations were terminated. However, once past this stage, other cost issues still had to be resolved.

The next major objective was to agree in advance on how to calculate the degree of his success and to convert that value into an acceptable fraction of the maximum price. Most projects involved partial clarification of the uncertainties, since any project involving the human species is at best a matter of chance. To accomplish this, at the end of the mission, Agent 00111 calculated the probabilities of each of the m disjoint events occurring, using both the *a priori* probabilities and any additional knowledge he had discovered. These probabilities yielded the *a posteriori* probabilities. By applying the entropy formula to the *a posteriori* probabilities, Agent 00111 obtained the remaining uncertainty and hence established the difference between the initial and the remaining uncertainty as the uncertainty his efforts had eliminated. This difference was the information gain for which he was paid.

This practice was capable of penalizing him if the ground rules changed too dramatically. For example, he had once been asked to establish which one of m candidates would be named head of the supersecret XP organization. The customer had persuaded Agent 00111 that one of the candidates, the beloved nephew of deputy party chief Z, was a clear favorite. Agent 00111 had concluded that there was not much uncertainty in the case; however, he not only failed to determine the true candidate, but established that Z's nephew had been shot for refusing a bribe from the party chief's favorite niece, with the *a posteriori* result that every remaining candidate was in fact equally likely. On the basis of this revision of probabilities, it was apparent that there was more uncertainty in the system than Agent 00111 had originally estimated, and he had apparently achieved negative success!

By and large, however, the strategy worked well. Agent 00111 had long since become very cynical and hard-nosed about accepting a customer's evaluation of *a priori* probabilities, erring as much as possible toward the equally likely hypothesis. (He smiled; his customer took his skepticism as the experience of an old pro while he saw it as maximizing his profit.) Normally, Agent 00111 was sufficiently competent in his field work to achieve a heavy bias toward one or more outcomes to ensure, on average, a positive reduction in uncertainty. Moreover, the idea of a refund in the event of a complete failure was a marketing concept that continued to leave his speechless rivals at a disadvantage.

Agent 00111 continued reflecting on his former assignment of determining who would be named head of the XP organization. That project was very typical of the murky waters of espionage: The chance of finding the information directly was almost impossible and fraught with peril. In fact, as in quantum mechanics, if detected, the act of direct inquiry could alter the outcome; therefore, the *a posteriori* probabilities almost always had to be based on indirect observation and indirect inference by correlating secondary, partially related, tangential concerns. The difficulty was taking a mass of indirect evidence (of which any one item could be false or irrelevant) and determining how much knowledge about the goal had been obtained and how much uncertainty eliminated.

In the case of the next head of *XP*, it might be impossible to establish the exact successor. More realistically, it might be possible for Agent 00111 to determine

Which (if any) of the m candidates is on the short list of acceptable candidates of certain appointment committee members.

The relative influence of committee members on the committee's decision.

Of course, as in the shooting of the deputy party chief's nephew, the probability distribution for the committee and members' influence on the selection outcome could change without warning, and Agent 00111 could hardly expect sympathy from his client, who had hired him for his results, not his methods.

More generally, Agent 00111 might negotiate with a client about which of m disjoint events in **M** will happen. In this scenario, Agent 00111 cannot directly observe or resolve the issue, and a more realistic goal would involve studying a related set of n discrete outcomes in a set **N**. Even if Agent 00111 determines which one of the n outcomes N_j will happen, there is still uncertainty about which one of the m outcomes in **M** will happen. If, as often happens, Agent 00111 succeeds in establishing a much more accurate assessment of the relative probabilities of occurrence for each of the n outcomes of **N**, it is necessary to compute the amount of information $I(\mathbf{M}; \mathbf{N})$ that has been supplied on the m outcomes in **M**. This is a fairly direct task (see Section 1.3.2) that essentially measures the degree to which events in **N** influence events in **M**, and vice versa. The quantity $I(\mathbf{M}; \mathbf{N})$ is known as the *average uncertainty reduction* or *mutual information* between **M** and **N**.

$$H(\mathbf{M}|\mathbf{N}) = H(\mathbf{M}) - I(\mathbf{M}; \mathbf{N})$$

is the *equivocation of* **M** *given* **N**, i.e., the remaining uncertainty about the outcome from **M** given Agent 00111's results in establishing the probabilities of elements in **N** and the influence of **N** on **M**.

Interestingly although Agent 00111 had on occasion been caught by a negative uncertainty reduction, on the average he had managed to reduce the uncertainty by studying **N**. In other words, the actual outcome of **N** could render a decision about the outcome of **M** even more uncertain, but a knowledge of the probability distribution on **N** largely eliminated uncertainty.

1.3.2. Mutual Information and Equivocation

Agent 00111 had agreed to reduce the uncertainty about which of several events in a disjoint set **M** of *m* events would occur. In the case of the deputy party chief Z's nephew being shot, the event N that Agent 00111 observed was outside of the agreed frame of reference on which the bargain was sealed, i.e., the charge system failed. The compromise that Agent 00111 suggested was

$$H(\mathbf{M}) = \text{uncertainty before mission} = \sum_{M \in \mathbf{M}} Pr(M) \log \frac{1}{Pr(M)} \quad (31)$$

$$H(\mathbf{M}|N) = \text{uncertainty after mission} = \sum_{M \in \mathbf{M}} Pr(M|N) \log \frac{1}{Pr(M|N)} \quad (32)$$

Notice that by this time, Agent 00111 had settled on a shortened notation for entropy computations. A fair payment should be proportional to the reduction in uncertainty accomplished by the mission, namely,

$$\text{uncertainty reduction} = H(\mathbf{M}) - H(\mathbf{M}|N) \quad (33)$$

This uncertainty reduction can be interpreted as *information gained about the event set* **M** *when observing the event N.*

The uncertainty reduction proved to be an ideal compromise for two reasons:

1. In the case when N was an element of **M**, the uncertainty reduction was equal to the *a priori* uncertainty $H(\mathbf{M})$, since the *a posteriori* uncertainty was zero.

$$H(\mathbf{M}|N) = 0 \text{ for } N \in \mathbf{M} \quad (34)$$

Hence, in comparable cases, the two pricing schemes were consistent.

2. Despite the fact that in particular cases the uncertainty reduction could be negative, *on the average the uncertainty reduction must be nonnegative.*

The positivity of average uncertainty reduction is verified as follows. Suppose that **M** denotes the set of disjoint events about which information is sought and suppose that **N** *denotes the set of disjoint events about which information is received.* The average uncertainty reduction $I(\mathbf{M}; \mathbf{N})$ is computed as

Introduction

$$I(\mathbf{M}; \mathbf{N}) = \sum_{N \in \mathbf{N}} Pr(N)[H(\mathbf{M}) - H(\mathbf{M}|N)] = H(\mathbf{M}) - H(\mathbf{M}|\mathbf{N}) \quad (35)$$

where

$$H(\mathbf{M}|\mathbf{N}) = \sum_{N \in \mathbf{N}} Pr(N) H(\mathbf{M}|N) \quad (36)$$

The quantity $H(\mathbf{M}|\mathbf{N})$, called the *equivocation* of $\mathbf{M}$ given $\mathbf{N}$ (that is, when observing an event in $\mathbf{N}$), indicates the average *a posteriori* uncertainty due to imperfect observations. Hence, we shall now show that the average *a posteriori* uncertainty is less than the *a priori* uncertainty. From Equations 31 and 32 and elementary probability manipulations, we have

$$\begin{aligned} I(\mathbf{M}; \mathbf{N}) &= \sum_{N \in \mathbf{N}} \sum_{M \in \mathbf{M}} Pr(M, N) \log \left[\frac{Pr(M|N)}{Pr(M)} \right] \\ &= \sum_{(M,N) \in \mathbf{M} \times \mathbf{N}} Pr(M, N) \log \left[\frac{Pr(M, N)}{Pr(M) Pr(N)} \right] \end{aligned} \quad (37)$$

Applying the logarithmic inequality (Equation 29) then gives

$$I(\mathbf{M}; \mathbf{N}) \geq \sum_{(M,N) \in \mathbf{M} \times \mathbf{N}} Pr(M, N) \left[1 - \frac{Pr(M) Pr(N)}{Pr(M, N)} \right] = 0 \quad (38)$$

which verifies the second result.

The average uncertainty reduction $I(\mathbf{M}; \mathbf{N})$ is usually called the *mutual information between the sets* $\mathbf{M}$ *and* $\mathbf{N}$ *of events*. An inspection of Equation 37 verifies the appropriateness of the term mutual, since it is obvious there that the roles of $\mathbf{M}$ and $\mathbf{N}$ may be interchanged without changing the mutual information. Notice that equality holds in Equation 38; i.e., the mutual information between $\mathbf{M}$ and $\mathbf{N}$ is zero if and only if $\mathbf{M}$ and $\mathbf{N}$ are statistically independent sets of events. Thus,

$$I(\mathbf{M}; \mathbf{N}) = 0 \text{ iff } Pr(M, N) = Pr(M) Pr(N)$$

$$\text{for all } M \in \mathbf{M}, N \in \mathbf{N} \quad (39)$$

Since the equivocation is also nonnegative (it is an average of nonnegative entropy computations), the following upper bounds follow directly from Equation 35 and the symmetry of $I(\mathbf{M}; \mathbf{N})$:

$$I(\mathbf{M}; \mathbf{N}) \leq H(\mathbf{M}) \qquad (40)$$

$$I(\mathbf{M}; \mathbf{N}) \leq H(\mathbf{N}) \qquad (41)$$

Equality in Equation 40 holds if and only if $H(\mathbf{M}|N) = 0$ for all $N \in \mathbf{N}$, or equivalently if and only if each N implies a specific event in $\mathbf{M}$

$I(\mathbf{M}; \mathbf{N}) = H(\mathbf{M})$ iff $Pr(M|N) N \in \mathbf{N} = 1$ or 0 for all $M \in \mathbf{M}$ and $N \in \mathbf{N}$

A similar statement can be made for the case of equality in Equation 41.

Example 1.1. Agent 11000 communicates with his source of information by phone, unfortunately in a foreign language over a noisy connection. Agent 11000 asks questions requiring only yes or no answers from the source. Due to the noise and language barrier, Agent 11000 hears and interprets the answer correctly only 80% of the time; he fails to understand the answer 15% of the time; and he misinterprets the answer 5% of the time. Before asking the question, Agent 11000 expects the answer yes 60% of the time.

The space of *true events* is $\mathbf{M} = \{Y, N\}$, where Y equals yes and N equals no. The event space corresponding to Agent 11000's interpretation of the answer is $\mathbf{N} = \{AY, AU, AN\}$, where AY, AU, and AN represent answer yes, answer unintelligible, and answer no. The *a priori* uncertainty for the space of true events is

$$H(\mathbf{M}) = H(0.6, 0.4) = 0.97095 \text{ bits}$$

Using Bayes' law, we can determine the following probabilities:

$$Pr(AY) = 0.8 \times 0.6 + 0.05 \times 0.4 = 0.50$$
$$Pr(AU) = 0.15 \times 0.6 + 0.15 \times 0.4 = 0.15$$
$$Pr(AN) = 0.05 \times 0.6 + 0.8 \times 0.4 = 0.35$$
$$Pr(Y|AY) = (0.8 \times 0.6)/0.5 = 0.96$$
$$Pr(N|AY) = (0.05 \times 0.4)/0.5 = 0.04$$
$$Pr(Y|AU) = 0.6$$
$$Pr(N|AU) = 0.4$$
$$Pr(Y|AN) = (0.05 \times 0.6)/0.35 \doteq 0.0857$$
$$Pr(N|AN) = (0.8 \times 0.4)/0.35 \doteq 0.9143$$

Introduction

The residual uncertainties conditioned on each answer are

$$H(\mathbf{M}|AY) = H(0.96, 0.04) \doteq 0.24229 \text{ bits}$$
$$H(\mathbf{M}|AU) = H(\mathbf{M}) \doteq 0.97095 \text{ bits}$$
$$H(\mathbf{M}|AN) = H(0.0857, 0.9143) \doteq 0.42195 \text{ bits}$$

Note that no information is gained if the answer is unintelligible, since

$$H(\mathbf{M}) - H(\mathbf{M}|AU) = 0$$

The average uncertainty about $\mathbf{M}$ after making an observation in $\mathbf{N}$ is

$$H(\mathbf{M}|\mathbf{N}) \doteq 0.5 \times 0.24229 + 0.15 \times 0.97095$$
$$+ 0.35 \times 0.42195 = 0.41447 \text{ bits}$$

Obviously, the line of communication between Agent 11000 and his source leaves something to be desired. The mutual information between $\mathbf{M}$ and $\mathbf{N}$ is

$$I(\mathbf{M}; \mathbf{N}) \doteq 0.97095 - 0.41447 = 0.55648 \text{ bits}$$

Hence, on the average, Agent 11000 gets 0.55648 bits out of a possible 0.97095 bits per answer.

Exercises

1. You are allowed to observe a randomly chosen card from a standard deck of 52 cards. How much information do you receive about the selected card when you observe:

    (a) The suit of the card?

    (b) The rank of the card?

    (c) Both the rank and suit of the card?

    (d) What conclusions can you draw from (a), (b), and (c) about the mutual information on the rank and suit of a card from a standard deck?

2. An information source has an alphabet consisting of the digits 1–8. The digits occur with equal probability.

   (a) How much information do you gain about a randomly selected digit when told that:

      (i) The selected digit is a multiple of 2?

      (ii) The selected digit is a multiple of 3?

   (b) What is the average information gained about a randomly selected digit when told the number of times 2 divides the selected digit and the number of times 3 divides the selected digit? (That is, you are told the highest power of 2 and the highest power of 3 that divide a particular digit.)

   (c) Given the information in (b), what is the average additional amount of information required to identify the selected digit?

   (d) What is the entropy of this source?

3. A spy is employed as a newscaster for a radio station to which his confederate listens. The spy uses his newscast to communicate one of two binary messages, "contact me" or "do not contact me." He sends the contact-me message with probability p. The method of communication is as follows: If the newscaster wishes to broadcast contact me, he reports all news from the Associated Press wire, a source of entropy $H(AP)$ bits per broadcast. For a do-not-contact-me message, the newscaster uses news from United Press International, having entropy $H(UPI)$ bits per broadcast. The news source used is obvious to the listener.

   (a) What is the *a priori* uncertainty about the newscast? Interpret your answer.

   (b) What is the mutual information between the spy's contact-me, do-not-contact-me alphabet and the newscast?

4. Two companies quite often submit proposals to a government agency for research grants. When Expressavision, Inc., submits a proposal, it wins a contract with a probability of 0.75, is asked to rewrite the proposal with a probability of 0.125, and is turned down with a probability of 0.125. When the Nocturnal Aviation Corp. submits a proposal, it is asked to rewrite it with a probability of 0.75; 0.125 of the time, it wins a grant; and 0.125 of the time, it loses outright. Generally, Nocturnal Aviation submits seven times as many proposals as Expressavision. A rewritten proposal is considered a new submission.

Introduction

(a) How much information do you gain about the identity of the applicant when told that

(i) A grant has been awarded?

(ii) The proposal is to be rewritten?

(iii) The proposal has been turned down?

(b) On the average, how much information do you gain about the applicant when informed of one of the three preceding events?

(c) On the average, how much information do you gain about the possibility of awarding the grant, i.e., the event set in (i), when informed of the identity of the applicant?

1.4. Handling Large Amounts of Information

1.4.1. The Problem for Agent 00111

Agent 00111 was comfortable and happy with pricing a project involving a reasonably small number of outcomes. However, his recollection of initial attempts at applying the same principles to determine the cost of supplying a 1000-page report to an agency were not happy, and although the problems were resolved, the solution still struck him as almost mystical in places.

The problem, of course, is size. Agent 00111 tried to imagine the uncertainty involved in determining the contents of a book when told only that it is 1000 pages long. He gave up, smiling at the likely reaction of requesting his department to list all the possibilities. The problem does not have to be that generic. If the book is about production quantities of concrete in Country X in the last 5 years, the content is considerably more defined than if it is about love and power in a hot house world of corporate finance, wealth, and greed, although that could have formula plots. Less flippantly, it was not obvious to Agent 00111 that a 1000-page book in Chinese held more or less information than a 1000-page book in French. Finally, Agent 00111 realized that the purpose of the document was relevant. A large competitive proposal might have a very small amount of information once the considerable amount of boilerplate material had been discarded. However, this did not help him; it was not his job to make subjective decisions about what to discard. He had to compute the amount of information contained and charge accordingly.

Many years earlier, he had given up the idea of direct computation. Estimations of uncertainty had to be made, so workable models were established that gave realistic approximate assessments of the uncertainty involved in projects with very large amounts of uncertainty. This method had required becoming involved with several technical and scientific types. In any case, the situation had been resolved, at least for English texts, by the appearance of several articles that, using techniques of no interest to 00111, cited the number of bits of information per letter in English. In many pricing exercises, this information allowed Agent 00111 to multiply these estimates by the approximate number of letters in a book of a particular length.

Occasionally, this was too rough an estimate of the amount of information. He would have to obtain several texts similar to the one he was trying to cost and give them to his scientists with instructions to obtain a model based on the documents' statistics. From this model, known in their parlance as an nth-order Markov chain, the scientists would eventually give him an estimate of the bits per symbol. The scientists needed careful guidelines on how accurate a model to generate. Less accurate models were much quicker and cost much less in computer bills; however, these overestimated the amount of information and consequently the price to his clients. Therefore, in competitive cost-sensitive situations, Agent 00111 used a more accurate and more complicated estimate.

1.4.2. A General Model for an Information Source

In this section, we look at n^{th}-order Markov chains and, when they are both homogeneous and irreducible, their entropy. The basic theory of Markov chains is outlined in Section 1.5. Here, we examine an nth-order homogeneous irreducible Markov chain as a model for obtaining an estimate of the entropy of English.

Let us examine a general model of an information source and see if simplifying assumptions can be made that will yield a useful estimation procedure for the amount of uncertainty. Let $M(1)$, $M(2)$, ..., $M(k)$ be a sequence of k events about which we are uncertain, i.e., for each choice of i, $M(i)$ is chosen from a collection $\mathbf{M}_i$ of possible events. The variable i may be thought of as a time or space variable or simply as an index. In Agent 00111's problem, the index may be thought of as a spatial variable, with $M(i)$ indicating the ith symbol (a random event) in the text and $\mathbf{M}_i$ indicating the alphabet from which $M(i)$ is selected. This is by no means the only possibility. $M(i)$ could be the ith word in the text and $\mathbf{M}_i$ the vocabulary from which it is selected, or the basic random event $M(i)$ could be a sentence, paragraph, page, or chapter, etc., selected from an appropriate collection $\mathbf{M}_i$. However, if the set $\mathbf{M}_i$ is not easily described because of its size or complexity, we immediately return to Agent 00111's enumeration problem. It is best to think of the set $\mathbf{M}_i$ as some convenient unit small enough to be workable and large enough to be interesting.

Introduction

Thinking abstractly now, the event sequence $M(1), \ldots, M(k)$ can be regarded as a single event from the collection $\mathbf{M}_1 \times \mathbf{M}_2 \times \cdots \times \mathbf{M}_k$ of all event sequences, and therefore the entropy of the sequence must be given by

$$H(\mathbf{M}_1 \cdots \mathbf{M}_k) = - \sum_{(M(1),\ldots,M(k)) \in \mathbf{M}_1 \times \cdots \times \mathbf{M}_k} Pr[M(1), \ldots, M(k)] \log Pr[M(1), \ldots, M(k)] \quad (42)$$

The grouping axiom (see Section 1.2.2) states that this entropy can be computed by summing the uncertainty about $M(1), \ldots, M(k-1)$ and the average uncertainty about $M(k)$ given the previous events in the sequence. (Each $\mathbf{S}_i$ in the statement of the grouping axiom corresponds to a particular compound event in $\mathbf{M}_1 \times \cdots \times \mathbf{M}_{k-1}$). Hence,

$$H(\mathbf{M}_1 \cdots \mathbf{M}_k) = H(\mathbf{M}_1 \cdots \mathbf{M}_{k-1}) + \sum_{[M(1),\ldots,M(k-1)] \in \mathbf{M}_1 \times \cdots \times \mathbf{M}_{k-1}} Pr[M(1), \ldots, M(k-1)] H[\mathbf{M}_k | M(1), \ldots, M(k-1)] \quad (43)$$

The sum in Equation 43 is easily recognized as an equivocation term of the form defined in Equation 36, giving

$$H(\mathbf{M}_1 \cdots \mathbf{M}_k) = H(\mathbf{M}_1 \cdots \mathbf{M}_{k-1}) + H(\mathbf{M}_k | \mathbf{M}_1 \cdots \mathbf{M}_{k-1}) \quad (44)$$

Equation 44 allows the grouping axiom to be rephrased as follows: The uncertainty about a sequence of k events is equal to the uncertainty of the first $k-1$ events plus the equivocation (average remaining uncertainty) of the kth event after observing the preceding events.

Iteration of the result (Equation 44) yields a further simplification

$$H(\mathbf{M}_1 \cdots \mathbf{M}_k) = H(\mathbf{M}_1) + \sum_{i=2}^{k} H(\mathbf{M}_i | \mathbf{M}_1 \cdots \mathbf{M}_{i-1}) \quad (45)$$

Notice that the nonnegative characteristic of mutual information, specifically $I(\mathbf{M}_i ; \mathbf{M}_1 \times \cdots \times \mathbf{M}_{i-1})$, implies

$$H(\mathbf{M}_i) \geq H(\mathbf{M}_i | \mathbf{M}_1 \times \cdots \times \mathbf{M}_{i-1}) \quad (46)$$

since uncertainty is reduced on the average by the observation process. Applying Equation 46 to Equation 45 gives us an upper bound on the uncertainty of the event sequence that is much easier to estimate. Only the marginal probability functions, defined on the individual event sets $\mathbf{M}_i$, $i = 1, \ldots, k$, are required to compute the bound

$$H(\mathbf{M}_1 \times \cdots \times \mathbf{M}_k) \leq \sum_{i=1}^{k} H(\mathbf{M}_i) \qquad (47)$$

Equality holds in Equation 47 if and only if

$$H(\mathbf{M}_i) = H(\mathbf{M}_i | \mathbf{M}_1 \times \cdots \times \mathbf{M}_{i-1}) \qquad \text{for all } i \qquad (48)$$

which by Equation 39 holds if and only if $\mathbf{M}_i$ and $\mathbf{M}_1 \times \cdots \times \mathbf{M}_{i-1}$ are statistically independent collections of events for all $1 < i \leq k$. If the event sequence generator produces independent event sequences, and hence Equation 47 holds with equality, the generator is called a *memoryless information source*.

One technique to tighten the upper bound shown in Equation 47 involves using tighter bounds on the individual terms in Equation 45. An nth-*order bound* is derived using the fact that

$$H(\mathbf{M}_i | \mathbf{M}_1 \times \cdots \times \mathbf{M}_{i-1}) \leq H(\mathbf{M}_i | \mathbf{M}_{i-n} \times \mathbf{M}_{i-n+1} \times \cdots \times \mathbf{M}_{i-1}) \quad (49)$$

for all n with $1 \leq n < i$. Verbally, we make the assumption that the correlation of events in $\mathbf{M}_i$ with earlier events is dominated by the n preceding events. If this assumption is not true, we are ignoring information, increasing the uncertainty in the events of $\mathbf{M}_i$, yielding the inequality in Equation 49. The tighter upper bound on event sequence uncertainty is given by

$$H(\mathbf{M}_1 \times \cdots \times \mathbf{M}_k) \leq H(\mathbf{M}_1) + \sum_{i=2}^{n} H(\mathbf{M}_i | \mathbf{M}_1 \times \cdots \times \mathbf{M}_{i-1})$$
$$+ \sum_{i=n+1}^{k} H(\mathbf{M}_i | \mathbf{M}_{i-n} \times \cdots \times \mathbf{M}_{i-1}) \quad (50)$$

The bound in Equation 50 requires knowledge or estimates of probability distributions defined on the $(n + 1)$-fold collections $\mathbf{M}_{i-n} \times \cdots \times \mathbf{M}_i$ for all appropriate choices of i, as opposed to the smaller amount of knowledge required for constructing Equation 47.

With the bounding technique used to verify the nonnegativity of mutual information, it is possible to verify that *equality* holds in Equation 49 and hence in Equation 50 if and only if

$Pr[M(i)|M(1),\ldots,M(i-1)]$
$$= Pr[M(i)|M(i-n),\ldots,M(i-1)] \quad (51)$$

for all i such that $i > n$ and for all choices of $[M(1), \ldots, M(i)]$ in $\mathbf{M}_1 \times \cdots \times \mathbf{M}_i$. An event sequence generator with the properties indicated in Equation 51 is called an nth-order Markov source. Hence, the nth-order bound on sequence entropy indicated in Equation 50 holds with equality if and only if the event sequence generator is an nth-order Markov source.

A final assumption is usually made when dealing with large amounts of information, namely, that the event sequence is a *stationary random process*. Roughly speaking, this means that any particular event subsequence will appear at any place (time or index) within the sequence with the same probability. In written text, this would mean, for example, that the probability that the ith word in the text is "help" is a constant independent of i. Furthermore, similar statements could be made for all possible words, phrases, sentences, etc. To be mathematically precise, a stationary random process must meet two conditions: First, the basic event collections must be the same. Hence

$$\mathbf{M}_i \text{ is identical to an } \mathbf{M} \text{ for all } i. \quad (52)$$

Second, for all choices of $[M(1), \ldots, M(i)]$, i, and τ,

$Pr[\mathbf{M}_1 = M(1), \ldots, \mathbf{M}_i = M(i)]$
$$= Pr[\mathbf{M}_{1+\tau} = M(1), \ldots, \mathbf{M}_{i+\tau} = M(i)] \quad (53)$$

The stationary source assumption reduces the bound in Equation 50 to

$$H(\mathbf{M}^k) \leq H(\mathbf{M}) + \sum_{i=2}^{n} H(\mathbf{M}|\mathbf{M}^{i-1}) + (k-n)H(\mathbf{M}|\mathbf{M}^n) \quad k \geq n \quad (54)$$

for k events using an nth-order approximation for the entropy functions. In Equation 54, the notation has been simplified, with $\mathbf{M}^k$ indicating the set of all k-tuples of elements from $\mathbf{M}$ and $H(\mathbf{M}|\mathbf{M}^n)$ indicating the equivocation of an event alphabet after observing the n immediately previous event alphabets.

In the case of stationary sources with their stable statistical structure, it seems logical to normalize the entropy of the event sequence by dividing by the sequence length. Define

$$H(\mathbf{M}|\mathbf{M}^\infty) = \lim_{k\to\infty} \frac{1}{k} H(\mathbf{M}^k)$$

Then, applying the normalization to Equation 54 yields the result

$$H(\mathbf{M}|\mathbf{M}^\infty) \le H(\mathbf{M}|\mathbf{M}^n)(\text{bits/event}) \qquad (55)$$

which holds with equality for n^{th}-order Markov sources. In any situation where the entropy of a stationary source is stated without reference to the computation performed, it is undoubtedly $H(\mathbf{M}|\mathbf{M}^\infty)$, as defined in Shannon, 1948.

Without performing all of the statistical work required to develop estimates of the entropy of English, Agent 00111 was able to obtain results from the open literature. Assuming that for $n = 100$, $H(\mathbf{M}|\mathbf{M}^n)$ is near its limiting value, a reasonable estimate of the entropy $H(\mathbf{M}|\mathbf{M}^\infty)$ of English is about 1 bit/letter, as shown in Table 1.1.

Exercises

1. Without using the grouping axiom, verify that

$$H(\mathbf{M} \times \mathbf{N}) = H(\mathbf{M}) + H(\mathbf{M}|\mathbf{N})$$

2. Verify that for any event space $\mathbf{P}$,

$$H(\mathbf{M}|\mathbf{N}) \ge H(\mathbf{M}|\mathbf{N} \times \mathbf{P})$$

and determine the conditions for equality.

Table 1.1. Estimates of the Entropy of English

	$H(\mathbf{M}|\mathbf{M}^n)$ in bits/letter		
n	Shannon (1951)	Burton and Licklider (1955)	Hurd (1965)
0 (Memoryless)	4.03	4.0–4.5	4.112
1	3.32	3.2–3.8	3.460
2	3.10	2.8–3.2	2.896
99	0.3–1.3	—	—
127	—	1.04–1.95	—

Introduction

3. Zipf's law asserts that if the words in a language are ordered in a list according to relative frequency of occurrence, then the probability of the nth word in the list occurring is proportional to $1/n$. (This law is empirical and approximate.)

   (a) Accepting this premise and assuming that language can be modeled by a memoryless word source, estimate the entropy of languages having the following vocabulary sizes: (i) 10, (ii) 100, (iii) 1000, (iv) 10,000.

   (b) Assuming that the average word length of a language is five letters and the language has a vocabulary of 10,000 words, estimate the entropy of the language in bits/letter.

   (c) Imagine a language that has infinitely many words, with the words in one-to-one correspondence with the positive integers. Can Zipf's law apply to such a language? Explain.

1.5. Tutorial on Homogeneous Markov Sources

This section contains a tutorial on Markov sources and the connection to entropy. It may be regarded as a generalization of the theory in Section 1.4.2. For those familiar with Markov chains, it is not essential reading.

An nth-order Markov source is completely described probabilistically by the distributions

$$Pr[M(1), \ldots, M(n)] \quad \text{and} \quad (56a)$$

$$Pr[M(i)|M(i-n), \ldots, M(i-1)] \quad \text{for } i = n+1, n+2, \cdots \quad (56b)$$

for all possible choices of the indicated events. If we define the *state* of the nth-order source at time i as the n most recent event observations

$$s(i) = M(i-n+1), M(i-n+2), \ldots, M(i) \quad (57)$$

then the source description given in Equations 56a–b reduces to

$$Pr[s(n)] \quad (58a)$$

$$Pr[s(i)|s(i-1)] \quad \text{for } i = n+1, n+2, \cdots \quad (58b)$$

for all possible choices of the indicated states. Since the sequence of states $s(n), s(n+1), \cdots$ contains the same information as $M(1), M(2), \ldots$,

Table 1.2. A Binary Second-Order Markov Source

$M(i)$	$M(i-1)$	$M(i-2)$	$s(i)$	$s(i-1)$	$Pr[s(i)\|s(i-1)]$
0	0	0	0 0	0 0	p_i
1	0	0	1 0	0 0	$1 - p_i$
0	1	0	0 1	1 0	$1 - q_i$
1	1	0	1 1	1 0	q_i
0	0	1	0 0	0 1	r_i
1	0	1	1 0	0 1	$1 - r_i$
0	1	1	0 1	1 1	$1 - s_i$
1	1	1	1 1	1 1	s_i

the state sequence is a valid representation of the nth-order Markov source from events in $\mathbf{M}_i$ as a first-order Markov source from states in $\mathbf{M}_{i-n+1} \times \mathbf{M}_{i-n+2} \times \cdots \times \mathbf{M}_i$.

The source description in Equation 58 can be displayed in a *trellis diagram*, which indicates the flow of probability from state to state as the index increases. Consider for example a binary second-order Markov source from the following statistical description in Table 1.2. The beginning of the trellis diagram for this source is shown in Figure 1.2. Since the *transition probabilities*

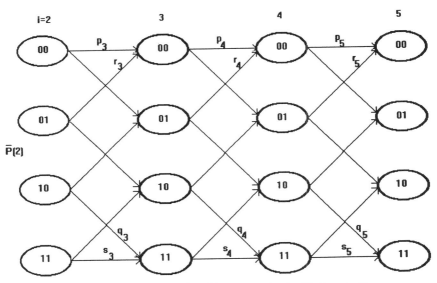

Figure 1.2. State diagram for a binary second-order Markov source.

Introduction

$$Pr[s(i)|s(i-1)]$$

sum over all possible $s(i)$ to unity,

$$\sum_i Pr[s(i)|s(i-1)] = 1$$

Figure 1.2 includes all the information about the source, including the initial state probabilities $\bar{P}(2)$.

In general, we define an ordering on the possible states at each time i by subscripting them $s_1(i), s_2(i), \ldots, s_{N_i}(i)$, where

$$N_i = |\mathbf{M}_{i-n+1} \times \cdots \times \mathbf{M}_i| \tag{59}$$

is the number of states at time i. Then it is possible to define a *state probability vector at time (index) i* as

$$\bar{P}(i) = \begin{bmatrix} Pr[s_1(i)] \\ Pr[s_2(i)] \\ \vdots \\ Pr[s_{N_i}(i)] \end{bmatrix} \tag{60}$$

and a transition probability matrix $T(i)$, which is an $N_i \times N_{i-1}$ matrix whose entry in the jth row and the kth column is given by

$$Pr[s_j(i)|s_k(i-1)]$$

Then the state probability vector at index i may be computed recursively using

$$\bar{P}(i) = T(i)\bar{P}(i-1) \tag{61}$$

or equivalently, given the initial state probability vector $\bar{P}(n)$,

$$\bar{P}(i) = \left[\prod_{j=n+1}^{i} T(j) \right] \bar{P}(n) \qquad \text{for } i \geq n \tag{62}$$

A major simplification occurs in the description of the nth-order Markov source when its transition probabilities are stationary, i.e., when

$$s_j(i) = s_j(n) \qquad \text{for all } i > n \tag{63a}$$

and

$$T(i) = T \quad \text{for all } i > n \quad (63b)$$

A source having the property of Equation 63 is said to be *homogeneous*. Then Figure 1.2 is periodic with the state names and transition probabilities identical for each value of the index i. In this case, the trellis diagram in Figure 1.2 is usually replaced by the *state diagram* of the source. For the binary second-order source described earlier, a homogeneity assumption gives the state diagram illustrated in Figure 1.3. In the state diagrams of homogeneous sources (see Figure 1.3), a state represents itself for all values of the index i, and since transition probabilities are independent of the index i, the state diagram and the initial state vector $\bar{P}(n)$ completely describe the source.

The entropy of a homogeneous Markov source is easily computed assuming (as we have in our n^{th}-order models) that the source output $M(1)$, $M(2), \ldots, M(k)$, $k > n$, may be mapped in a one-to-one fashion into a state sequence $s(n), s(n+1), \ldots, s(k)$. Then letting $\mathbf{S}_i$ represent the collection $\{s(i)\}$ of possible states at index i, the entropy of k symbols from the source is reduced to

$$H(\mathbf{M}_1 \times \cdots \times \mathbf{M}_k) = H(\mathbf{S}_n \times \cdots \times \mathbf{S}_k) \quad (64)$$

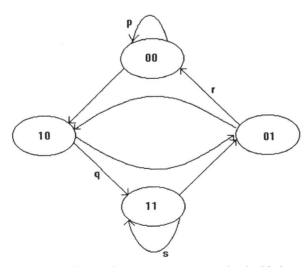

Figure 1.3. A state diagram for a homogeneous second-order Markov source.

Introduction

Since the state sequence is first-order Markov,

$$H(\mathbf{M}_1 \times \cdots \times \mathbf{M}_k) = H(\mathbf{S}_n) + \sum_{i=n+1}^{k} H(\mathbf{S}_i | \mathbf{S}_{i-1}) \qquad (65)$$

Now

$$H(\mathbf{S}_i | \mathbf{S}_{i-1}) = \bar{H}^t \bar{P}(i-1) \qquad (66)$$

where the vector $\bar{H}^t$ is the transpose of the conditional entropy column vector $\bar{H}$

$$\bar{H} \triangleq \begin{bmatrix} H[\mathbf{S}_i | s_1(i-1)] \\ H[\mathbf{S}_i | s_2(i-1)] \\ \vdots \\ H[\mathbf{S}_i | s_N(i-1)] \end{bmatrix} \qquad (67)$$

and $\bar{P}(i-1)$ is the state probability distribution at time $i-1$. Equivalently, the nth entry in $\bar{H}$ is the entropy function of the probability distribution of the transitions, leaving state $s_n(i-1)$, i.e., the nth column of the state transition matrix $T(i)$. For a homogeneous source, $T(i)$ is not a function of i, and hence $\bar{H}$ is not a function of i. This results in a major simplification of Equation 65, namely,

$$H(\mathbf{M}_1 \times \cdots \times \mathbf{M}_k) = H(\mathbf{S}_n) + \bar{H}^t \left[\sum_{i=n}^{k-1} \bar{P}(i) \right]$$

$$= H(\mathbf{S}_n) + \bar{H}^t \left[\sum_{j=0}^{k-1-n} T^j \right] \bar{P}(n) \qquad (68)$$

where $H(\mathbf{S}_n)$ is the entropy of the state probability distribution at time n, namely, $\bar{P}(n)$.

Example 1.2. Consider the homogeneous binary second-order Markov source defined in Figure 1.3, with $p = 1 - q = 1 - r = s = 3/4$. Ordering the states $s_1 = 00$, $s_2 = 01$, $s_3 = 10$, and $s_4 = 11$ gives

$$T = \begin{bmatrix} 3/4 & 1/4 & 0 & 0 \\ 0 & 0 & 3/4 & 1/4 \\ 1/4 & 3/4 & 0 & 0 \\ 0 & 0 & 1/4 & 3/4 \end{bmatrix}$$

Then from Equation 68,

$$H(\mathbf{M}_1 \times \cdots \times \mathbf{M}_k) = H[\bar{P}(2)] + H(3/4, 1/4)\underline{j}^t\left[\sum_{i=n}^{k-1} \bar{P}(i)\right]$$

$$= H[\bar{P}(2)] + (k - n)H(3/4, 1/4)$$

Here $\underline{j}$ denotes the all 1s column vector, and we have reverted to a probability distribution argument for the entropy function. An added simplification occurred in this computation because the transition probability distributions, i.e., the T columns, all involved the same numbers, namely, 3/4, 1/4, 0, and 0.

Example 1.3. Consider the homogeneous binary second-order Markov source from transition probabilities as labeled in the state diagram in Figure 1.3, with $s_1 = 00$, $s_2 = 01$, $s_3 = 10$, and $s_4 = 11$; then,

$$T = \begin{bmatrix} p & r & 0 & 0 \\ 0 & 0 & \bar{q} & \bar{s} \\ \bar{p} & \bar{r} & 0 & 0 \\ 0 & 0 & q & s \end{bmatrix}$$

where for any x, $\bar{x} = 1 - x$. If

$$\bar{P}(2) = c \begin{bmatrix} r\bar{s} \\ \bar{p}s \\ \bar{p}s \\ \bar{p}q \end{bmatrix}$$

where $c^{-1} = r\bar{s} + 2\bar{p}s + \bar{p}q$, then $\bar{P}(3) = T\bar{P}(2) = \bar{P}(2)$. That is, $\bar{P}(2)$ is an eigenvector of the matrix T from eigenvalue 1. By iteration, it is obvious that $\bar{P}(i) = \bar{P}(2)$ for $i \geq 2$. Applying these results to Equation 68 gives the entropy of k symbols from this source

$$H(\mathbf{M}_1 \times \cdots \times \mathbf{M}_k) = H(\bar{P}(2)) + (k - n)\bar{H}^t\bar{P}(2)$$

where

$$\bar{H} = \begin{bmatrix} H(p, 1 - p) \\ H(r, 1 - r) \\ H(q, 1 - q) \\ H(s, 1 - s) \end{bmatrix}$$

Introduction

Equation 68 indicates that entropy computations for homogeneous Markov sources are easily carried out if the state probability vector $\bar{P}(i)$ is known for all i. In Example 1.2, $\bar{P}(i)$ did not enter the computation for $i > n$ because all states had similar transition probabilities. In Example 3 $\bar{P}(i)$ does not change with i, and the source is in fact *stationary*. Now we will consider generalizations of Example 3.

Let a *stationary state probability vector* $\bar{P}_{stat}$ be defined as any vector satisfying the relations

$$T\bar{P}_{stat} = \bar{P}_{stat} \tag{69a}$$

$$P_{stat\ n} \geq 0 \quad \text{for all } n \tag{69b}$$

$$\underline{j}^t \bar{P}_{stat} = 1 \tag{69c}$$

where $P_{stat\ n}$ is the nth component of $\bar{P}_{stat}$. In fact, Equations 69b and 69c follow from Equation 69a:

1. If T has eigenvalue 1, then there always exists at least one nonzero solution $(P_1, P_2, \ldots, P_N)^t$.

2. If $(P_1, P_2, \ldots, P_N)^t$ is a solution to Equation 69a, then so is $(|P_1|, |P_2|, \ldots, |P_N|)^t$. Hence Equation 69b is satisfied.

3. By scaling the preceding vector in (2), it is possible to satisfy the probability distribution constraint in Equation 69c.

Therefore a stationary state probability vector can be found for any homogeneous Markov source. It can further be shown that

4. If $\bar{P}(n) = \bar{P}_{stat}$, then the source output sequence is a stationary random process. The verification of these results is left to the reader.

The connections in the state diagram of a Markov source contain a great deal of information about the asymptotic properties of $\bar{P}(i)$ for large i. We say that *state s_j can be reached from state s_i* and write $s_i \rightarrow s_j$ if there exists a sequence of transitions beginning at s_i and ending at s_j in the state diagram (as a convention, we assume $s_i \rightarrow s_i$, i.e., s_i can be reached from s_i in zero transition). The state diagram can always be completely partitioned into disjoint sets (or clusters) of states with the property that s_i and s_j are in the same cluster if and only if $s_i \rightarrow s_j$ and $s_j \rightarrow s_i$.

Example 1.4.

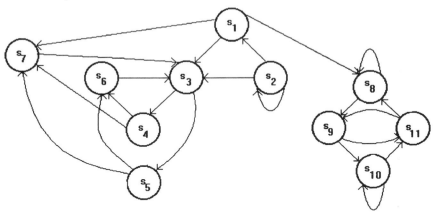

Figure 1.4. Example of a Markov source.

Consider the state diagram for the Markov source shown in Figure 1.4. The four clusters C_i, $i = 1, 2, 3, 4$, for this diagram are

$$C_1 = \{s_1\}$$
$$C_2 = \{s_2\}$$
$$C_3 = \{s_3, s_4, s_5, s_6, s_7\}$$
$$C_4 = \{s_8, s_9, s_{10}, s_{11}\}$$

The equivalent cluster diagram is shown in Figure 1.5.

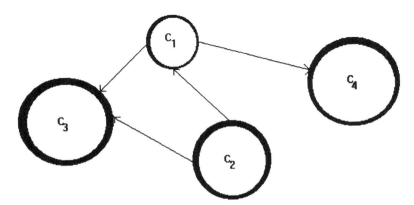

Figure 1.5. Cluster diagram for the Markov source in Figure 1.4.

Introduction

Roughly speaking, transition lines in a state diagram indicate the direction of flow of probability between states as the index i increases; arrows in the equivalent cluster diagram indicate this same effect. However, the cluster diagram indicates an *irreversible* flow of probability, since by the definition of a cluster, there is no transition sequence leaving a cluster C_i, entering another cluster C_j, and returning to C_i. Hence, it is intuitively obvious that as i increases, the probability will ultimately reside in the *terminal clusters,* i.e., those clusters having no exiting transitions to other clusters in the diagram.

When used to model languagelike information sources, Markov chains generally consist of a single (terminal) cluster. If there were two terminal clusters, a transition into one terminal cluster would forever prevent the occurrence of state sequences within a different cluster. This would *forever* exclude the use of certain words or phrases in the information source that would have appeared if a different transition sequence occurred. Obviously, the objection to two terminal clusters in a language model is due to the finality of the word forever. We will assume from now on that Markov models of information sources consist of exactly one terminal cluster, i.e., that $s_i \rightarrow s_j \rightarrow s_i$ for all choices of s_i and s_j in the state diagram. Such a source, consisting of exactly one cluster, is said to be *irreducible.*

If a homogeneous Markov source is turned on and allowed to pass through a large number of transitions, we might expect the state probability vector $\bar{P}(i)$ to approach some constant value. If in fact $\bar{P}(i)$ approaches a limit as i increases, we define the *steady-state state probability vector* $\bar{P}_{ss}$ as that limit:

$$\bar{P}_{ss} = \lim_{i \to \infty} \bar{P}(i) \tag{70}$$

If $\bar{P}_{ss}$ exists, it must be a stationary-state probability vector, since

$$T\bar{P}_{ss} = \lim_{i \to \infty} T\bar{P}(i) = \lim_{i \to \infty} \bar{P}(i+1) = \bar{P}_{ss} \tag{71}$$

The following results may be verified:

5. The stationary probability vector $\bar{P}_{stat}$ of a homogeneous Markov source is unique if and only if the source contains exactly one terminal cluster and equals $\bar{P}_{ss}$.

6. If the source is irreducible, then $\bar{P}_{stat}$ contains no zero entries.

Since $\bar{P}_{stat}$ is a function of only the transition probability matrix T and since (5) states that $\bar{P}_{stat}$ is unique for irreducible sources, the following hypothesis is plausible: If $\bar{P}_{ss}$ exists for an irreducible homogeneous Markov source, then it is independent of the initial state probability vector $\bar{P}(n)$.

Example 1.5. Consider the second-order binary Markov source with transition probability matrix

$$T = \begin{bmatrix} 2/3 & 2/3 & 0 & 0 \\ 0 & 0 & 5/6 & 5/6 \\ 1/3 & 1/3 & 0 & 0 \\ 0 & 0 & 1/6 & 1/6 \end{bmatrix}$$

T is diagonalizable to

$$T = EDE^{-1}$$

where

$$E = \begin{bmatrix} 10 & 4 & 1 & 0 \\ 5 & -5 & -1 & 0 \\ 5 & 2 & 0 & 1 \\ 1 & -1 & 0 & -1 \end{bmatrix}$$

$$D = \begin{bmatrix} 1 & 0 & 0 & 0 \\ 0 & -1/6 & 0 & 0 \\ 0 & 0 & 0 & 0 \\ 0 & 0 & 0 & 0 \end{bmatrix}$$

$$E^{-1} = \frac{1}{21} \begin{bmatrix} 1 & 1 & 1 & 1 \\ -6 & -6 & 15 & 15 \\ 35 & 14 & -70 & -70 \\ 7 & 7 & -14 & -35 \end{bmatrix}$$

Using Equation 62 and the preceding expansion, it is possible to compute $\bar{P}(i)$ for all i as

$$\bar{P}(i) = T^{i-n}\bar{P}(n) = ED^{i-n}E^{-1}\bar{P}(n)$$

As i increases,

$$\lim_{i \to \infty} D^{i-n} = \begin{bmatrix} 1 & 0 & 0 & 0 \\ 0 & 0 & 0 & 0 \\ 0 & 0 & 0 & 0 \\ 0 & 0 & 0 & 0 \end{bmatrix}$$

Introduction

and

$$\lim_{i \to \infty} \bar{P}(i) = \frac{1}{21} \begin{bmatrix} 10 \\ 5 \\ 5 \\ 1 \end{bmatrix} [1 \quad 1 \quad 1 \quad 1] \bar{P}(n)$$

$$= \frac{1}{21} \begin{bmatrix} 10 \\ 5 \\ 5 \\ 1 \end{bmatrix} \triangleq \bar{P}_{ss}$$

Using the results from Example 1.4, it is also easily verified that $\bar{P}_{ss}$ is the stationary state probability vector for the source.

The noteworthy point in Example 1.5, namely, that $\bar{P}_{ss}$ exists and is independent of the initial state probability vector $\bar{P}(n)$, demonstrates the following result:

7. If the transition probability matrix T of a homogeneous Markov source has the property that for some value of k, all the entries of T^k are nonzero, then

$$\lim_{j \to \infty} T^k = \bar{P}_{\text{stat}} \underline{j}^t = \bar{P}_{ss} \underline{j}^t \tag{72}$$

For an nth order Markov source with all possible transition probabilities nonzero, T^n consists entirely of nonzero entries.

Example 6 illustrates an irreducible homogeneous Markov source for which $\bar{P}_{ss}$ does not exist.

Example 1.6. Consider the Markov source defined in Figure 1.6.

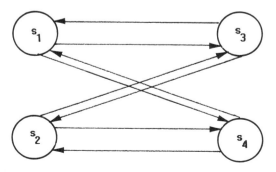

Figure 1.6. State diagram for Example 1.4. All transition probabilities are 1/2.

Then

$$T = \begin{bmatrix} 0 & 0 & 1/2 & 1/2 \\ 0 & 0 & 1/2 & 1/2 \\ 1/2 & 1/2 & 0 & 0 \\ 1/2 & 1/2 & 0 & 0 \end{bmatrix} = T^{2k+1} \qquad k = 0, 1, 2, \cdots$$

$$T^2 = \begin{bmatrix} 1/2 & 1/2 & 0 & 0 \\ 1/2 & 1/2 & 0 & 0 \\ 0 & 0 & 1/2 & 1/2 \\ 0 & 0 & 1/2 & 1/2 \end{bmatrix} = T^{2k} \qquad k = 1, 2, \cdots$$

Since T^i does not approach a fixed matrix as i increases, the limit of $T^{(i-n)}\bar{P}(i)$ cannot be expected to exist, except in special cases, e.g., when all states are equally likely initially. The problem indicated in Example 6 is obviously caused by a periodic structure in the state diagram. The states of the diagram may be divided into distinct *phase sets,* namely,

$$\Phi_0 = \{s_1, s_2\}, \ \Phi_1 = \{s_3, s_4\}$$

Any observation from this source consists of a sequence of states alternately selected from Φ_0 and Φ_1.

The following discussion is helpful in determining if T^j ever has all nonzero entries. Consider a path through the state diagram that begins at state s_i and first returns to s_i r transitions later. The number r is called a *first-recurrence time* of s_i. Let R_i be the collection of first-recurrence times of the state s_i. Either all elements of R_i have a common divisor greater than 1 or they do not. The greatest common divisor of the elements in R_i, say, h_i, corresponds to the minimum period obeyed by the first-recursion times; i.e., if s_i recurs after r steps, we know that h_i divides r. For this reason, the greatest common divisor of the R_i is known as the *period of the state s_i*. In general, a Markov chain may have states with different periods. To simplify,

8. In an irreducible homogeneous Markov source, the greatest common divisor of the first-recurrence times of a state s_i is independent of the choice of state. With the preceding assumptions, we define

$$\eta = \gcd\{R_i\} \text{ for all } i, \tag{73}$$

where η is called the *period* of the source. In a more general setting, the period of the source can be defined as the least common multiple of the (possibly different) state periods. We will continue to look at irreducible homogeneous periodic Markov sources.

Introduction

The states of a periodic source may be separated into η phase sets, Φ_0, Φ_1, ..., $\Phi_{\eta-1}$ as follows: Put s_1 in Φ_0. If s_i may be reached from s_1 in m steps, then s_i is in $\Phi_{m \bmod \eta}$. ($m \bmod \eta$ is the remainder when m is divided by η.)

9. If the states are ordered according to the index of their phase set, then the transition probability matrix T in partitioned form appears as a cyclic permutation matrix

$$T = \begin{bmatrix} 0 & 0 & 0 & \cdots & 0 & T_0 \\ T_1 & 0 & 0 & \cdots & 0 & 0 \\ 0 & T_2 & 0 & \cdots & 0 & 0 \\ 0 & 0 & T_3 & \cdots & 0 & 0 \\ \vdots & \vdots & \vdots & \cdots & & \vdots \\ & & & \cdots & & \\ 0 & 0 & 0 & 0 & T_{\eta-1} & 0 \end{bmatrix} \quad (74)$$

where T_i is a $|\Phi_i| \times |\Phi_{i-1 \bmod \eta}|$ matrix.

10. A limit of $[T^\eta]^i$ exists as i increases

$$\lim_{i \to \infty} (T^\eta)^i = \begin{bmatrix} \bar{P}_{ss_0} \underline{j}^t & 0 & \cdots & 0 \\ 0 & \bar{P}_{ss_1} \underline{j}^t & \cdots & 0 \\ \vdots & \vdots & \cdots & \vdots \\ 0 & 0 & \cdots & \bar{P}_{ss_{\eta-1}} \underline{j}^t \end{bmatrix} \quad (75)$$

where $\bar{P}_{ss_i}$ is the steady-state (or stationary-state) state probability vector of the matrix $T_{i+\eta} \cdots T_{i+2} T_{i+1}$. Note that the matrix subscript arithmetic is mod η.

11. The stationary-state probability vector of the source is given by

$$\bar{P}_{\text{stat}} = \frac{1}{\eta} \begin{bmatrix} \bar{P}_{ss_0} \\ \bar{P}_{ss_1} \\ \vdots \\ \bar{P}_{ss_{\eta-1}} \end{bmatrix} \quad (76)$$

Obviously, when η equals 1, the results in (74–76) reduce to (72).

Thus we see that for an irreducible homogeneous Markov source of period 1, the entropy of the source is given by

$$H(\mathbf{M}|\mathbf{M}^\infty) = \lim_{k \to \infty} \frac{1}{k} H(\mathbf{M}_1 \times \cdots \times \mathbf{M}_k)$$
$$= \bar{H}^t \lim_{k \to \infty} \bar{P}(k)$$
$$= \bar{H}^t \bar{P}_{ss} = \bar{H}^t \bar{P}_{\text{stat}} \qquad (77)$$

where $\bar{H}$ is defined in Equation 67. As a word of caution, remember that the discussion here has assumed a *finite* number of states in the source. Generalizing to infinite state sources may lead to difficulties.

A final result for homogeneous irreducible Markov sources involves the relationship between long sample sequences generated by the information source and the statistical description of the source. Let $N^k_{j_0 j_1 \ldots j_m}$ be the number of values of i, $1 \leq i \leq k - m$, for which

$$s(i) = s_{j_0}, s(i+1) = s_{j_1}, \ldots, s(i+m) = s_{j_m} \qquad (78)$$

That is, $N^k_{j_0 \ldots j_m}$ is the number of times that the consecutive subsequence $s_{j_0}, s_{j_1}, \ldots, s_{j_m}$ appears in the source output state sequence $s(1), s(2), \ldots, s(k)$. Notice that $N^k_{j_0 \ldots j_m}$ is a random variable whose value depends on the randomly generated state sequence.

Example 1.7.

i	1	2	3	4	5	6	7	8	9	10
$s(i)$	0	0	1	0	1	0	1	1	0	1

Let $s_1 = 1$, $s_2 = 0$. If $j_0 = 1$, $j_1 = 2$, $j_2 = 1$, then $s_{j_0} s_{j_1} s_{j_2} = 101$. The sequence 101 occurs for i equaling 3, 5, and 8 in the preceding sequence, and hence $N^{10}_{121} = 3$ for the preceding sequence.

Theorem 1.1 can be verified.

Theorem 1.1. For any finite-order homogeneous irreducible Markov source, the random variable $(1/k) N^k_{j_0 j_1 \ldots j_m}$ converges in probability to P, where

$$P = \left[\prod_{i=1}^{m} Pr(s_{j_i} | s_{j_{i-1}}) \right] P_{\text{stat } j_0}$$

Introduction

for all choices of m and $j_0, j_1, \ldots, j_m$; that is,

$$\lim_{k \to \infty} Pr\left\{\left|\frac{1}{k} N^k_{j_0 j_1 \cdots j_m} - P\right| > \epsilon\right\} = 0 \tag{79}$$

Here $Pr(s_{j_i} | s_{j_{i-1}})$ is the transition probability of moving from state $s_{j_{i-1}}$ to state s_{j_i}, and $P_{\text{stat} j_0}$ is the stationary probability of being in state j_0. Hence the theorem states that the fractional number of times that a specified state subsequence occurs in a long sequence of states converges in probability to the probability of occurrence of the specified subsequence at a given position in the output of an irreducible stationary Markov source having the same transition probabilities. The preceding theorem, which resembles a law of large numbers, has its roots in an elementary form of *ergodic* theory. Sources satisfying this theorem are referred to as *ergodic* sources.

Problems

1. (a) Draw the state diagram of a binary third-order stationary Markov source.
   (b) Simplify (a) under the following two conditions:
      (i) Two successive changes in symbol (i.e., $\cdots$ 010 $\cdots$ or $\cdots$ 101 $\cdots$) are not allowed.
      (ii) All allowable transitions from any state are equiprobable.

2. What is the entropy of a ternary, fifth-order stationary Markov source with equally likely transitions?

3. Find the average information obtained from observing n symbols if the first-order stationary Markov source has the following transition probability matrix:

$$\begin{bmatrix} \frac{1}{3} & \frac{1}{2} & \frac{1}{4} \\ \frac{1}{3} & \frac{1}{4} & \frac{1}{4} \\ \frac{1}{3} & \frac{1}{4} & \frac{1}{2} \end{bmatrix}$$

4. Assuming that all transitions from any given state of an n-th order Markov

source give rise to distinct messages from the source and that no two states can be identified with the same sequence of messages, identify the alphabet size and order of the sources with the state diagrams (not necessarily complete) shown in Figure 1.7 and label the states.

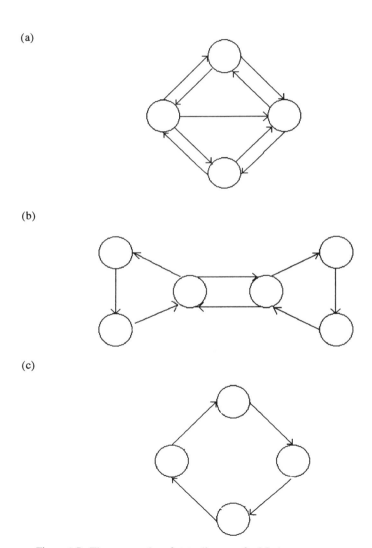

Figure 1.7. Three examples of state diagrams for Markov sources.

Introduction

5. The answers to Problem 4 are not necessarily unique. Can the state diagram of a binary second-order source always be interpreted as the diagram of a quaternary first-order source? Is this true of the reverse implication? Can you generalize these results?

6. What is the source entropy of the binary stationary Markov source having the following defining characteristic: Given $M(t-1)$, $M(t-2)$, ..., the event at time t will be identical to the event at time $t-2$ with probability 0.75. What is the entropy of a sequence of n symbols from this source?

7. Consider a Markov source where each state has the same number of outgoing transitions. Show that the entropy of the source is maximized when these transitions are equiprobable.

8. Figure 1.8 defines the state diagram of a ternary first-order Markov source.

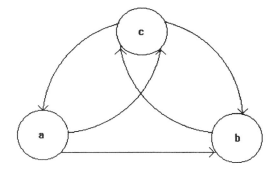

Figure 1.8. A ternary first-order Markov source.

Assign transition probabilities to the diagram to maximize the entropy of the source.

9. If you were allowed to remove two transition arrows from the state diagram of a third-order binary Markov source, indicate how you would do so to create a state diagram with the following properties:

   (a) Two terminal clusters and one nonterminal cluster (one way)

   (b) One terminal cluster and one nonterminal cluster (many ways)

10. Characterize the set of possible stationary distributions for the source with the following transition probability matrix:

$$\begin{bmatrix} 1/36 & 0 & 0 & 0 & 0 & 0 \\ 1/18 & 1/3 & 1/2 & 0 & 1/4 & 0 \\ 5/12 & 1/3 & 1/4 & 0 & 1/4 & 0 \\ 5/12 & 0 & 0 & 1/2 & 0 & 3/4 \\ 1/18 & 1/3 & 1/4 & 0 & 1/2 & 0 \\ 1/36 & 0 & 0 & 1/2 & 0 & 1/4 \end{bmatrix}$$

11. Consider the binary Markov source with transition probability matrix given by

$$\begin{bmatrix} 3/4 & 1/4 \\ 1/4 & 3/4 \end{bmatrix}$$

Assume the initial state probability vector is [1, 0].

(a) Compute T^i and $\bar{P}(i)$, $i = 2, 3,$ and 4.

(b) Find the general form for T^i and $\bar{P}(i)$.

12. Consider the source with the state diagram defined in Figure 1.9.

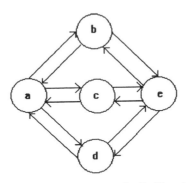

Figure 1.9. Markov source for Problem 12.

(a) Does this source have steady-state probabilities? Why?

(b) If the initial state is state a and all transitions are equally likely, derive the state probability vector $\bar{P}(i)$ as a function of i.

(c) Referring to (b), what is the entropy of 500 symbols from this source?

Introduction

13. Prove that any irreducible source having at least one state s_1 with a transition directly back to itself has a steady-state probability distribution.

14. How would you maximize the entropy of a periodic Markov source over the set of all possible transition probabilities given the condition that all states in the same cycle position have the same number of outgoing transitions?

15. Assume that the set of first-recurrence times for states in an irreducible Markov source is given by $\{462, 1188, 1650\}$. What is the period of the source?

16. Derive an expression for the entropy of a stationary periodic Markov source.

17. Consider the Markov source having the following transition probability matrix:

$$\begin{bmatrix} 0 & 0 & 0 & 1/3 & 0 & 1/3 & 0 & 0 \\ 1/8 & 0 & 0 & 0 & 1/8 & 0 & 0 & 0 \\ 1/2 & 0 & 0 & 0 & 0 & 0 & 0 & 3/4 \\ 0 & 1 & 3/4 & 0 & 0 & 0 & 0 & 0 \\ 0 & 0 & 0 & 1/3 & 0 & 1/3 & 0 & 0 \\ 0 & 0 & 1/4 & 0 & 0 & 0 & 1 & 0 \\ 3/8 & 0 & 0 & 0 & 7/8 & 0 & 0 & 1/4 \\ 0 & 0 & 0 & 1/3 & 0 & 1/3 & 0 & 0 \end{bmatrix}$$

(a) Find the period of the source.

(b) Find its stationary distribution.

(c) What is its entropy?

1.6. The Number of Typical Sequences

1.6.1. Agent 00111 Uncovers a Puzzle

Recalling his efforts to estimate information contained in a large work written in, say, English, led Agent 00111 to one of the truly remarkable, and at least to him, surprising aspects of his methods. In his early days, faced with estimating the information in a written text, Agent 00111 had ignored all those theoretical types who confused him with strange concepts and turned to something as physical as a list of phrases. Agent 00111 had attempted to write down every possible string of n symbols from the Roman alphabet. This counting exercise, when spaces were included, had 27^n entries, so he soon abandoned this method. In a second effort, he tried to write down only the n-tuples that made sense in English; that is, only words, parts of words and parts of text were listed. He continued this task (with the aid of several complaining staff members) for many values of n. Once the lists were complete, Agent 00111 ordered a count of the number of n-tuples. He found out that the lists of n-tuples were all approximately 2^{nH} long, where H was a number that did not change from one value of n to another when n was large; H was in fact a constant. This had mildly surprised Agent 00111, but in truth, he had not known quite what to expect. However, the next result gave him a shock. The value of H seemed familiar. It took him a little while to remember, but then he had it: The value of H was exactly the same as what the literature said was the entropy of English in bits per symbol!

At first, Agent 00111 assumed this was a coincidence. Without confiding in his staff—he had learned something in his career—he repeated the exercise for Chinese and French. If his staff had complained about n-tuples in English, the Chinese-inspired complaints nearly deafened him. In fact, as when Agent 00111 required English civil servants to count French n-tuples, questions of his sanity were raised. Still Agent 00111 persevered. He found that there were approximately $2^{n \cdot H_C}$ entries on the Chinese list and $2^{n \cdot H_F}$ entries on the French list. Independently, he asked his scientists to estimate the entropy of French and Chinese. Some time later, they gave him the answers: H_F and H_C.

Agent 00111 swallowed his pride; there had to be a connection. He had started with the most pragmatic listing approaches only to find that list lengths were directly related to his scientists' Holy Grail of abstraction, the entropy of a language.

From that point on, Agent 00111 made the idea of entropy the cornerstone of his textual and other large-volume pricing. His scientists, who, he grudgingly conceded, were not so other-worldly as he had first thought, assured him that entropy was the cornerstone of information theory. They also rambled on about asymptotic equipartition properties and Shannon–McMillan theorems, which had not helped him too much. His concerns, as ever, were more practical.

1.6.2. List Length and Entropy

In Section 1.6.2, we prove the results that surprised Agent 00111; i.e., we demonstrate the connection between the entropy of a language and the correlation between rate of increase in the number of possible sequences and the length of the sequence. The notation in Section 1.4.2 and Theorem 1.1 are used.

Introduction

We formulate the relationship between list length and entropy in a precise manner for n^{th}-order homogeneous irreducible Markov sources, as described in Section 1.6.1. Define the random variable I_k by

$$I_k = -\frac{1}{k} \log Pr[M(1), \ldots, M(k)] \tag{80}$$

Obviously, I_k is a function of the source's random output symbol sequence $M(1), \ldots, M(k)$. The expected value of I_k is given by

$$E\{I_k\} = -\frac{1}{k} \sum_{M(1), \ldots, M(k) \in \mathbf{M}} Pr[M(1), \ldots, M(k)] \log Pr[M(1), \ldots, M(k)]$$

$$= \frac{1}{k} H(\mathbf{M}_1 \times \cdots \times \mathbf{M}_k) \tag{81}$$

As k increases, this mean value approaches the entropy of the source.

$$\lim_{k \to \infty} E\{I_k\} = H(\mathbf{M} | \mathbf{M}^\infty) \tag{82}$$

If the value of I_k for a particular sequence $M(1), \ldots, M(k)$ is within ϵ of its asymptotic average value $H(\mathbf{M}|\mathbf{M}^\infty)$, i.e.,

$$|I_k - H(\mathbf{M}|\mathbf{M}^\infty)| < \epsilon \tag{83}$$

then $M(1), \ldots, M(k)$ is said to be ϵ-*typical*. We now prove the following relations for homogeneous irreducible Markov sources.

Theorem 1.2a. Asymptotic Equipartition Property (*Shannon–McMillan Theorem*) The sequence of random variables I_k, $k = 1, 2, \ldots$, converges in probability to the source entropy $H(\mathbf{M}|\mathbf{M}^\infty)$; that is, for any $\epsilon > 0$,

$$\lim_{k \to \infty} Pr\{|I_k - H(\mathbf{M}|\mathbf{M}^\infty)| \geq \epsilon\} = 0$$

Proof: The random variable I_k can be rewritten in state notation for the nth-order Markov source as

$$I_k = -\frac{1}{k} \log Pr[s(n), s(n+1), \ldots, s(k)]$$

$$= -\frac{1}{k}\left\{\log Pr[s(n)] + \sum_{i=n+1}^{k} \log Pr[s(i)|s(i-1)]\right\} \quad (84)$$

Let us denote the collection of possible states by **S**, which contains states s_1, $s_2, \ldots, s_{|\mathbf{S}|}$. When

$$s(i) = s_j \quad \text{and} \quad s(i-1) = s_m \quad (85a)$$

we will write

$$Pr[s(i)|s(i-1)] = Pr(s_j|s_m) \quad (85b)$$

where it is understood that s_j is the state immediately following s_m in the sequence. The index i is dropped because the source is assumed to be homogeneous, making transition probabilities independent of i. Furthermore, let $N_{jm}^{(\alpha)}$ denote the number of times that state s_j immediately follows state s_m in an α length sequence of states. Then

$$I_k = -\frac{1}{k}\left\{\log Pr[s(n)] + \sum_{j=1}^{|\mathbf{S}|} \sum_{m=1}^{|\mathbf{S}|} N_{jm}^{(k-n)} \log Pr(s_j|s_m)\right\} \quad (86)$$

Obviously, I_k depends on the random state sequence only through the randomly selected initial state $s(n)$ and the state pair counts $N_{jm}^{(k-n)}$ for all possible j and m.

In the previous section, we found that for an nth-order homogeneous Markov source,

$$H(\mathbf{M}|\mathbf{M}^\infty) = \bar{H}^t \bar{P}_{stat}$$

$$= \sum_{m=1}^{|\mathbf{S}|} P_{stat\,m} \sum_{j=1}^{|\mathbf{S}|} Pr(s_j|s_m) \log \frac{1}{Pr(s_j|s_m)} \quad (87)$$

where $P_{stat\,m}$ is the stationary probability of being in state s_m. Substituting Equations 86 and 87 into Equation 84 gives

$$|I_k - H(\mathbf{M}|\mathbf{M}^\infty)|$$

$$= \left| -\frac{\log Pr[s(n)]}{k} - \sum_{j=1}^{|\mathbf{S}|} \sum_{m=1}^{|\mathbf{S}|} \left[\frac{N_{jm}^{(k-n)}}{k} - P(s_j|s_m)P_{stat\,m}\right] \log Pr(s_j|s_m) \right| \quad (88)$$

Introduction

Using { } to denote events, it follows that

$$\{|I_k - H(\mathbf{M}|\mathbf{M}^\infty)| \geq \epsilon\} \subset \left\{-\frac{\log Pr(s(n))}{k} \geq \frac{\epsilon}{|\mathbf{S}|^2 + 1}\right\}$$

$$\cup \left[\bigcup_{j=1}^{|\mathbf{S}|}\bigcup_{m=1}^{|\mathbf{S}|}\left\{\left|\frac{N_{jm}^{(k-n)}}{k} - P(s_j|s_m)P_{\text{stat}m}\right||\log Pr(s_j|s_m)| \geq \frac{\epsilon}{|\mathbf{S}|^2 + 1}\right\}\right] \quad (89)$$

since for the left side of Equation 89 to be greater than ϵ, at least one of the $|\mathbf{S}|^2 + 1$ terms on the right side of Equation 89 must exceed $\epsilon/(|\mathbf{S}|^2 + 1)$. Applying inclusion and union bounds of probability to Equation 89 yields

$$Pr\{|I_k - H(\mathbf{M}|\mathbf{M}^\infty)| \geq \epsilon\} \leq Pr\left\{-\frac{\log Pr[s(n)]}{k} \geq \frac{\epsilon}{|\mathbf{S}|^2 + 1}\right\}$$

$$+ \sum_{j=1}^{|\mathbf{S}|}\sum_{m=1}^{|\mathbf{S}|} Pr\left\{\left|\frac{N_{jm}^{(k-n)}}{k} - P(s_j|s_m)P_{\text{stat}m}\right||\log Pr(s_j|s_m)| \geq \frac{\epsilon}{|\mathbf{S}|^2 + 1}\right\} \quad (90)$$

At this point, if $Pr(s_j|s_m)$ and hence $N_{jm}^{(k-n)}$ are zero, the corresponding probability term on the right side is zero and can therefore be dropped from the sum. It is now possible to find conditions under which all remaining terms on the right side of Equation 90 go to zero as k increases.

1. Let P_0 be the smallest nonzero value of $Pr[s(n)]$ for $s(n) \in \mathbf{S}$; then,

$$Pr\left\{-\frac{\log Pr(s(n))}{k} \geq \frac{\epsilon}{|\mathbf{S}|^2 + 1}\right\} = 0 \quad \text{for } k > \frac{(|\mathbf{S}|^2 + 1)|\log P_0|}{\epsilon} \quad (91)$$

2. Notice that

$$\left\{\left|\frac{N_{jm}^{(k-n)}}{k} - P(s_j|s_m)P_{\text{stat}m}\right||\log Pr(s_j|s_m)| \geq \frac{\epsilon}{|\mathbf{S}|^2 + 1}\right\} \subset \left\{\left|\frac{N_{jm}^{(k-n)}}{(k-n)}\right.\right.$$

$$\left.\left.- Pr(s_j|s_m)P_{\text{stat}m}\right| \geq \frac{\epsilon}{(|\mathbf{S}|^2 + 1)|\log Pr(s_j|s_m)|} - \frac{n}{k-n}Pr(s_j|s_m)P_{\text{stat}m}\right\} \quad (92)$$

In more detail, Equation 92 is true because if

$$\left|\frac{a}{k} - p\right|c > d$$

where c, a and d are strictly positive real numbers, $p \geq 0$, and k is a positive integer, then, for any nonnegative integer n for which $k > n \geq 0$, it follows by elementary manipulation that

$$\left|\frac{a}{k-n} - p\right| c > d$$

$$\left|\frac{a}{k-n} - p\right| > \frac{d}{c}$$

$$\left|\frac{a}{k-n} - p\right| + \left(\frac{n}{k-n}\right) p \geq \frac{d}{c}$$

$$\left|\frac{a}{k-n} - p\right| \geq \frac{d}{c} - \left(\frac{n}{k-n}\right) p$$

Equation 92 follows by setting

$$a = N_{jm}^{k-m}$$

$$p = Pr(s_j | s_m) P_{\text{stat}m}$$

$$c = |\log Pr(s_j | s_m)|$$

$$d = \frac{\epsilon}{|\mathbf{S}|^2 + 1}$$

The probability of the event on the right side of Equation 92 goes to zero (and likewise for the event on the left side) if the random variable $N_{jm}^{(k-n)}/(k-n)$ converges in probability to $Pr(s_j|s_m)P_{\text{stat}m}$, i.e., for any $\epsilon' > 0$

$$\lim_{k \to \infty} Pr\left\{ \left| \frac{N_{jm}^{(k)}}{k} - Pr(s_j|s_m) P_{\text{stat}m} \right| > \epsilon \right\} = 0 \tag{93}$$

Hence the proof of the complex asymptotic equipartition property will be complete if Equation 93 holds for all choices of j and m. This requirement is satisfied when the Markov source is *ergodic*. Since irreducible homogeneous n^{th}-order Markov sources are ergodic, the proof of Theorem 1.2a is complete.

Theorem 1.2b. The Number of Typical Sequences: For any $\epsilon > 0$ and $\gamma > 0$, there exists a positive real number $K(\epsilon, \gamma)$ such that the number N_k of ϵ-typical event sequences of length k is bounded by

Introduction

$$(1 - \gamma)e^{k[H(\mathbf{M}|\mathbf{M}^\infty)-\epsilon]} < N_k < e^{k[H(\mathbf{M}|\mathbf{M}^\infty)+\epsilon]}$$

whenever $k > K(\epsilon, \gamma)$, and the entropy is in nats.

Briefly, for large k, the probability that a randomly selected k-tuple is ϵ-typical approaches 1, and the number of ϵ-typical k-tuples is on the order of $e^{kH(\mathbf{M}|\mathbf{M}^\infty)}$.

Proof: Every typical sequence $M(1), \ldots, M(k)$ has a value of I_k satisfying

$$H(\mathbf{M}|\mathbf{M}^\infty) - \epsilon < I_k < H(\mathbf{M}|\mathbf{M}^\infty) + \epsilon \tag{94}$$

or equivalently, using the definition of I_k, in equation 80,

$$e^{-k[H(\mathbf{M}|\mathbf{M}^\infty)+\epsilon]} < Pr[M(1), \ldots, M(k)] < e^{-k[H(\mathbf{M}|\mathbf{M}^\infty)-\epsilon]} \tag{95}$$

where entropy is measured in nats/symbol. Let N_k be the number of ϵ-typical sequences of length k. Summing terms in Equation 95 over the set $(\mathbf{M}^k)_{\text{typ}}$ of ϵ-typical sequences gives

$$N_k e^{k[H(\mathbf{M}|\mathbf{M}^\infty)+\epsilon]} < \sum_{[M(1), \ldots, M(k)] \in (\mathbf{M}^k)_{\text{typ}}} Pr[M(1), \ldots, M(k)]$$

$$< N_k e^{-k[H(\mathbf{M}|\mathbf{M}^\infty)-\epsilon]} \tag{96}$$

The asymptotic equipartition property states that for $\epsilon > 0$, $\gamma > 0$, when $k > K(\epsilon, \gamma)$, the probability that a length k sequence is not ϵ typical is less than γ. Hence

$$1 - \gamma < \sum_{[M(1), \ldots, M(k)] \in (\mathbf{M}^k)_{\text{typ}}} Pr[M(1), \ldots, M(k)] \leq 1 \tag{97}$$

Combining the inequalities in Equations 96 and 97 produces the desired result

$$(1 - \gamma)e^{k[H(\mathbf{M}|\mathbf{M}^\infty)-\epsilon]} < N_k < e^{k[H(\mathbf{M}|\mathbf{M}^\infty)+\epsilon]} \tag{98}$$

Hence, Theorem 1.2b is proved.

Since ϵ and γ may be made arbitrarily small, there is a tendency to say that for large k,

$$N_k \sim e^{kH(\mathbf{M}|\mathbf{M}^\infty)} \quad \text{(not correct)} \tag{99}$$

But it should be noted that the allowable values of k, namely, $k > K(\epsilon, \gamma)$, are functions of ϵ and γ. Thus, we cannot arbitrarily assume that $e^{\pm K(\gamma, \epsilon)\epsilon}$ goes to 1 as ϵ vanishes. However, by taking logarithms and dividing by k, Equation 98 reduces to

$$\frac{\ln(1-\gamma)}{k} + H(\mathbf{M}|\mathbf{M}^\infty) - \epsilon < \frac{\ln N_k}{k} < H(\mathbf{M}|\mathbf{M}^\infty) + \epsilon \tag{100}$$

Then it is legitimate to say that for large k,

$$\frac{\ln N_k}{k} \sim H(\mathbf{M}|\mathbf{M}^\infty) \qquad \text{(correct)} \tag{101}$$

since for arbitrarily small ϵ and γ, k can be increased to make $[\ln(1-\gamma)]/k$ disappear, leaving the arbitrarily small error ϵ in Equation 101. Thus the proper interpretation of the bounds on N_k is that the number of ϵ-typical sequences N_k has an asymptotic exponential rate of growth $[\ln(N_k)]/k$ with increasing k, given by $H(\mathbf{M}|\mathbf{M}^\infty)$.

Exercises

1. Each sequence of k messages has a probability assigned to it that can be used to determine its information content I_k. Is the most probable sequence always an ϵ-typical sequence? Explain your answer.

2. Using Chebychev's inequality from probability theory, develop your own proof of the asymptotic equipartition property for memoryless information sources having finite symbol alphabets.

3. Consider the following functions

$$f_1(k) = e^{ak} + k^b$$
$$f_2(k) = e^{ak} + e^{bk}$$
$$f_3(k) = e^{a(k+b)}$$
$$f_4(k) = k^b e^{ak}$$
$$f_5(k) = b^k e^{ak}$$

where a and b are constants. What are the asymptotic exponential rates of growth of these functions as k increases?

Introduction

4. Clarify the role of the choice of information units in the asymptotic equipartition property and the bounds on a typical sequence count by indicating what changes would have to be made if the entropy $H(\mathbf{M}|\mathbf{M}^\infty)$ were specified in bits.

1.7. The Utility of Information Source Models

1.7.1. Agent 00111 and Language Generation

Despite his reservations about scientists, Agent 00111 was very good at languages and their seemingly infinite subtleties; indeed, these were his trade. Agent 00111 was genuinely interested in the list length connection to entropy but not necessarily for the most honorable of motives: He intended to reverse the process by using computer models to churn out totally bogus reports that appeared to have meaning. If the reports were not too unacceptable, Agent 00111 could claim that the odd phraseology was due to codenames and imperfect decryption and suggest that it would surely profit his customer to purchase the document at a reduced fee to study further. In addition, he had a residual skepticism. If he were pricing on the basis of a model, he wanted to see the model's output.

In any case, Agent 00111 persuaded his scientists to use their models to generate text (see the examples in Section 1.7.2). The results were interesting but a bit inconclusive. Agent 00111 certainly got what he paid for: The simplest models generated text that looked hopeless; the higher models (which cost dearly in computer time) did in fact generate patches of almost plausible text. At that point in his career (the 1960s), Agent 00111 had concluded that computer capability and economics would have to improve before he could embark on such a scam.

Agent 00111 pointed out to his scientists that their models worked only at the lexical level. Could they use other models starting at a higher grammatical level and gradually work down to the choice of words and/or letters?

1.7.2. Language Models and Generation

In Section 1.7.2, Agent 00111's instructions and orders are explored. Some of the implications are very deep and go beyond the scope of this book. We start out by examining the models that have been used to date.

In previous sections, we used n^{th}-order Markov source models having finite numbers of states. Can such a source be used to model a natural language, as, say, English? The answer depends on the use of the model. Suppose, for example, someone wishes to observe alphabet sequences in English. Even a memoryless source would satisfactorily model this aspect of the language. In any n^{th}-order Markov model of English ($n > 2$) the frequency of occurrence of single letters, digraphs, and trigraphs is used. For this purpose, a second-order Markov source based on the transition probabilities of the language should prove adequate for testing. However, a memoryless source could not possibly provide the digraph and trigraph statistics to imitate the language properly.

Section 1.7.1 showed that approximately $2^{nH(\mathbf{M}|\mathbf{M}^\infty)}$ typical sequences of length n occur in a Markov source having entropy $H(\mathbf{M}|\mathbf{M}^\infty)$ bits/symbol. This information is useful in determining the number of distinct n-tuples in the language that an n-tuple processor must be prepared to handle. It takes very little intuition to guess that in testing communication channels as natural language processors, it is necessary for the test source entropy $H(\mathbf{M}|\mathbf{M}^\infty)$ to match the entropy of the natural language. Such a test source (mathematical or real) might be a high-order Markov source.

How does output from a Markov source modeling a natural language look? Suppose we examine the following Markov imitations of English:

Memoryless with English letter frequencies

  RAEEPAO_PA_ADGANS_SEBNSE_CDSIA_MF_

  IL_DMIU_NEAA_MSES_IBIFEOEAOK_NOE

First-order Markov with English digraph frequencies

  FARAIN_T_S_BLS_WAS_WANTHE_

  E_TEN_FRN_D_CTOFANDROGRLE

Second-order Markov with English trigraph frequencies

  IMMIL_DIENTION_NOT_SOCRAR_

  THATERERY_THE_NEW_STATUGHEILIC

In the first case, the single-symbol frequency distribution resembles English as is indicated by the large number of vowels and lack of such low-frequency letters as X, Q, J, etc. In the second example, all digraphs are statistically similar to those of English, since they virtually eliminate vowel strings and tend to set up alternating vowel-consonant structures. In the third example using common English trigraphs, the nonsense words can be pronounced;

Introduction

short English words tend to occur; and single-letter nonsense words are definitely eliminated.

Practical difficulties prevent the use of significantly higher order models: Either the mechanism for generating or describing the output sequence is too cumbersome, or the time required is too large. It may be possible to treat words in a language as symbols and model English, for example, as a low-order Markov source with short-word sequences being typical of English. While this artifice guarantees that the observed words are present in an English dictionary, it still does not produce grammatically correct sentences.

There are at least three levels at which information is generated by a language:

1. Lexicographic information—ordering letters to produce words (spelling)

2. Syntactic information—ordering words to make sentences (grammar)

3. Semantic information—using words and sentences to translate a train of thought

A Markov source of order n will produce letter strings of all lengths. In this sense, the Markov source adequately models the lexical information in a language. Syntactic information modeling is the subject of the formal theory of languages, which we will briefly introduce in this section. Semantic information has not yet been effectively modeled and perhaps, philosophically speaking, a semantic information source cannot be imitated. At the very least, it is an objective of Artificial Intelligence that has not yet been realized.

The formal theory of languages is most simply described as the study of *finite* descriptions of languages and the study of algorithms answering certain questions about the language. This theory has immediate application in compiler design for formal computer languages.

What is grammar? Grammar is the correct ordering of words to make a sentence. To decide whether or not a sequence of words is grammatically correct, we can parse the sequence, i.e., break it down into subsequences that must satisfy grammatical rules, e.g., rules of combination. Symbolically, a sentence might be parsed as shown in Figure 1.10. Without continuing further, we see that a grammatical structure can be thought of as a sequence of substitution operations called *productions*. We started initially with a *nonterminal* symbol (one not present in the final structure called **sentence**). We substituted the concatenation **subject phrase, verb phrase, object phrase** for **sentence**. Then **adjective** and **noun** were substituted for **subject phrase**, *etc.*, and finally, terminal symbols were substituted, such as *brilliant* for **adjective**, *00111* for **noun**, *etc.* It is easy to imagine other productions that could lead to much more elaborate structures.

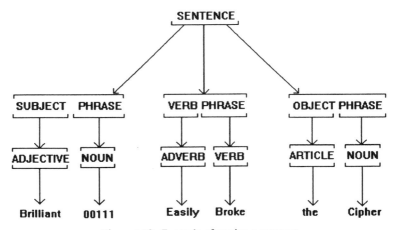

Figure 1.10. Example of parsing a sentence.

An elementary example demonstrates the utility of the grammar concept. Let T be the terminal alphabet, N the nonterminal alphabet, and S the set of starting productions.

Example 1.8. $T = \{0, 1\}$, $N = \{S, S_0, S_1\}$, $S = \{S\}$. P contains the following productions:

1. $S \rightarrow 0SS_0$

2. $S \rightarrow 1SS_1$

3. $S \rightarrow 0S_0$

4. $S \rightarrow 1S_1$

5. $S_0 \rightarrow 0$

6. $S_1 \rightarrow 1$

The language generated by this grammar is the set of all palindromic binary sequences. In this type of grammar, called a context-free grammar, production outputs can always be put in the form of a nonterminal mapped into a terminal followed by a (possibly empty) string of nonterminals. One possible sentence in the language is shown in Figure 1.11.

Introduction

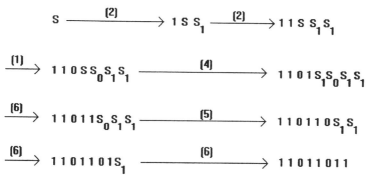

Figure 1.11. Example of the grammar approach.

Example 1.8 illustrates the beauty of the grammar approach to modeling information sources. The palindromic property (all sentences have reversal symmetry) could not be imitated by any finite-order Markov source, yet the grammar is completely described by the *finite* sets T, N, S, and P. Example 1.8 is neither the simplest nor the most complicated conceptually. But it lacks one indispensable property in our study of the information content of source outputs: No probability distribution is defined on the elements of the language. One special method of describing a distribution involves assigning next-production probabilities. Table 1.3 demonstrates this method for Example 1.8.

Assuming that the initial distribution of productions involving the starting variable S is uniform, the probability that the information source will generate the sentence 11011011 is $(0.25)(0.4)(0.3)(0.05)(1.0)(1.0)(1.0)(1.0) = 0.0015$. Here, it is assumed that the next production is always a substitution for the leftmost nonterminal in a partially constructed sentence.

While this description is reminiscent of a Markov chain computation with states corresponding to productions, this is not generally the case. For a context-free grammar to yield finite-length sentences, some production(s) (such as (5) and (6) in Example 1.8) must remove a nonterminal without adding another to the string. When these productions are applied, the next nonterminal to be replaced by substitution cannot always be remembered by any *bounded* memory. Hence next-production distributions are specified for all possible leftmost nonterminals in these situations.

A natural language cannot be perfectly cast in the framework of formal languages. Regardless of how liberally the grammatical rules are written, the adaptivity of a human receiver makes it possible to contrive nongrammatical sentences that convey thoughts. In fact, it is this very property that causes

Table 1.3. Example of Sentence Production

Production	Next nonterminal	Next production	From probability
(1) $S \to 0SS_0$	S	(1)	0.7
		(2)	0.2
		(3)	0.05
		(4)	0.05
(2) $S \to 1SS_1$	S	(1)	0.3
		(2)	0.4
		(3)	0.2
		(4)	0.1
(3) $S \to 0S_0$	S_0	(5)	1.0
(4) $S \to 1S_1$	S_1	(6)	1.0
(5) $S_0 \to 0$	S_0	(5)	1.0
	S_1	(6)	1.0
(6) $S_1 \to 1$	S_0	(5)	1.0
	S_1	(6)	1.0

language structure to change over a period of time. As an example, in English, the statements (1) the spy was handed the codebook, and (2) the codebook was handed to the spy, are both grammatically correct *and have the same meaning!* A *formal* grammatical theory would require the roles of spy and codebook to be interchanged from one sentence to the other.

Natural languages also preserve certain expressions that embody archaic (or extinct) grammatical constructions. How would you parse the sentence, "Oh, woe is me."? The sentence does *not* express identity between woe and me, since archaic English requires I (nominative case) rather than me for such an interpretation. In fact, me is a *dative of reference,* and the sentence means, "Alas, woe is mine."

Exercises:

1. Examples of nth-order Markov source outputs imitating English text are generated as follows:

    (a) Obtain a large text written in English.

    (b) Select an *n*-tuple $m_1 m_2 \cdots m_n$ at random from *n*-tuples in the text.

    (c) Turn to a part of the text far enough away from those previously

Introduction

selected to assure the independence of the new location's contents from those previously used.

(d) Assuming that the output constructed thus far ends in $m_{i+1}m_{i+2} \cdots m_{i+n}$, locate a copy of this n-tuple in the text.

(e) Let the letter immediately following the copy equal m_{i+n+1} in your output stream.

(f) Return to (c) as long as you wish to generate new output.

Generate a set of examples similar to those in this section using a foreign language text. Can your classmates identify the language from the output that you generated?

2. Find three different syntactic interpretations of the sentence "May flies like the wind."

Can you find two semantic interpretations corresponding to one of these syntactic interpretations?

3. In the example of palindromic binary sequences, the production sequence that generated the sentence 11011011 had probability 0.0015.

(a) How much information is gained about the production sequence when 11011011 is obtained and known to be a complete sentence?

(b) How much information is gained about the production sequence when 11011011 is obtained but it is not known whether or not it is a complete sentence?

(c) Is information about the grammatical structure (i.e., production sequence) what the receiver really desires, or does the receiver simply desire the source output sequence? *Hint:* Is there a one-to-one relationship between the paths through an auxiliary Markov chain and sentences in the grammar?

1.8. Notes

Using the logarithm of the number of possible messages to measure information (or uncertainty) was first suggested by Hartley (1928). Shannon (1948) constructed a sound mathematical theory of communication based on using logarithmic information measures and probabilistic source descriptions. At the same time, Wiener (1948, 1949) was developing a theory that deserves major credit for moving the synthesis of communication systems into the domain of statistical theory. Shannon's theory seems most easily adaptable to discrete (digital) systems of communication while Wiener concentrated his work on problems in time series analysis and time-continuous stochastic processes. Undoubtedly, the research of these two eminent scholars is the foundation of modern communication theory.

Shannon deserves credit for introducing the entropy concept into communication theory. This basic concept is treated by most texts on information theory, e.g., those listed in the References.

For the interpretation of entropy in statistical hypothesis-testing situations, see Kullback (1959).

In addition to the axiomatic derivations of entropy expressions by Shannon (1948), Khintchin (1953), and Lee (1964), other derivations have been supplied by Feinstein (1958) and Tverberg (1958). Fano (1961, chap. 2) derives a mutual information measure from a basic set of axioms. Renyi (1961, 1965) gives an axiomatic derivation of the entropy expression and furthermore investigates other possible methods of measuring information.

The treatment of properties of finite Markov chains is considered in many texts on probability theory, including Feller (1950), Parzen (1962), Kemeny and Snell (1960), Prabhu (1965), Karlin and Taylor (1975, 1981), and Kleinrock (1975). An *information-theoretic* proof of the convergence of the state probability vector to a steady-state distribution was given by Renyi (1961). The state diagram for the nth-order binary Markov process is known as the de Bruijn Graph B_n. For more on the properties of this graph, see Golomb (1967).

For further textbook discussion of ergodicity and the asymptotic equipartition property of information sources, see Ash (1965, chap. 6), Gallager (1968, chap. 3), and Wolfowitz (1961). Comprehensive and more general treatments are found in Billingsley (1965), Pinsker (1964), and Csiszar and Korner (1981). The fundamental result, namely, the asymptotic equipartition property, is due to McMillan (1953), and Breiman (1957) generalized the result to finite-alphabet ergodic sources.

The statistical properties of natural languages have interested cryptanalysts for centuries. Early studies of English have been carried out by Shannon (1951), Burton and Licklider (1955), and Hurd (1965).

The concept of transformational grammars has arisen from the work of Chomsky (1956, 1959). For a review of the theory of formal languages and its connection to automata theory, see Aho and Ullman (1968). Problems in modeling natural languages formally were first discussed in Chomsky (1969) and Fodor and Katz (1964).

References

N. M. Abramson. 1963. *Information Theory and Coding.* McGraw-Hill, New York.

A. V. Aho and J. D. Ullman. 1968. "The Theory of Languages." *Math. Systems Theory* **2**: 97–125.

R. L. Adler, D. Coppersmith, M. Hasner. 1983. "Algorithms for Sliding Block Codes—An Application of Symbol Dynamics to Information Theory." *IEEE Trans. Inform. Theory:* IT-29: 5–22, 1983: 5–22.

R. B. Ash. 1965. *Information Theory.* Interscience, New York.

P. Billingsley. 1965. *Ergodic Theory and Information.* Wiley, New York.

R. E. Blahut. 1987. *Principles and Practice of Information Theory.* Addison Wesley, New York.

L. Breiman. 1957. "The Individual Ergodic Theory of Information Theory." *Ann. Math. Stat.* **28**: 809–11; errata in *Ann. Math. Stat.* **31**: 809–10.

G. Burton and J. C. R. Licklider. 1955. "Long-Range Constraints in the Statistical Structure of Printed English." *Amer. Jour. Psych.* **68**: 650–53.

N. Chomsky. 1956. "Three Models for the Description of Languages." *IEEE Trans. Inform. Theory* **2**: 113–24.

———. 1959. "On Certain Formal Properties of Grammars." *Inf. Contr.* **2**: 137–67.

———. 1969. *Aspects of the Theory of Syntax.* MIT Press, Cambridge, MA.

T. M. Cover and J. A. Thomas. 1991. *Elements of Information Theory.* Wiley, New York.

I. Csiszar and T. Korner. 1981. *Information Theory: Coding Theorems for Discrete Memoryless Systems.* Academic Press, New York.

R. M. Fano. 1961. *Transmission of Information: A Statistical Theory of Communication,* MIT Press and Wiley, New York.

A. Feinstein. 1958. *Foundations of Information Theory.* New York: McGraw-Hill.

W. Feller. 1950. *An Introduction to Probability Theory and Its Applications.* Vol. 1. Wiley, New York.

J. A. Fodor and J. Katz. 1964. *The Structure of Language.* Prentice-Hall.

R. G. Gallager. 1968. *Information Theory and Reliable Communication.* Wiley, New York.

S. W. Golomb. 1967. *Shift Register Sequences.* Holden-Day, San Francisco. Revised edition, Aegean Park Press, Laguna Hills, California, 1982.

S. Guiasu. 1976. *Information Theory with Applications.* McGraw-Hill, New York.

R. V. L. Hartley. 1928. "Transmission of Information," *Bell Sys. Tech. J.* **7**: 535–63.

W. J. Hurd. 1965. "Coding for English As a Second-Order Markov source with New Trigram Statistics." Report No. 24. Electrical Engineering Dept., University of Southern California.

F. Jelinek. 1968. *Probabilistic Information Theory: Discrete and Memoryless Models.* McGraw-Hill, New York.

S. Karlin and H. M. Taylor. 1975. *A First Course in Stochastic Processes.* Academic Press, San Diego, CA.

———. 1981. *A Second Course in Stochastic Processes.* Academic Press, San Diego, CA.

J. G. Kemeny, J. L. Snell. 1960. *Finite Markov Chains.* Van Nostrand, Princeton, NJ.

A. L. Khintchin. 1953. "The Entropic Concept of Probability." *Uspekhi Mat. Nauk.* **8**: 3–20.

———. 1957. *Mathematical Foundations of Information Theory.* Dover, New York.

L. Kleinrock. 1975. *Queueing Systems.* Wiley, New York.

S. Kullback. 1959. *Information Theory and Statistics.* Wiley, New York.

H. J. Larson and B. O. Shubert. 1979. *Probabilistic Models in Engineering Sciences.* Vol. 2. Wiley, New York.

P. M. Lee. 1964. "On the Axioms of Information Theory." *Ann. Math. Stat.* **35**: 415–18.

B. Mcmillan. 1953. "The Basic Theorems of Information Theory." *Ann. Math. Stat.* **24**: 196–219.

M. Mansuripur. 1987. *Introduction to Information Theory.* Prentice Hall, Englewood Cliffs, NJ.
R. J. McEliece. 1977. *The Theory of Information and Coding.* Addison Wesley, Reading, MA.
A. Papoulis. 1965. *Probability, Random Variables, and Stochastic Processes.* McGraw-Hill, New York.
E. Parzen. 1962. *Stochastic Processes.* Holden-Day, San Francisco, CA.
M. S. Pinsker. 1964. *Information and Information Stability of Random Variables and Processes.* Holden-Day, San Francisco, CA.
N. U. Prabhu. 1965. *Stochastic Processes.* Holden-Day, San Francisco, CA.
A. Renyi. 1961. "On Measures of Entropy and Information." *Fourth Berkeley Symposium on Math. Stat. and Prob.* **1**: 547–61.
———. 1965. "On the Foundations of Information Theory." *RISI* **33**, no. 1: pp. 1–14.
F. M. Reza. 1961. *An Introduction to Information Theory.* McGraw-Hill, New York.
C. E. Shannon. 1948. "A Mathematical Theory of Communication." *BSTJ* **27**: 379–423, 624–56.
———. 1951. "Prediction and Entropy of Printed English." *BSTJ* **30**: 50–65.
G. Strang. 1988. *Linear Algebra and Its Applications.* 3d ed. Harcourt Brace Jovanovich, Orlando, FL.
H. Tverberg.,1958. "A New Derivation of the Information Function." *Math. Scand.* **6**: 297–98.
N. Wiener. 1948. *Cybernetics.* Wiley.
———. 1949. *Extrapolation, Interpolation, and Smoothing of Stationary Time Series.* MIT Press, Cambridge, MA, and Wiley, New York.
J. Wolfowitz. 1961. *Coding Theorems of Information Theory.* Springer: Berlin-Heidelberg.

2

Coding for Discrete Noiseless Channels

2.1. The Problem

2.1.1. Agent 00111's Problem

Agent 00111 was a legendary master of espionage because he had found answers (sometimes only partial answers) to several espionage dilemmas. One answer, discussed in Chapter 1, was an accounting and budgeting system for the amount of delivered information. However, the same principles could also be applied to other problem areas, such as communicating the information he received.

On one hand, the more relevant the information Agent 00111 sent to his client(s), the more money he made, since he was paid per unit of uncertainty eliminated and the more his reputation was enhanced. However, his chances of being detected behind enemy lines also increased with the quantity of data he sent. For example, Agent 00111 often wrote messages in code on the back of postage stamps that he used on postcards to his poor ailing mother in Hampstead, England. Of course, you can only write so much on a stamp, even if microdot-writing technology were readily available. If he was too successful in his mission, Agent 00111 had to send more postcards, and there were limits to how dutiful a son could be without attracting attention. If nothing else, it was imperative for Agent 00111 to restrict his communications to the absolutely bare essentials. Thus, Agent 00111 had to find a way of extracting information from his source material and placing it into a format that minimized the quantity of material that had to be sent. This problem is the subject of Chapter 2. However, Agent 00111 also had to protect transmitted data from channel corruption. This was rarely a major concern when using postal stamps, since his postcards were not frequently lost, trampled underfoot, or chewed by dogs, although ink from the cancellation mark once rendered some of his message illegible. In other scenarios, for example, jammed radio transmissions, restoring degraded text was of fundamental importance to him, and this is the subject of Chapters 5–7.

Agent 00111 was temporarily happy to forget about data corruption. The immediate problem was how to condense his communications to the absolute minimum, yet enable his client to extract the data unambiguously. A secondary problem arose if it were impossible to meet the latter requirement. How could Agent 00111 transmit as much information as possible with a minimum of ambiguity or distortion? Solutions to these problems involve *entropy*.

2.1.2. Problem Statement

Many real-life communication channels are relatively free of noise but frequently break down. One useful criterion for determining whether or not a channel is free of noise involves comparing channel interference to equipment reliability. We can legitimately maintain that a channel is free of noise if the expected time between errors made determining the transmitted symbol at the receiver is much greater than the average operation time before catastrophic transmitter or receiver failure. In Chapter 2, we assume that Agent 00111 communicates through a noiseless channel in this sense, since coded information arrives at his headquarters exactly as it was sent, providing he is not discovered.

A *discrete communication channel* is modeled by specifying (1) an *input alphabet* $\mathbf{X}$, (2) an *output alphabet* $\mathbf{Y}$, and (3) a conditional probability distribution $Pr(\bar{y}|\bar{x})$ defined for all n-tuples $\bar{x} \in \mathbf{X}^n$, all n-tuples $\bar{y} \in \mathbf{Y}^n$, and for all choices of n. The channel is said to be *noiseless* if for each output n-tuple $\bar{y} \in \mathbf{Y}^n$, $Pr(\bar{x}|\bar{y})$ is either 0 or 1 for all $\bar{x} \in \mathbf{X}^n$. Equivalently, a channel is noiseless if the input n-tuple $\bar{x}$ can be determined accurately from the output n-tuple $\bar{y}$ in all cases. Thus, the receiver can perform the inverse $\bar{y} \rightarrow \bar{x}$ of the noiseless channel, mapping $\bar{x} \rightarrow \bar{y}$, and the combined transmitter–

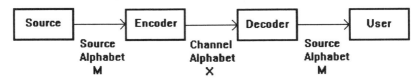

Figure 2.1. A communication system containing a noiseless channel.

Coding for Discrete Noiseless Channels

Table 2.1. Example of a Binary Block Code

Source message	Code word
M_1	0
M_2	1
M_3	10

channel–receiver processing is an identity mapping. Since the identity mapping can be absorbed into the encoder or decoder, we can consider an equivalent system, such as the one in Figure 2.1. The role of the encoder–decoder combination in this situation is to map message sequences into sequences from the permitted set of channel symbols and then back to message sequences. To convey the source message sequence to the user, the encoder mapping must be constructed so that the decoder can always perform the inverse mapping without ambiguity.

As we have implied, a *code* is a mapping of source message sequences into code-symbol sequences. The simplest type of mapping that translates a source message into a code-symbol sequence is a *block code*. Table 2.1 shows a binary code for a source alphabet of three symbols. If the code is a one-to-one mapping (as in Table 2.1), the code is said to be *nonsingular*. Obviously, a single transmitted source message can always be determined from the code received if the code is nonsingular.

Suppose the decoder has been asked to decode a message consisting of two source symbols. The new code formed by placing two words from the original code in succession is called the *second extension of the code*. If the second extension of the code is nonsingular, then decoding is unique. Second and third extensions of the code shown in Table 2.1 are presented in Tables 2.2 and 2.3, respectively. In this example, the second extension is nonsingular, but the third extension contains two identical sequences corresponding to $M_2 M_1 M_3$ and $M_3 M_2 M_1$. Thus, if the decoder is told only that 1010 represents a sequence of three code words, the inverse mapping cannot be performed without ambiguity. In fact, most communication systems are required to work with *less* information than Table 2.3 contains. For example, the receiver is seldom fortunate enough to know the number of transmitted code words and might not even know exactly where the stream starts; Agent 00111, for instance, would not write on his postcard, "This is a 22-word coded message starting exactly 0.4 cm from the top left of the stamp."

Table 2.2. Second Extension of the Binary Code in Table 2.1

Symbol from M^2	Second-extension code word
M_1M_1	00
M_1M_2	01
M_1M_3	010
M_2M_1	10
M_2M_2	11
M_2M_3	110
M_3M_1	100
M_3M_2	101
M_3M_3	1010

Code dictionaries that can always perform the inverse mapping without being told the number of code words are called *uniquely decodable codes*. A uniquely decodable code is called:

A U_F *dictionary* if all *finite* code-word sequences can be uniquely decoded.

A U_S *dictionary* if all *semiinfinite* code word sequences can be uniquely decoded.

A U_I *dictionary* if all *doubly infinite* code word sequences can be uniquely decoded.

The structure of these dictionaries differs, since the sequences on which their decoders perform inverse mappings contain different amounts of synchronization information.

Table 2.3. Third Extension of the Binary Code in Table 2.1

Symbol from S^3	Third-extension code word	Symbol from S^3	Third-extension code word	Symbol from S^3	Third-extension code word
$M_1M_1M_1$	000	$M_2M_1M_1$	100	$M_3M_1M_1$	1000
$M_1M_1M_2$	001	$M_2M_1M_2$	101	$M_3M_1M_2$	1001
$M_1M_1M_3$	0010	$M_2M_1M_3$	1010	$M_3M_1M_3$	10010
$M_1M_2M_1$	010	$M_2M_2M_1$	110	$M_3M_2M_1$	1010
$M_1M_2M_2$	011	$M_2M_2M_2$	111	$M_3M_2M_2$	1011
$M_1M_2M_3$	0110	$M_2M_2M_3$	1110	$M_3M_2M_3$	10110
$M_1M_3M_1$	0100	$M_2M_3M_1$	1100	$M_3M_3M_1$	10100
$M_1M_3M_2$	0101	$M_2M_3M_2$	1101	$M_3M_3M_2$	10101
$M_1M_3M_3$	01010	$M_2M_3M_3$	11010	$M_3M_3M_3$	101010

Coding for Discrete Noiseless Channels

In a *finite sequence* of words, the location of the first symbol in the first code word and the last symbol of the last code word is known to the decoder. These positions are known as *points of word synchronization* in the sequence. In a *semi-infinite sequence* of codewords, the decoder knows *a priori* only the initial point of word synchronization. In a *doubly infinite sequence* of codewords, no synchronization information is initially available. For example, Agent 00111 might switch on a radio receiver in the middle of a lengthy broadcast. The most fundamental problem in coding for the noiseless channel is designing nonsingular codes (U_F, U_S, and U_I dictionaries, as the case may be) that allow a simple method for determining all points of word synchronization (i.e., all code word beginnings) in a code-word sequence.

Exercise

1. Is the statement, "A dictionary is U_F if and only if all finite extensions of the code are nonsingular" true? Explain.

2.2. An Algorithm for Determining Unique Decodability in the U_F and U_S Cases

A reasonable approach to studying unique decodability is to begin by asking what problems can occur in decoding a semi-infinite sequence of codewords. Table 2.4 gives a finite sequence of code words and the code used to construct the sequence. Suppose we try to decode this sequence and see where problems arise. Since we know the location of the beginning of the first word and in this example, that it starts with zero, we deduce that the source message sequence begins with a. Likewise, the second word in the sequence is uniquely decoded as c. At this point, the next symbols can be interpreted in several ways, including the two ways shown in Figure 2.2.

Table 2.4. A Code Example

Source message	Code word	
a	0	A sequence of code words:
b	10	01101010110...
c	110	↑ point of word synchronization
d	101	

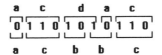

Figure 2.2. Two interpretations of a coded sequence.

Exercise

1. Can you find other interpretations?

This is not a U_F dictionary code, since we have located a finite sequence that is not uniquely decodable. The decoding problem stems from the fact that the code word for b, 10, is a *prefix* of the code word for d, 101; and in turn, when the prefix 10 is deleted, the *suffix* for d, namely 1, is the prefix of several code words—10, 110, and 101. Hence, 1010, 10110, and 10101 are all ambiguous. If we continue this procedure, deleting 1 as a prefix, we obtain 0, 10, and 01 as possible code-word beginnings, which would still render decoding ambiguous. For the sequence we constructed, the code word 0 turned out to be a suffix. It is obvious that when a code word also appears as a suffix in this analysis, two different decodings are possible for a finite sequence of symbols.

The following algorithm (usually referred to as the Sardinas–Patterson algorithm) systematically checks all possible ambiguous sequence beginnings to establish the unique decodability of finite sequences. The algorithm works by (1) sequentially generating a new table from an old table, (2) testing the table, and if the test is passed, (3) repeating the procedure. Finite code word sequences are uniquely decodable if the test is always passed. In this algorithm, we use the term *suffix* to mean the remainder of a word or word segment when a specified prefix is removed. Quantities designated Seg are sets of words or word segments. We first describe how to generate the tables and then the testing method.

1. Let Seg 0 be the collection of all code words, i.e., words in the code.

2. In Seg 1, list the suffixes of code words with members of Seg 0 (code words) as prefixes.

3. Construct Seg k, $k > 1$, using the following inductive process:

    a. In Seg k, list suffixes of members of Seg $k - 1$ having members of Seg 0 as prefixes.

    b. In Seg k, list suffixes of members of Seg 0 having members of Seg $k - 1$ as prefixes.

Coding for Discrete Noiseless Channels

Test: If for any $k > 0$,

Seg k contains a complete code word, then the code is not uniquely decodable in any sense.

If no Seg k other than Seg 0 contains a code word, then the code forms a U_F dictionary.

Figures 2.3 and 2.4 show the logic for the algorithm.

The Seg table for our example is shown in Figure 2.5. Seg table calculations could have been terminated as soon as code word 0 appeared in Seg 2. However, by continuing the table, we can enumerate all basic ambiguously decodable finite sequences. Arrows in Figure 2.5, indicating which element of Seg($k - 1$) was used to obtain an element in Seg k, aid in the enumeration process. Figure 2.6 shows ambiguously decodable finite sequences through Seg 4. Both the Seg table and Figure 2.6 could be continued indefinitely if desired. It is obvious that a nonuniquely decodable finite sequence of words exists for a code if and only if some Seg k, $k \neq 0$, contains a code word.

Suppose we are given a code table and wish to determine in what sense (if any) the code is uniquely decodable. We can use the Seg table approach to enumerate all possibilities. In doing so, we observe three possible outcomes:

1. A code word appears in Seg k, $k > 0$.

2. No code word appears, but the construction terminates with an empty Seg.

3. No code word appears, but the construction continues indefinitely.

If Possibility 1 is true, then the code is not a U_F dictionary, since a nonuniquely decodable sequence can be constructed from the Seg table. If Possibility 2 is true, the code is U_F, since the existence of any finite ambiguously decodable sequence would have caused a code word to appear in some Seg k, $k > 0$, in the code table. In addition, since Possibility 2 terminates in an empty Seg, a semi-infinite ambiguous sequence cannot exist, and the code under consideration is also a U_S code. By reconstructing the sequences from the Seg table, it is obvious that Possibility 3 is due to the existence of a semiinfinite ambiguously decodable sequence whose two interpretations never agree on a terminal point of word synchronization. Thus, Possibility 3 implies that the code is U_F, but not U_S.

A very simple example of Possibility 3 is shown in Table 2.5. The Seg table for Table 2.5 is given in Figure 2.7a, and by reconstruction, the nonuniquely decodable semi-infinite sequence is shown in Figure 2.7b. If the sequence in Figure 2.7 were finite in length, word synchronization could be established by decoding backward from the end of the sequence.

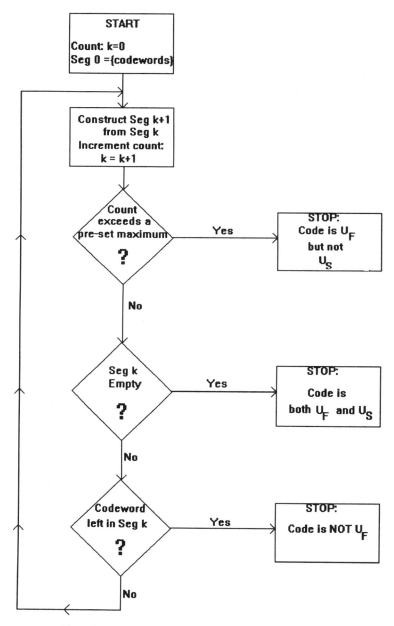

Figure 2.3. Flow chart for the Sardinas–Patterson algorithm.

Coding for Discrete Noiseless Channels

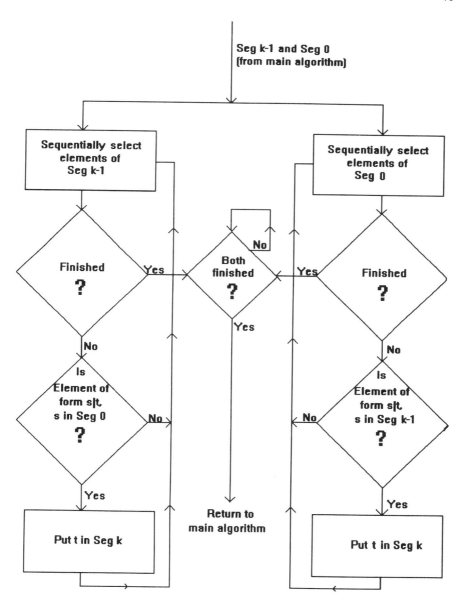

Note: if x = (a,b,.......,c) and y = (r,s,.........,t), x|y = (a,b,....,c,r,s,.......t)

Figure 2.4. Flow chart for Seg k construction.

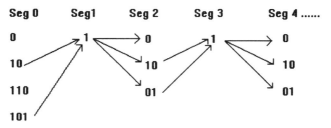

Figure 2.5. Example of a Seg table.

The number of Segs required to complete the Sardinas–Patterson algorithm is bounded, since only a finite number of Segs, say, K, can be formed from code-word suffixes. This means that if an empty Seg does not occur before K Segs have been calculated, one Seg must be repeated twice, and thus Seg k = Seg($k + n$) for some k and n. By induction, Seg($k + i$) = Seg($k + n + i$) for $i > 0$, and the Seg table is periodic. Thus, it takes at most K Segs to determine whether Possibility 1, 2, or 3 holds.

Exercises

1. Which of the codes in Table 2.6 are U_F? U_S? Give counterexamples when possible. Are any of these codes U_I dictionaries?

2. Can you derive an upper bound for K, the number of distinct Segs that can be formed from a U_F code, in terms of appropriate code parameters?

3. (a) If the order of all symbols in all words of a U_F code are reversed, is the new reversed code a U_F code?

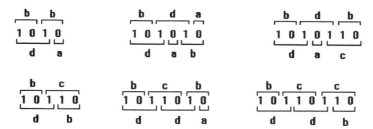

Figure 2.6. Ambiguous sequences in the Seg table example.

Coding for Discrete Noiseless Channels

Table 2.5. Example of Code That Is U_F but Not U_S

Source symbol	Code word
a	0
b	01
c	11

(b) If a U_S code is reversed as in (a), how would you classify the new code? Explain.

4. Can either of the following two dictionaries produce finite ambiguous sequence? If so, exhibit an ambiguous sequence.

(a) AA   (b) AA
    AAB   AB
    ABB   ABB
    BAB   BA
    BABA   BBBA

(a)

```
Seg 0      Seg 1        Seg 2        Seg 3
  0          1 ——————→ 1 ——————→ 1 ——————→
  0 1      ↗
  0 1 1
```

(b)

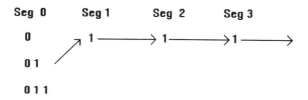

Figure 2.7. Another example of a Seg table.

Table 2.6. Code Examples for Exercise 1

Code A	Code B	Code C	Code D
0	1	0	100
10	01	01	110
01	010	011	1001
11	100	0111	11000

5. Segment tables look very much like trellis diagrams for Markov sources, with table elements as states. Ignoring for the moment the assignment of probabilities to the transitions, convert the trellislike diagram of a Seg table into the equivalent *state* diagram (a directed graph) of the Seg table. You will probably have to include an empty state.

   (a) Given such a segment state diagram, how would you decide whether a code is a U_F or U_S dictionary?

   (b) Sketch the Segment state diagram for the following code:

   $$
   \begin{array}{l}
   0\ 1 \\
   1\ 0\ 0 \\
   1\ 0\ 0\ 0 \\
   1\ 0\ 1\ 0 \\
   1\ 0\ 0\ 0\ 0 \\
   0\ 0\ 0\ 1\ 1\ 0 \\
   1\ 0\ 1\ 1\ 1\ 1 \\
   1\ 1\ 1\ 1\ 0\ 1
   \end{array}
   $$

   (c) Apply your decision method in (a) to the diagram in (b).

   (d) Are there any obvious advantages to the Segment state diagram approach?

2.3. A Simple Coding Theorem for Fixed-Rate Sources

The Sardinas–Patterson algorithm has one obvious corollary; namely, codes with all words of equal length are U_S dictionaries (no code words are

Coding for Discrete Noiseless Channels

prefixes or suffixes to any other code word). These simple codes are used when the information source emits messages at a given rate, say, one message every T_S seconds, and the channel accepts a symbol every T_c seconds. Distinct sequences of k source messages are mapped into distinct sequences of n channel symbols so that

$$nT_c = kT_S = m_o \text{lcm}(T_c, T_S) \qquad (1)$$

where m_o is an integer, and lcm stands for least common multiple. Such sources are called *fixed rate*. Fixed rate sources equate time durations for the following two reasons: (1) If a code word had a shorter duration than its corresponding message sequence, then the channel would at some time lack an input and be forced to transmit a pause. But this implies that we have pauses as part of the channel input alphabet and there exists one more input symbol available than we had assumed. Instead, we assume pause is already counted in the alphabet of channel symbols. (2) If a code word is longer in duration than its corresponding message, then a sufficiently long sequence of short-duration messages will cause any fixed finite amount of storage built into the encoder to overflow.

Under what conditions are there enough code words (i.e., sequences of n channel symbols) to ensure that every sequence of k source messages is represented in the channel by a *distinct* code word? Since the number of messages in the source alphabet is $|\mathbf{M}|$ and the number of symbols in the channel alphabet is $|\mathbf{X}|$, then distinct representation of message sequences is possible if

Number of message sequences

$$= |\mathbf{M}|^k \leq |\mathbf{X}|^n = \text{number of code words.} \qquad (2)$$

Taking the log of Equation 2 and applying Equation 1 yields

$$|\mathbf{M}|^k \leq |\mathbf{X}|^n \Leftrightarrow \frac{\log|\mathbf{M}|}{T_S} \leq \frac{\log|\mathbf{X}|}{T_c} \qquad (3)$$

Obviously, every sequence of k messages can be encoded into a distinct sequence of n code symbols when the right side of Equation 3 is satisfied.

This settles the question of when there are enough code words. In Agent 00111's example, the question is settled when his information rate, i.e., his intelligence rate, is less than his capacity, measured in terms of the number of letters per week and the amount of data that can be placed on a stamp.

However, Agent 00111 lives in an imperfect world: What happens when there are more message sequences than code words available in a given interval or to put it more bluntly, when his intelligence exceeds his ability to transmit it? The answer is deceptively simple: Assign all but one of the available code words to the most frequently used sequences and assign the remaining code word to a universal "cannot encode" symbol denoted as $\bar{x}_a$.

For example, Agent 00111 may be asked to supply answers to 20 related questions about a country's internal politics. Each answer may have one of 10 preassigned values, giving 10^{20} possible answer sets, whereas Agent 00111 has only 10^6 possible messages. Fortunately for Agent 00111, all answer sequences are seldom equally likely. For instance, suppose there are two questions—what is the status of the deputy party chief and who has been appointed head of project Z? If the answer to the first question is the deputy party chief has been shot, it is unlikely that he has been appointed head of project Z. In other words, not every answer sequence is equally likely and relationships between questions work in Agent 00111's favor. Note, however, that receipt of the cannot encode signal is news in itself—something unexpected has happened. Note secondly that genuinely unrelated questions and equally likely answer sequences really pose a problem to Agent 00111.

The real question for Agent 00111, however, is how much information does he lose by having to discard some answer sequences? To answer this, we must return to the Asymptotic Equipartition theorem in Section 1.6 to reinterpret results for the set of possible answer sequences. With this strategy in mind, we again ask what happens when there are more message sequences than code words available in a given interval? One approach is to encode as many message sequences as possible into distinct code words, then transmit the code word $\bar{x}_a$ when any of the remaining ambiguously coded messages occur. If the most likely messages are encoded distinctly, this method yields the least probability of transmitting $\bar{x}_a$ and having the decoder confront an ambiguous code word. Let us require the probability of an ambiguous message to be less than α. The asymptotic equipartition property states that for any $\alpha > 0$ and $\epsilon > 0$, there exists a number $K(\alpha, \epsilon)$ such that a source k-tuple of messages will be ϵ-typical with probability greater than $1 - \alpha$ when $k > K(\alpha, \epsilon)$. Hence, if we encode the N_k ϵ-typical sequences in *distinct* code words, and map all nontypical sequences into the n-tuple $\bar{x}_a$, the probability of transmitting $\bar{x}_a$ will be at most α for large enough k. The total number of code words required for this task is $N_k + 1$. Furthermore N_k is upper bounded by

$$N_k < e^{k[H(\mathbf{M}|\mathbf{M}^\infty)+\epsilon]} \tag{4}$$

where $H(\mathbf{M}|\mathbf{M}^\infty)$ is the entropy of the source in nats/source symbol. The desired encoding can then be accomplished whenever

Coding for Discrete Noiseless Channels

$$e^{k[H(\mathbf{M}|\mathbf{M}^\infty)+\epsilon]} + 1 \leq |\mathbf{X}|^n = e^{n \log_e |\mathbf{X}|} \tag{5}$$

and $k > K(\alpha, \epsilon)$. This is equivalent to requiring the *exponential growth rate of the set of ϵ-typical sequences to be less than the exponential growth rate of the set of code words*. Taking logarithms in Equation 5 and choosing epsilon judiciously gives the equivalent requirement that

$$kH(\mathbf{M}|\mathbf{M}^\infty) < n \log_e |\mathbf{X}| \tag{6}$$

Hence, ϵ-typical sequence encoding is a valid and reliable technique whenever

$$\frac{H(\mathbf{M}|\mathbf{M}^\infty)}{T_S} < \frac{\log|\mathbf{X}|}{T_c} \tag{7}$$

when the base of the logarithm on the right side corresponds to information units of $H(\mathbf{M}|\mathbf{M}^\infty)$. When the source is memoryless and uses equally likely messages, Equation 7 is reduced to Equation 3.

From experience, Agent 00111 was aware of Equations 5, 6, and 7. He was surprised at first but then found it increasingly logical and useful to be able to take the statistics or entropy of a language and to translate these and a given set of communication constraints into a given amount of information lost to his client. Equations 5–7 allowed Agent 00111 to cut his losses in an imperfect environment.

However, a final question remains to be answered: How well can we encode if

$$\frac{\log|\mathbf{X}|}{T_c} < \frac{H(\mathbf{M}|\mathbf{M}^\infty)}{T_S}?$$

The answer, which we will derive from information theoretic considerations, is not very well at all, a fact known only too well by Agent 00111. The formal answer can be derived as follows.

Suppose we examine the mutual information between the set $\mathbf{M}^k$ of source message sequences and the set $\mathbf{X}^n$ of code words.

$$I(\mathbf{M}^k; \mathbf{X}^n) = H(\mathbf{M}^k) - H(\mathbf{M}^k|\mathbf{X}^n) \tag{8}$$

Obviously, $I(\mathbf{M}^k; \mathbf{X}^n)$ can be bounded from above by the entropy $H(\mathbf{X}^n)$ of one of the extension alphabets under consideration. In turn, the entropy of the code is upper bounded by $n \log|\mathbf{X}|$; thus,

$$I(\mathbf{M}^k; \mathbf{X}^n) \leq n \log|\mathbf{X}| \tag{9}$$

The mutual information can also be bounded from below, using the following inequality in conjunction with Equation 8:

$$H(\mathbf{M}^k) \geq kH(\mathbf{M}|\mathbf{M}^\infty) \tag{10}$$

This yields

$$kH(\mathbf{M}|\mathbf{M}^\infty) - H(\mathbf{M}^k|\mathbf{X}^n) \leq I(\mathbf{M}^k; \mathbf{X}^n) \leq n \log|\mathbf{X}| \tag{11}$$

Now let $\mathbf{A}$ denote the set of messages encoded ambiguously in the code word $\bar{x}_a$; then

$$H(\mathbf{M}^k|\mathbf{X}^n) = Pr(\mathbf{A})H(\mathbf{M}^k|\bar{x}_a) + \sum_{\bar{x} \in \mathbf{X}^n - \{\bar{x}_a\}} Pr(\bar{x})H(\mathbf{M}^k|\bar{x}) \tag{12}$$

Since all code words that do not equal $\bar{x}_a$ represent distinct message sequences for which there is no uncertainty, $H(\mathbf{M}^k|\bar{x})$ is zero for all $\bar{x} \neq \bar{x}_a$. Of course, $H(\mathbf{M}^k|\bar{x}_a)$ is bounded from above in the following manner:

$$H(\mathbf{M}^k|\bar{x}_a) \leq k \log|\mathbf{M}| \tag{13}$$

Substituting Equations 9–13 into Equation 8 gives

$$n \log|\mathbf{X}| \geq k[H(\mathbf{M}|\mathbf{M}^\infty) - P(\mathbf{A}) \log|\mathbf{M}|] \tag{14}$$

Writing this as an inequality on $P(\mathbf{A})$ gives

$$P(\mathbf{A}) \geq \left[\frac{H(\mathbf{M}|\mathbf{M}^\infty)}{T_S} - \frac{\log|\mathbf{X}|}{T_c}\right]\frac{T_S}{\log|\mathbf{M}|} \tag{15}$$

Obviously, when the entropy

$$\frac{H(\mathbf{M}|\mathbf{M}^\infty)}{T_S} > \frac{\log|\mathbf{X}|}{T_c}$$

the probability of ambiguous encoding is bounded away from zero and arbitrarily high reliability in communication cannot be achieved.

Figure 2.8 summarizes results from this section by indicating where the probability $Pr(\mathbf{A})$ of decoding ambiguity as a function of $(1/T_c) \log|\mathbf{X}|$, i.e.,

Coding for Discrete Noiseless Channels

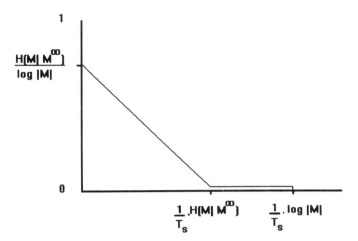

Figure 2.8. Performance bounds for well-designed codes. Probability of decoding ambiguity. Note: $\frac{1}{T_c} \log |X|$ is the channel capacity.

the amount of information per second that Agent 00111 can send, must lie for a *well-designed* code. The significance of the quantity $(1/T_c) \log |X|$, often referred to as the *capacity of the noiseless channel,* is discussed in Section 2.4.

Exercises

1. Consider the problem of communicating information through a noiseless binary channel that accepts one symbol every second. The memoryless information source emits two messages every 3 seconds; its message alphabet is ternary, with message probabilities 0.75, 0.125, and 0.125. For $k = 2, 4, 6, 8$, tabulate (a) the total number of message k-tuples, (b) the total number of code words, and (c) the minimum probability of ambiguous encoding.

2. One of the problems involved in computing code performance for fixed-rate sources is enumerating the number of message sequences occurring with a given probability. Suppose we define the following generating function for a memoryless information source

$$g(x) = \sum_{M \in \mathbf{M}} x^{\log_b Pr(M)}$$

where x is an indeterminate, **M** is the message alphabet, and $Pr(M)$ is the *a priori* probability of the message M. If $g(x)$ is raised to the nth power, $g^n(x)$ is a polynomial of the form

$$g^n(x) = \sum_i k_i x^{h_i}$$

(a) Assuming $h_i \neq h_j$ for $i \neq j$, identify what n, k_i, and h_i represent.

(b) If x is set equal to base b of the logarithm used, what is $g^n(b)$?

(c) Evaluate the number of sequences of four messages as a function of the message sequence probability for the memoryless source with message probabilities 1/3, 1/3, 1/9, 1/9, 1/9.

(d) What is the quantity $(d/dx)g(x)|_{x=b}$?

2.4. The Significance of Information Theory

In Section 2.3, we proved a simple coding theorem stating that it is possible to communicate information from a fixed-rate source through a noiseless channel with arbitrarily low (but not necessarily zero) probability of error if

$$\frac{H(\mathbf{M}|\mathbf{M}^\infty)}{T_S} < \frac{\log|\mathbf{X}|}{T_c} \qquad (16)$$

and it is not possible to do so if the inequality is reversed. Most quantities in Equation 16 have information-theoretic interpretations. The quantity $H(\mathbf{M}|\mathbf{M}^\infty)/T_S$ simply computes the *source information rate* R. Assuming logarithm base 2, R is given by

$$R = \frac{H(\mathbf{M}|\mathbf{M}^\infty)}{T_S} \text{ bits per second} \qquad (17)$$

where $H(\mathbf{M}|\mathbf{M}^\infty)$ is the source entropy in bits per message and T_S is the duration of a message in seconds per message.

The quantity $\log|\mathbf{X}|/T_c$ also has an information-theoretic interpretation. In the problem under consideration, we are trying to communicate through a channel like the one depicted in Figure 2.1. The average information gained

Coding for Discrete Noiseless Channels

by the decoder about a sequence of n channel input symbols by observing the channel output sequence of n symbols is given by the mutual information $I(\mathbf{X}^n; \mathbf{Y}^n)$. If mutual information is truly a valid measure of information transmitted through the channel, the obvious design procedure is to maximize the channel's mutual information per unit time over all parameters within the designer's control. In this case, the designer has control only over the coding operation and hence over the input probability distribution $Pr(\bar{x}_T)$ of channel input sequences $\bar{x}_T$ with duration T and over the choice of T. We define this maximum value of mutual information to be the *capacity* C of the channel. Thus,

$$C = \max_{T, Pr(\bar{x}_T)} \frac{I(T)}{T} \qquad (18)$$

where $I(T)$ denotes mutual information between channel inputs and outputs of duration T.

Suppose we now compute the capacity of the specific noiseless channel discussed in Section 2.3. Since the time duration of a sequence of n symbols was simply nT_c, the capacity expression reduces to

$$C = \frac{1}{T_c} \max_{n, Pr(\bar{x})} \frac{I(\mathbf{X}^n; \mathbf{Y}^n)}{n} \qquad (19)$$

where $\mathbf{X}$ and $\mathbf{Y}$ are the channel input and output alphabets, respectively. From the definition of a noiseless channel, it is easily shown that $H(\mathbf{X}^n|\mathbf{Y}^n)$ is zero and thus

$$I(\mathbf{X}^n; \mathbf{Y}^n) = H(\mathbf{X}^n) \qquad (20)$$

Furthermore, our knowledge of the entropy function has shown that

$$H(\mathbf{X}^n) \le nH(\mathbf{X}) \qquad (21)$$

with equality if and only if the source is memoryless. Equations 20 and 21 are enough to reduce the capacity expression to

$$C = \frac{1}{T_c} \max_{P(x)} H(\mathbf{X}) \qquad (22)$$

But the entropy of an alphabet $\mathbf{X}$ is maximized if all of its $|\mathbf{X}|$ symbols are equally likely; thus,

$$C = \frac{\log|\mathbf{X}|}{T_c} \tag{23}$$

The coding theorem in Section 2.3 can now be interpreted strictly in terms of information theory, using R as defined in Equation 17 and C as defined in Equation 18. The theorem states that a fixed-rate source with information rate R can be communicated reliably through a noiseless channel of capacity C if $R < C$ but reliable communication is impossible if $R > C$.

The significance of information theory is that it truly supplies an abstract measure of information that can be used consistently to predict (with rate R and capacity C) when reliable communication can be established through a wide variety of channels. Thus, coding theorems that relate rate R and capacity C to our ability to design good encoders and decoders justify our study of information theory.

Terminology suggests the following true analogy. Suppose you moved 1000 ft^3 of belongings from Los Angeles to London, with a loading time of one day. Obviously, you must enter your belongings in a transportation channel at the *rate* of 1000 ft^3/day. The vans and boats leaving Los Angeles for London must have an available *capacity* of at least 1000 ft^3/day or reliable transportation is not available.

The analogy can be extended. A cubic volume of 1000.1 ft^3 might be available in a moving van. Unfortunately, even though the volume of goods is less, you may still be unable to pack them in the available volume if your 40-ft ham radio antenna does not fit, or miscellaneous chair legs protrude, etc. To accomplish your objective, you must transform (encode) your goods into more easily packed pieces by dismantling the antenna and unscrewing chair legs. Of course, we hope that you have coded each piece for easy assembly (decoding) at your destination rather than reducing all the furniture to sawdust for efficient packing.

As the rate approaches capacity, the packing (coding) problem becomes extremely difficult unless your belongings are small cubes (or other space-filling objects) all the same size. A solution to the odd-shaped object problem is to cut all your belongings into infinitesimal uniform solid cubes. Of course, the decoding problem then becomes unbelievably difficult. Using large rectangular solids called containers has recently revolutionized the transportation industry by simplifying loading (encoding) and unloading (decoding) problems for cargo from the viewpoint of a shipping firm. Individual containers must be loaded and unloaded, which are the encoding and decoding problems of exporters and importers, respectively.

The difficulty with reducing everything to sawdust is that the *physical entropy,* which has essentially the same mathematical expression as our *statistical entropy,* is extremely high, and it takes prodigious efforts to reduce physical entropy. The revered second law of thermodynamics, which states that physical entropy constantly increases in any closed physical system to which no external energy has been supplied, has been neatly paraphrased in the expression, "you can't unscramble an egg." Decoding, however, is a form of unscrambling, and (at best) it requires data-processing energy to be supplied to the system.

Exercise

1. If it is impossible to achieve highly reliable communication for long message sequences ($k > K(\alpha, \epsilon)$), does this preclude the possibility of highly reliable communication of short message sequences? Answer this question based on the following considerations:

   (a) Let the short message sequence contain k_S messages, $k_S M = k$ for large values of M, and assume the probability of failure in a single short message transmission is α_S. Assuming successive transmissions are independent, how is α_S related to α, the probability of failures in transmitting M consecutive short sequences?

   (b) Now answer the main question given the two following definitions: Assume highly reliable communication of short message sequences means

   i. High probability of *no failures* in long sequences of Mk_S messages

   ii. High probability of *no failures* in a single sequence of k_S messages

   (c) Which of the preceding definitions of reliable communication is used in the interpretation of the coding theorem in Section 2.3?

2.5. Tree Codes

Although Agent 00111 had discovered the Information Theory results, these would have been of little use to him if they were merely theoretical. He had to find specific codes that were U_S dictionaries, then assign sequences to them. Section 2.5 considers this problem, which is essentially simple.

We can use the insight gained from the Sardinas–Patterson algorithm to construct U_S dictionaries. Recall that the simplest and quickest exit from the algorithm (Figure 2.3) occurred when Seg 1 was empty, in which case, the dictionary is both U_S and U_F. A code of this type is said to satisfy the *prefix property*, or alternatively, it is said to be *prefix free*. Explicitly, a prefix-free code has the property that no code word is the prefix of any other code word. Of course, any block code with n symbols per code word possesses this property. However, these codes were often inefficient for Agent 00111. For example, if the deputy party chief has been shot and this was the first information he had to communicate, it would be ridiculous to waste valuable stamp area answering subsequent questions about the poor fellow's potential for blackmail, promotion, and fame. Some answers precluded others, and Agent 00111 had to exploit this fact to communicate more efficiently. A more general type of prefix-free code allowed this to happen.

Constructing U_S codes satisfying the prefix property is most simply shown by studying the decoding process. Suppose that $\mathbf{M} = \{M_i, i = 1, 2, \ldots, |\mathbf{M}|\}$ represents a source alphabet encoded in a binary code over a channel alphabet $\mathbf{X} = \{0, 1\}$ having the prefix property. The receiver can perform symbol-by-symbol tree decoding of any semiinfinite coded sequence of source messages as follows: The first received symbol is either 1 or 0. While in general this is not enough information to identify the corresponding source message M_i unambiguously, it at least serves to distinguish between those M_i whose code words begin with 0 and those whose code words begin with 1. If indeed the received symbol (1 or 0) is an entire code word, the prefix property guarantees that this is the only interpretation and the next symbol begins a *new* code word. If the received symbol is not an entire code word, the next symbol further restricts source symbols that could be involved, and this continues until the received sequence of symbols is itself a code word. Given the prefix property, this code word uniquely corresponds to a source message, and the next received symbol begins a new code word, so that the decoding cycle is ready to begin again. Thus, given the initial synchronization, the prefix property guarantees *instantaneous* decoding of each code word as soon as its last symbol is observed. This, on occasion, had been an added bonus for Agent 00111.

Suppose at some stage of the decoding process, the only symbol that could follow the binary digits already received is a 1, for example. This means that the point has been reached where the next preordained binary symbol conveys no information and could therefore just as well have been omitted. This would certainly be the case in any optimum (minimum word length) coding scheme. Thus, it is reasonable to impose full *binary tree* conditions on the decoder: The decoder can be represented as a tree with one *initial* node, $|\mathbf{M}| - 2$ *interior* nodes representing proper code-word prefixes, and $|\mathbf{M}|$ *terminal* nodes representing code words (see Figure 2.9). Two arrows emerge from the initial node and each interior node. One arrow enters each terminal node. By convention, the nodes are given binary labels as follows: The initial node is unlabeled. Its immediate successor on the left is labeled 0, and its immediate successor on the right is labeled 1. The remaining nodes

Coding for Discrete Noiseless Channels

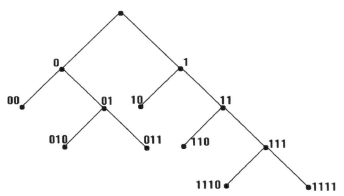

Figure 2.9. An Order 7 binary tree code.

are given the labels of their immediate predecessors with an extra symbol at the right end of the label: 1 for right-side descendants and 0 for left-side descendants. In the example shown in Figure 2.9, the initial node is unlabeled; the interior nodes are 0, 1, 01, 11, and 111; and the terminal nodes labeled 00, 10, 010, 011, 110, 1110, and 1111 are code words. Let us interpret this tree according to Agent 00111's problem with the deputy party chief. Either the deputy party chief has been shot or he has not been shot. If he has, we indeed have a terminal node!

Given any semiinfinite binary sequence to decode, a binary tree decoder will always assign a decoding to it, and the decoding assigned will be unique and instantaneous. Specifically, starting at the top of the tree, each symbol will act as an instruction (go right or go left) until a terminal node is reached. At this point, one decoding cycle has been completed and a code word determined, and the decoder is then reset at the top of the tree.

The number of branches traversed in passing through the code tree from the initial node in a tree to the terminal node is called the *length* of the code word that the terminal node represents. Equivalently, the length of a code word is the number of symbols used in its construction. The following is a simple test to determine if a tree code with specified code-word lengths can be constructed.

Theorem 2.1 (Kraft). A tree code with code word lengths $l_1, l_2, \ldots, l_{|\mathbf{M}|}$ can be constructed if and only if

$$\sum_{i=1}^{|\mathbf{M}|} |\mathbf{X}|^{-l_i} \leq 1 \qquad (24)$$

where $|\mathbf{X}|$ is the number of symbols in the channel alphabet $\mathbf{X}$. The inequality in Equation 24 may also be written as

$$\sum_{l=1}^{L} N_l |\mathbf{X}|^{-l} \leq 1 \tag{25}$$

where N_l is the number of words of length l and L is the maximum word length.

Proof a (tree existence $\Rightarrow$ inequality): Let I_l be the number of interior nodes in the tree of length l. Since the total number of nodes of length l is the result of at most $|\mathbf{X}|$ branches from each interior node of length $l-1$,

$$N_l + I_l \leq |\mathbf{X}| I_{l-1} \tag{26}$$

for $l \geq 1$. Note equality occurs in Equation 26 if and only if I_{l-1} nodes each produce $|\mathbf{X}|$ branches. Since there is one initial interior node,

$$I_0 = 1 \tag{27}$$

Evaluating Equation 26 for $l = 1, 2, \ldots$, gives

$$l = 1: N_1 + I_1 \leq |\mathbf{X}| \tag{28a}$$

$$l = 2: N_2 + I_2 \leq |\mathbf{X}| I_1 \leq |\mathbf{X}|^2 - N_1 |\mathbf{X}| \tag{28b}$$

$$l = 3: N_3 + I_3 \leq |\mathbf{X}| I_2 \leq |\mathbf{X}|^3 - N_1 |\mathbf{X}|^2 - N_2 |\mathbf{X}| \tag{28c}$$

In general,

$$I_l \leq |\mathbf{X}|^l \left(1 - \sum_{l'=1}^{l} N_{l'} |\mathbf{X}|^{-l'} \right) \tag{29}$$

Since no interior nodes are required for the longest word length L, I_L is zero, and the Kraft inequality follows directly from Equation 29, with l set equal to L.

Proof b (inequality $\Rightarrow$ tree existence): Consider the following method for constructing a tree with N_l terminal nodes of length l, $l = 1, 2, \ldots, L$.

Coding for Discrete Noiseless Channels

1. Construct a tree with $|\mathbf{X}|$ terminal nodes of length 1.
2. Set $l = 1$; $A_1 = |\mathbf{X}|$.
3. Reserve N_l nodes of the A_l available nodes of length l for terminal nodes.
4. Add $|\mathbf{X}|$ branches to each of the remaining $A_l - N_l$ nodes to create A_{l+1} nodes of length $l + 1$.
5. If $N_{l'} > 0$ for some $l' > l$, then increase l by one and return to Step 3; otherwise, stop.

Steps 3 and 4 imply that the number of available nodes of length $l + 1$ is given by

$$A_{l+1} = |\mathbf{X}|(A_l - N_l), \, l = 1, 2, 3, \ldots \tag{30}$$

For the preceding method to lead to a tree, the following must hold:

$$N_l \leq A_l, \, l = 1, 2, \ldots, L \tag{31}$$

Kraft's inequality (Equation 25) can be used to verify Equation 31 in the following manner.

For $l = 1$: Since all the terms in the sum in Equation 25 are nonnegative, the $l = 1$ term does not exceed 1 when the inequality is satisfied.

$$N_1 |\mathbf{X}|^{-1} \leq 1 \tag{32}$$

which implies that

$$N_1 \leq |\mathbf{X}| = A_1 \tag{33}$$

For $l = 2$: When Equation 25 is satisfied, the sum of the first two terms in Equation 25 does not exceed 1.

$$N_1 |\mathbf{X}|^{-1} + N_2 |\mathbf{X}|^{-2} \leq 1 \tag{34}$$

which implies that

$$N_2 \leq |\mathbf{X}|(|\mathbf{X}| - N_1) \tag{35}$$

Using Equation 30 yields

$$N_2 \leq A_2 \tag{36}$$

For *l in general:* The first *l* terms in the Kraft inequality imply that

$$\sum_{l'=1}^{l} N_{l'} |\mathbf{X}|^{-l'} \leq 1 \tag{37}$$

or

$$N_l \leq |\mathbf{X}| \{ |\mathbf{X}| [|\mathbf{X}| (\cdots) - N_{l-2}] - N_{l-1} \} \tag{38}$$

which, applying Equation 30 to each parenthesized term, gives

$$N_l \leq A_l \tag{39}$$

By iteration, the proof of Kraft's result is now complete.

Note that codes with the prefix property are a subclass of all the possible U_F or U_S codes. It would therefore seem reasonable that the best of all possible codes might or might not be a code with the prefix property; i.e., imposing the prefix requirement might force the code to be suboptimal on occasion. In fact, this is not so. An optimal U_F or U_S code can always be found with the prefix property, as the following result indicates.

Theorem 2.2 (McMillan). A necessary and sufficient condition for the existence of a U_F or U_S dictionary with word lengths $l_1, l_2, \ldots, l_{|\mathbf{M}|}$ is that the set of word lengths satisfy the Kraft inequality in Equation 24.

The surprising result for arbitrary U_F and U_S dictionaries is the absence of a penalty in terms of increased word length for demanding instantaneous (tree) decoding.

Proof: The sufficiency proof of the preceding statement could be borrowed from the proof of inequality for tree codes, since tree codes are U_S and U_F. The necessity of the Kraft inequality for U_F and U_S dictionaries can be demonstrated by investigating the uniqueness of long code-word sequences. Suppose we examine the structure of all possible sequences of *m* code words. We can generate the composite word lengths involved by performing the following calculation:

$$\left(\sum_{l=1}^{L} N_l r^{-l} \right)^m = \sum_{j=m}^{mL} n_j r^{-j} \tag{40}$$

Coding for Discrete Noiseless Channels

By raising $\sum_{l=1}^{L} N_l r^{-l}$ to the mth power, the resulting polynomial in the variable r is composed of terms whose coefficients n_j denote the number of distinct sequences of m code-words of length j.

For example, if $N_1 = 2$ and $N_2 = 1$, the possible sequences of three code words can be represented by

$$(2r^{-1} + r^{-2})^3 = 8r^{-3} + 12r^{-4} + 6r^{-5} + r^{-6}$$

If a denotes the length 2 code word and b and c denote length 1 code words in this example, then the following enumerates the possible sequences of three code words as a function of the number of symbols in the sequence:

Three symbols: *bbb, bbc, bcb, cbb, ccb, cbc, bcc, ccc*

Four symbols: *abb, abc, acb, acc, bab, bac, cab, cac, bba, bca, cba, cca*

Five symbols: *aab, aac, aba, aca, baa, caa*

Six symbols: *aaa*

For a uniquely decodable U_S or U_F code dictionary, regardless of the value of m, the n_j sequences of m code words totaling j code symbols must be distinct. This implies that n_j must be less than or equal to the number of distinct sequences $|\mathbf{X}|^j$ of j code symbols. This reduces the inequality in Equation 40 to the following:

$$\left(\sum_{l=1}^{L} N_l r^{-l}\right)^m \leq \sum_{j=m}^{mL} |\mathbf{X}|^j r^{-j} \tag{41}$$

Evaluating Equation 41 for $r = |\mathbf{X}|$ gives

$$\left(\sum_{l=1}^{L} N_l |\mathbf{X}|^{-l}\right)^m \leq mL \tag{42}$$

Since the left side of Equation 42 varies exponentially with m and the right varies only linearly in m, Equation 42 can hold for very large m if and only if

$$\sum_{l=1}^{L} N_l |\mathbf{X}|^{-l} \leq 1 \tag{43}$$

Since unique decodability implies that Equation 43 is true for all m, we have shown that every set of uniquely decodable code-word lengths must satisfy the Kraft inequality.

Without going through an elaborate search procedure, it is now possible to verify the existence of U_S dictionaries by simply examining dictionary-word lengths.

Example 2.1. Is it possible to construct a U_S dictionary using the 10 Arabic numerals 0, 1, ..., 9 as code symbols containing six code words of length 1, 35 code words of length 2, 34 code words of length 3, and 162 code words of length 4? The answer is *no,* since by Theorem 2.2 (McMillan)

$$6 \times 10^{-1} + 35 \times 10^{-2} + 34 \times 10^{-3} + 162 \times 10^{-4} = 1.0002 > 1$$

However, if two words of length 4 are eliminated, the synthesis of a U_S dictionary is possible, and in fact the dictionary can be a tree code. So, by using both Theorem 2.1 (Kraft) and Theorem 2.2 (McMillan), Agent 00111 can determine whether or not he may structure a dictionary around a set of constraints.

Exercises

1. Suppose $|\mathbf{X}| = q > 2$. Then it is possible to define *full q-ary* tree conditions as requiring (1) all *initial* and *interior* nodes of the tree each to emit q branches and (2) all exterior nodes to be code words.

   (a) What is the relationship between the number of exterior nodes $|\mathbf{M}|$ and the number of interior nodes?

   (b) How many exterior nodes can exist in a full q-ary tree?

2. In many cases, a code tree is adequately described by a list of its code-word lengths, with the lengths placed in ascending order. Two codes are said to be *distinct* if their *list descriptions* (as described in the preceding sentence) are not identical. Construct a table of the number of distinct code trees for as many values of $|\mathbf{M}|$ as possible when (a) $|\mathbf{X}| = 2$ and (b) $|\mathbf{X}| = 3$.

3. Show that when

$$\sum_{l=1}^{L} N_l |\mathbf{X}|^{-l} < 1$$

Coding for Discrete Noiseless Channels

then there exists a semiinfinite sequence of symbols from X that cannot be decoded into a sequence of words from the prefix code with N_l words of length l, $l = 1, \ldots, L$.

4. Consider the following code:

$$1, 01, 0011, 0010, 100001, 100101, 010.$$

   (a) How many code words of length 12 symbols are there in the fourth extension of the preceding code?

   (b) Is this code uniquely decodable in any sense?

   (c) Can you construct a binary code that is a U_F dictionary and has the same word lengths as the preceding code?

5. Agent 00111 has decided to use a code with three words of length 1, five words of length 2, and 17 words of length 3. How large a code alphabet does he need?

2.6. A Coding Theorem for Controllable Rate Sources

In Section 2.6, we assume that Agent 00111 is equipped with a radio that enables him to send one message of exactly 1000 bits. He cannot send more data for fear of detection, and he cannot send less data due to the radio's construction. However, he may send the message at any time; there are no deadlines to meet. His goal is to include as much genuine information in the 1000 bits as possible. He has a packing problem: Like furniture, different messages have different lengths, and it is inefficient to make all the lengths the same. Agent 00111 wishes to give high-probability messages a shorter length than low-probability messages, i.e., he must determine *message* probabilities. Section 2.6 exploits these probabilities to give optimal packings or encodings.

The first coding theorem for fixed-rate sources (from Section 2.3) indicated that high-performance coders for communication systems were possible if and only if the source entropy in bits/second was less than the channel capacity in bits/second. Since time is critical in communications, the required comparison was in terms of rates: a source information rate and a channel capacity (rate). Many situations do not require careful attention to time, e.g., information storage problems, communication situations where infinite or very large buffer storage is available to the encoder, etc. When the encoder does not have to match a fixed-duration message to a code word of the same duration, we refer to the source as having a *controllable rate*.

With their variety of word length, tree codes are naturally suited to controllable rate sources. Since time is no longer a consideration, a reasonable figure of merit for a code from a given source $\mathbf{M}$ is the *average code-word length* $\bar{l}$ given by

$$\bar{l} = \sum_{M \in \mathbf{M}} Pr(M) l(M) \tag{44}$$

where $l(M)$ is the length of the code word representing source message M and $Pr(M)$ is the *a priori* probability that message M will be sent. The immediate question is how small can $\bar{l}$ be made for a U_S, U_F, or tree code?

Consider a code for $\mathbf{M}$ with known *a priori* probabilities for each $M \in \mathbf{M}$, constructed by choosing $l(M)$ to satisfy

$$|\mathbf{X}|^{-l(M)+1} > Pr(M) \geq |\mathbf{X}|^{-l(M)} \tag{45}$$

for all M. Such a code can be uniquely decodable, since by summing each side of the right side inequalities in Equation 45, we have

$$1 \geq \sum_{M \in \mathbf{M}} |\mathbf{X}|^{-l(M)} \tag{46}$$

and hence word lengths satisfy the Kraft–McMillan inequality. Taking logarithms in Equation 45 gives

$$[-l(M) + 1] \log |\mathbf{X}| > \log Pr(M) \geq -l(M) \log |\mathbf{X}| \tag{47}$$

which yields the following bounds on $l(M)$:

$$\frac{-\log Pr(M)}{\log |\mathbf{X}|} \leq l(M) < \frac{-\log Pr(M)}{\log |\mathbf{X}|} + 1 \tag{48}$$

Multiplying Equation 48 by $Pr(M)$ and summing over $M \in \mathbf{M}$ yields

$$\frac{H(\mathbf{M})}{\log |\mathbf{X}|} \leq \bar{l} < \frac{H(\mathbf{M})}{\log |\mathbf{X}|} + 1 \tag{49}$$

These bounds on $\bar{l}$ hold for any U_S code construction satisfying Equation 45.

In fact, the lower bound on $\bar{l}$ must hold for any U_F code. To see this, note that for a U_F code

Coding for Discrete Noiseless Channels

$$\bar{l} - \frac{H(\mathbf{M})}{\log|\mathbf{X}|} \geq \bar{l} + \frac{\log\left[\sum_{M \in \mathbf{M}} |\mathbf{X}|^{-l(M)}\right]}{\log|\mathbf{X}|} - \frac{H(\mathbf{M})}{\log|\mathbf{X}|}$$

$$= \frac{1}{\log|\mathbf{X}|} \sum_{M \in \mathbf{M}} Pr(M) \log\left\{\frac{Pr(M)}{\left[\frac{|\mathbf{X}|^{-l(M)}}{\sum_{M' \in \mathbf{M}} |\mathbf{X}|^{-l(M')}}\right]}\right\} \geq 0 \quad (50)$$

The left-side inequality in Equation 50 follows directly from the Kraft–McMillan inequality and the right-side inequality can be recognized as the divergence inequality.

With adequate storage, it is possible to block encode the kth extension $\mathbf{M}^k$ of the information source, with each code word representing a distinct k-tuple of messages. Since $\mathbf{M}^k$ is itself a source, Equation 49 applies and

$$\frac{H(\mathbf{M}^k)}{\log|\mathbf{X}|} \leq \bar{l}_k < \frac{H(\mathbf{M}^k)}{\log|\mathbf{X}|} + 1 \quad (51)$$

where $\bar{l}_k$ is the average word length of a code word representing a sequence of k messages from $\mathbf{M}$. To compare Equations 51 and 49, we can divide Equation 51 by k, since $\bar{l}_k/k$ and $\bar{l}$ are both measures of the average number of code symbols required to represent a source symbol. As k increases, $H(\mathbf{M}^k)/k$ approaches the source entropy $H(\mathbf{M}|\mathbf{M}^\infty)$ for stationary sources. Therefore, lower and upper bounds on $\bar{l}_k/k$ derived from Equation 51 must approach $H(\mathbf{M}|\mathbf{M}^\infty)$ for large k. In summary, we have proved Theorem 2.3.

Theorem 2.3 (Shannon). For any $\epsilon > 0$, there is a finite integer $K(\epsilon)$ such that the minimum average number $\bar{l}_k$ of channel symbols from $\mathbf{X}$ necessary to encode a source message k-tuple from $\mathbf{M}^k$ into a U_F or U_S code is bounded by

$$\frac{H(\mathbf{M}|\mathbf{M}^\infty)}{\log|\mathbf{X}|} \leq \frac{\bar{l}_k}{k} < \frac{H(\mathbf{M}|\mathbf{M}^\infty)}{\log|\mathbf{X}|} + \epsilon \quad (52)$$

whenever $k \geq K(\epsilon)$.

Assuming that logarithms are base 2, the units of $H(\mathbf{M}|\mathbf{M}^\infty)$ are bits/message, and the units of $\bar{l}_k/k$ are (channel) symbols/message. Since Theorem 2.3 compares $\bar{l}_k/k$ to $H(\mathbf{M}|\mathbf{M}^\infty)/\log|\mathbf{X}|$, this indicates that the units of

$\log_2|\mathbf{X}|$ are bits/symbol. This can also be seen by noticing that $\log_2|\mathbf{X}|$ is a capacity C defined by

$$C = \max_{n,p(\bar{x})} \frac{I(\mathbf{X}^n; \mathbf{Y}^n)}{n}$$

$$= \log_2|\mathbf{X}| \tag{53}$$

The maximization is performed as in Section 2.4, with the difference that mutual information per unit time was maximized in Section 2.4, while here mutual information per channel symbol is maximized (time not being important). In effect, the channels discussed in Sections 2.4 and 2.6 are identical, and hence the capacities measured in Sections 2.4 and 2.6 differ only in units: bits/symbol as opposed to bits/second.

In favorable situations, $\bar{l}$ can achieve its lower bound $H(\mathbf{M})/\log|\mathbf{X}|$. This occurs in Equation 50 when most inequalities are in fact equalities. This is achieved when

$$\sum_{M \in \mathbf{M}} |\mathbf{X}|^{-l(M)} = 1 \tag{54a}$$

$$Pr(M) = |\mathbf{X}|^{-l(M)} \tag{54b}$$

The condition for equality could have been deduced heuristically when we calculated channel capacity for the noiseless channel. Recall that the mutual information $I(\mathbf{X}^n; \mathbf{Y}^n)$ is maximized when the code-symbol sequence from the encoder represented a memoryless source of equally likely symbols. Suppose we restrict the type of code we wish to use to uniquely decodable (U_S) codes having the prefix property. We can guarantee that the encoded message will appear to be from a memoryless source of equally likely symbols if for any position in the symbol sequence *and hence for any position in the code tree,* the symbol to follow is any one of the $|\mathbf{X}|$ equally likely code symbols. A code tree having this characteristic assigns the probability $1/|\mathbf{X}|$ of taking any path from one node to the next node. Thus, the tree must be full, each node having $|\mathbf{X}|$ branches, and the probability of a message M corresponding to a code word of length $l(M)$ symbols must be

$$\left(\frac{1}{|\mathbf{X}|}\right)^{l(M)}$$

Unfortunately, many *a priori* probability distributions do not satisfy Equation 54b, so the equally likely branching condition cannot be met exactly.

Example 2.2. Consider a memoryless information source with message probabilities

$$Pr(M_1) = Pr(M_2) = 1/3$$
$$Pr(M_3) = Pr(M_4) = 1/9$$
$$Pr(M_5) = Pr(M_6) = Pr(M_7) = 1/27$$

Define a trinit to be the unit information when using base 3 logarithms. With a code dictionary of 3 symbols, the preceding message source can be optimally coded using the tree code shown in Figure 2.10.

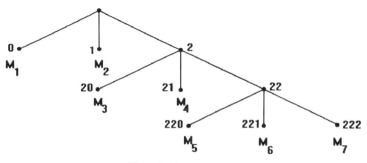

Figure 2.10. A tree code.

Note,

$$H = 2 \cdot 1/3 \log_3 3 + 2 \cdot 1/9 \log_3 9 + 3 \cdot 1/27 \log_3 27$$

$$= 1\frac{4}{9} \text{ trinits per message}$$

$$\bar{l} = 2 \cdot 1/3 \cdot 1 + 2 \cdot 1/9 \cdot 2 + 3 \cdot 1/27 \cdot 3 = 1\frac{4}{9} \text{ channel symbols per message.}$$

2.7. Huffman's Coding Procedure

Section 2.6 gave Agent 00111 hope for an accurate assessment of how much information he could ideally send within a fixed-size parcel. However,

Agent 00111 still has to find specific sequences for specific messages that form a U_F block code giving the smallest possible value for $\bar{l}$, the average number of channel symbols per message. Section 2.7 solves this problem for Agent 00111. In fact, the heuristic argument at the end of Section 2.6 gives a very strong indication of how the forthcoming algorithm works.

More formally, it has so far been determined only that the average word length $\bar{l}$ of a U_F block code for M cannot be less than

$$\frac{H(\mathbf{M})}{\log|\mathbf{X}|}$$

where $\mathbf{X}$ is the channel alphabet and in fact, having $\bar{l}$ within one unit of

$$\frac{H(\mathbf{M})}{\log|\mathbf{X}|}$$

is achievable. We now develop an algorithm for finding a U_F block code for $\mathbf{M}$ with the smallest possible value of $\bar{l}$. Such a code is said to be *compact* for the source $\mathbf{M}$. Based on Theorems 2.1 (Kraft) and 2.2 (McMillan), the code design can be restricted to tree codes.

To begin, consider having $|\mathbf{M}|$ terminal nodes as yet unconnected in any tree form but with known message probabilities $Pr(M_i)$, $i = 1, \ldots, |\mathbf{M}|$ assigned to the nodes. The following two properties of compact codes indicate how to begin combining the terminal nodes to form a tree.

Property 1. If $Pr(M_i) > Pr(M_j)$, then word lengths for M_i and M_j in the optimal tree code satisfy the inequality $l(M_i) \leq l(M_j)$. Colloquially, the more probable messages have shorter lengths in an optimal code.

Proof: Assume that the optimal code has $l(M_i) > l(M_j)$ and average word length $\bar{l}_{\text{opt}}$. Interchanging message assignments by assigning M_i to the word of length $l(M_j)$ and M_j to the word of length $l(M_i)$, generates another code with average word length $\bar{l}$, where

$$\bar{l}_{\text{opt}} - \bar{l} = P(M_i)l(M_i) + P(M_j)l(M_j) - P(M_i)l(M_j) - P(M_j)l(M_i)$$
$$= [P(M_i) - P(M_j)][l(M_i) - l(M_j)]$$
$$> 0 \qquad (55)$$

Since $\bar{l}_{\text{opt}}$ cannot exceed $\bar{l}$, the assumption that $l(M_i) > l(M_j)$ must be false.

Coding for Discrete Noiseless Channels

Property 2. Let $L = \max_i l(M_i)$. There must be at least two words of length L in the optimal code

Proof: Suppose that a code has only one word of maximum length L. Then the interior node just prior to the length L terminal node must have only one departing branch. This branch could be pruned, thereby lowering $\bar{l}$ without changing the number of available terminal nodes or the prefix property of the code.

At this point, it is obvious that in every optimal binary tree the two least probable (or two of the least probable if there are many) messages are assigned to messages of length L and their terminal nodes can be assumed to merge at a common interior node of length $L - 1$. Hence, for the two combined terminal nodes, only the length $L - 1$ prefix (L is still unknown) has yet to be determined. As will now be seen, the problem of choosing a code tree with $|\mathbf{M}|$ terminal nodes to minimize $\bar{l}$ can be replaced by an equivalent problem requiring only $|\mathbf{M}| - 1$ terminal nodes, with the length $L - 1$ prefix representing two messages in the modified procedure. We can now produce a construction by the inductive method.

Assume that messages are renumbered so that their probabilities are ordered

$$Pr(M_1) \geq Pr(M_2) \geq \cdots \geq Pr(M_{|\mathbf{M}|}) \tag{56}$$

Applying the rule of combination indicated by Properties 1 and 2 gives

$$\bar{l} = \sum_{i=1}^{|\mathbf{M}|} Pr(M_i) l(M_i)$$

$$= Pr(M_{|\mathbf{M}|-1}) + Pr(M_{|\mathbf{M}|}) + \sum_{i=1}^{|\mathbf{M}'|} Pr(M_i') l'(M_i') \tag{57}$$

where

$$M_i' = \begin{cases} M_i, & i = 1, 2, \ldots, |\mathbf{M}| - 2 \\ M_{|\mathbf{M}|} \cup M_{|\mathbf{M}|-1} & i = |\mathbf{M}| - 1 \end{cases}$$

$$l'(M_i') = \begin{cases} l(M_i), & i = 1, 2, \ldots, |\mathbf{M}| - 2 \\ l(M_{|\mathbf{M}|}) - 1 & i = |\mathbf{M}| - 1 \end{cases} \tag{58}$$

and

$$|\mathbf{M'}| = |\{M'_i\}| = |\mathbf{M}| - 1 \tag{59}$$

Equation 57 indicates that the minimization of $\bar{l}$ is now equivalent to the minimization of $\bar{l'}$ for a binary block code for the source $\mathbf{M'}$, where

$$\bar{l'} = \sum_{i=1}^{|\mathbf{M}|} Pr(M'_i) l'(M'_i) \tag{60}$$

Obviously, the number of terminal nodes required by the code for $\mathbf{M'}$ is less than the number required by the code for $\mathbf{M}$, yet *the problem format remains the same*. Hence, the rule of combination can be applied again to the smaller source $\mathbf{M'}$, and by induction, the algorithm will yield a compact binary tree code.

In summary, we have the Huffman algorithm:

Huffman Algorithm. To design a compact binary tree code, use a rule of combination that at each stage combines the two nodes of least probability. The sum of the two least probabilities is assigned to the interior node so formed and the compound message represented by the interior node replaces the two terminal node messages in the message list. The rule of combination is reapplied until one compound message of probability 1 exists and the initial node of the tree has been reached.

The following example illustrates Huffman's algorithm.

Example 2.3. With the four source messages A, B, C, and D and

$$Pr(A) = 0.4 \qquad Pr(B) = 0.3 \qquad Pr(C) = 0.2 \qquad Pr(D) = 0.1$$

we obtain the nodes shown in Figure 2.11a. The first step is to combine C and D to form a new event $C \cup D$ of probability 0.3.

$$Pr(A) = 0.4 \qquad Pr(B) = 0.3 \qquad Pr(C \cup D) = 0.3$$

yields the tree diagram shown in Figure 2.11b. The next step is to combine the two events (B and $C \cup D$) of probabilities 0.3 each to obtain the new event $B \cup C \cup D$ of probability 0.6.

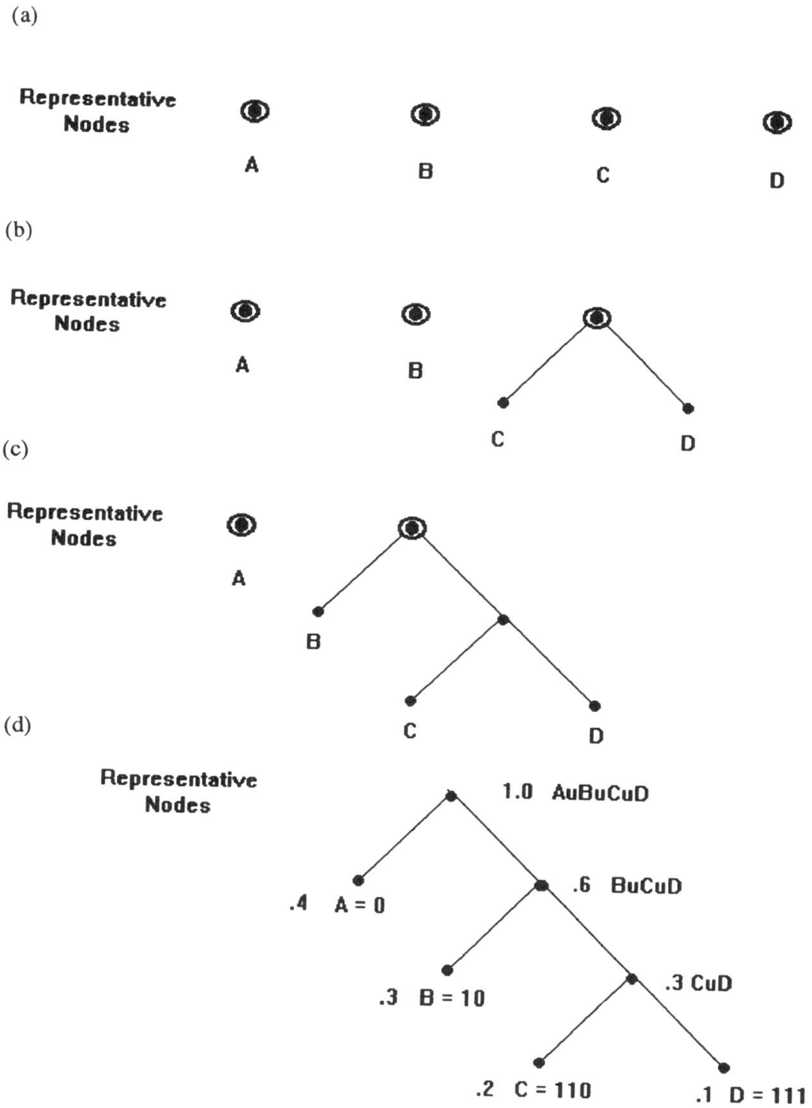

Coding for Discrete Noiseless Channels

(a)

Representative Nodes: A B C D

(b)

Representative Nodes: A B C D (with C and D joined below a node)

(c)

Representative Nodes: A, and a tree joining B with (C,D)

(d)

Representative Nodes:
- 1.0 AuBuCuD
- .4 A = 0
- .6 BuCuD
- .3 B = 10
- .3 CuD
- .2 C = 110
- .1 D = 111

Figure 2.11. Producing a tree code.

$$Pr(A) = 0.4 \qquad Pr(B \cup C \cup D) = 0.6$$

This gives the tree shown in Figure 2.11c. The final step is to combine the two remaining events A and $B \cup C \cup D$ into the universal event $A \cup B \cup C \cup D$ of probability 1. The tree that arises is shown in Figure 2.11d. All the nodes are shown with their associated code words and probabilities. Moreover, each interior node is labeled to identify the step in the combination process that produced it.

To compare a code to the lower bound on $\bar{l}$, we define the *efficiency* γ of a block code by

$$\gamma = \frac{H(\mathbf{M})}{\bar{l}C} \qquad (61)$$

where C is the capacity of the noiseless channel. The efficiency of the code in Example 2.3 is 0.972. From Equations 49 and 53, $\bar{l}$ is bounded from below by $H(\mathbf{M})/C$ for a block code for $\mathbf{M}$. Hence, the efficiency γ as defined in Equation 61 is at most 1. Care must be used in interpreting γ. An efficiency of 100% *does not imply* that we have found the general coding scheme that minimizes the average number of channel symbols per source message. As demonstrated by Theorem 2.3 (Shannon), for large k, the average number $\bar{l}_k/k$ of channel symbols per source message can be made arbitrarily close to $H(\mathbf{M}|\mathbf{M}^\infty)/C$, which may be less than $H(\mathbf{M})/C$. $\gamma = 1$ *does imply* that the best *block code for* $\mathbf{M}$ has been found.

Example 2.4. The effectiveness of the coding scheme in Example 2.3 can be measured by computing the average word length

$$\sum_{M \in \mathbf{M}} Pr(M)l(M) = 1 \times 0.4 + 2 \times 0.3 + 3 \times 0.2 + 3 \times 0.1$$

$$= 1.9 \text{ symbols/message}$$

which exceeds the source entropy

$$H(\mathbf{M}) = 0.4 \log(2.5) + 0.3 \log(10/3) + 0.2 \log 5 \doteq 1.51425 \text{ bits/message}$$

But it is *less* than the 2 symbols/message that the *other* binary tree of order 4 would require if assigning the code words 00, 01, 10, and 11, to A, B, C, and D in some order.

Exercises

1. Show that for each of the distinct binary trees having $|\mathbf{M}|$ terminal nodes, there is always a set of probabilities $\{Pr(M), M \in \mathbf{M}\}$ for which a given tree provides the optimum (Huffman) encoding.

2. Find a set of probabilities $Pr(M_1)$, $Pr(M_2)$, $Pr(M_3)$, and $Pr(M_4)$ for which both binary trees of order 4 provide equally good encodings. What is the general condition for this state of affairs to occur? What is the analogous result for trees of order 5? (See S. W. Golomb, 1980.)

3. Encode the source having message probabilities 0.2, 0.15, 0.15, 0.14, 0.10, 0.08, 0.08, 0.04, 0.02, 0.02, 0.01, 0.01 into a binary code. What is the code efficiency?

4. Generalize and prove Huffman's algorithm for a channel with input alphabet containing $|\mathbf{X}|$ symbols. *Hint:* Notice that unless the number of messages $n = k(|\mathbf{X}| - 1) + 1$ for some integer k, not every branch of a full tree can be identified with a message.

5. Find a ternary code for the eight message source having message probabilities 0.3, 0.2, 0.1, 0.1, 0.1, 0.1, 0.06, 0.04.

6. It is possible to encode a source *more efficiently* with

    (a) A ternary code instead of a binary code?

    (b) A quaternary code instead of a binary code?

    Justify your answers.

2.8. Efficiently Encoding Markov Sources

In several of the preceding discussions, we referred to Agent 00111's difficulty in sending several answers to related questions, since the answer to the first question could radically alter the probabilities of subsequent answers. The example of the deputy party chief being shot is an extreme case. If he were only in a mild state of disgrace, he could still possibly be made head of project Z. However, given the information about his death, the probabilities have changed. Prior to a mission, Agent 00111 would like to compute all his code books in advance and note the related probabilities. He can do this by extending Huffman coding to a Markov source model.

A Markov source has the property that the amount of information contained in the next message depends on the present state of the source. If a source is in state s_1, the next message may contain 1 bit of information; if the source is in state s_2, the next message may contain no information at all. Obviously, using one code word to represent this message in most of these situations is very inefficient.

Suppose we consider the situation where the decoder has somehow gained knowledge of the initial state $s(0)$ of the source. The probabilities of the next messages M_i, $i = 1, \ldots, |\mathbf{M}|$, are given by $P(M_i | s(0) = s_j)$, depending on the particular state s_j that initially occurs. A compact prefix code C_{s_j} can now be specified for the messages M_i, $i = 1, \ldots, |\mathbf{M}|$, based on this conditional probability distribution. As soon as the decoder determines the first transmitted message $M(1)$ to be M_i, it can use its knowledge of the source state diagram with $s(0) = s_j$ and $M(1) = M_i$ to determine the next state $s(1)$. The procedure can now be repeated. Obviously, prefix codes must be written for each state of the Markov source for this coding scheme to work.

Example 2.5. Consider the ternary source with the state diagram shown in Figure 2.12.

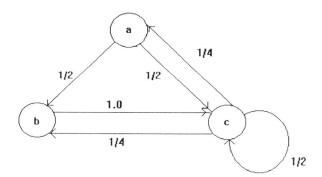

Figure 2.12. State diagram for a ternary source.

The stationary distribution of the source is given by

$$P(a) = \frac{2}{13} \qquad P(b) = \frac{8}{13} \qquad P(c) = \frac{3}{13}$$

If the usual Huffman coding procedure were applied to generate code words for a binary channel according to the stationary probabilities, the code would be

$$a \sim 10 \quad b \sim 0 \quad c \sim 11$$

with average word length

$$\bar{l} = 1\frac{5}{13} \text{ binary symbols per message.}$$

If we encode transitions, the required codes are

$$\begin{array}{ccc} \mathbf{C}_a & \mathbf{C}_b & \mathbf{C}_c \\ a - & a \sim 10 & a - \\ b \sim 0 & b \sim 0 & b - \\ c \sim 1 & c \sim 11 & c - \end{array}$$

No code at all is necessary when the source is in state c, since the output is deterministic. The average word length for this coding scheme is given by

$$\bar{l} = \bar{l}_a P(a) + \bar{l}_b P(b) + \bar{l}_c P(c)$$
$$= 1 \times \frac{2}{13} + \frac{3}{2} \times \frac{8}{13} + 0 \times \frac{3}{13} = 1\frac{1}{13} \text{ binary symbols per message.}$$

Exercises

1. Under what conditions is the encoding of transition probabilities 100% efficient?

2. Efficiently encode the Markov source having the state diagram shown in Figure 2.13 for communication through a binary noiseless channel. What is the efficiency of the code?

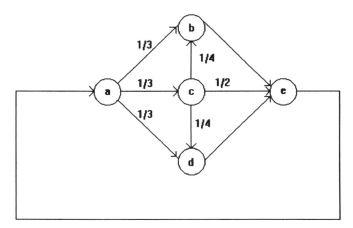

Figure 2.13. State diagram for a Markov source.

3. Encode the Markov source with the state diagram shown in Figure 2.14. Assume transition probabilities are chosen to maximize entropy and the source always starts at state a. Is there a 100% efficient binary coding scheme for this source that requires only one code?

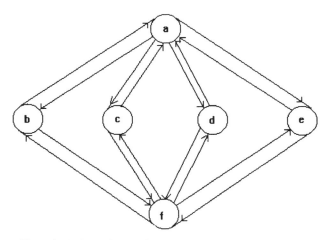

Figure 2.14. State diagram for the Markov source in Problem 3.

4. A very simple language consists entirely of 16 words and can be described by the first-order Markov source with the state diagram shown in Figure 2.15.

Coding for Discrete Noiseless Channels

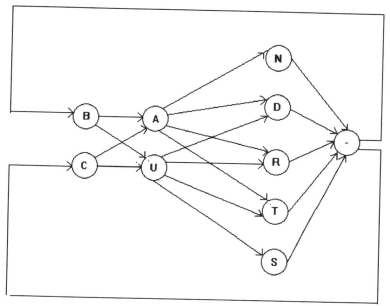

Figure 2.15. Markov source for Problem 4.

(a) List the 16 words in this language.

(b) Choose transition probabilities to maximize the entropy of the source.

(c) Starting in state i, the output of this source is to be communicated over a binary noiseless channel. Assume transition probabilities are given by your answer to (a). Find a simple 100% efficient coding scheme for this situation.

5. Show that next-state encoding of any Markov source **M** is at least as efficient as the simple Huffman code for **M**.

2.9. Variable Symbol Duration Channels

In many simple situations, input symbols to a noiseless channel are of unequal time duration. For example, Morse code uses dots and dashes, a dot taking much less time than a dash. The natural criterion to use in optimizing a communication system having symbols of unequal time duration is to maximize the information transmitted per unit time.

Let us consider a channel whose input alphabet **X** contains distinct symbols of time durations $T_1, \ldots, T_{|\mathbf{X}|}$, with T_1 the shortest of these time durations. Assuming that transmission begins at time zero, we consider the collection $\mathbf{C}_T$ of all possible distinct transmissions up to time T. A typical form for a sequence in $\mathbf{C}_T$ is a finite sequence of complete symbols followed by some fraction of a symbol, with the missing part of the last symbol having occurred after time T. Assuming that an observation of a fraction of a symbol is *not* sufficient to determine the symbol, we form the collection $\mathbf{D}_T$ composed of finite sequences in $\mathbf{C}_T$ with any fractional parts of final symbols removed. Hence, the number $|\mathbf{D}_T|$ of distinct sequences in $\mathbf{D}_T$ is in effect equal to the number of different messages that could be communicated over the channel.

The quantity $\log|\mathbf{D}_T|$ indicates the number of information units that can be specified in time T. It follows that the attainable transmission rate R_T during T seconds of communication through a noiseless channel is

$$R_T = \frac{\log|\mathbf{D}_T|}{T} \tag{62}$$

The *capacity* C of the variable symbol duration channel is defined as the channel's long-term attainable information rate

$$\begin{aligned}C &= \limsup_{T \to \infty} R_T \\ &= \limsup_{T \to \infty} \frac{\log|\mathbf{D}_T|}{T}\end{aligned} \tag{63}$$

We now evaluate C.

The number of sequences in $\mathbf{D}_T$ is difficult to evaluate as a function of T; however, a number with the same exponential growth rate as $|\mathbf{D}_T|$ is computable. Let $\mathbf{F}_t$ be the set of finite sequences of symbols whose duration is *exactly* t seconds. Notice that every element in $\mathbf{D}_T$ is a member of $\mathbf{F}_t$ for some $t \leq T$. Similarly, every element of $\mathbf{F}_t$ for any $t \leq T$ is either in $\mathbf{D}_T$ or is a prefix of an element of $\mathbf{D}_T$. Defining $(\mathbf{D}_T)_P$ as the set of all prefixes of $\mathbf{D}_T$, we have

$$\mathbf{D}_T \subset \left(\bigcup_{t \leq T} \mathbf{F}_t\right) \subset (\mathbf{D}_T)_P \tag{64}$$

Including itself, each member of $\mathbf{D}_T$ can have at most T/T_1 prefixes, since T_1 is the shortest duration channel symbol. Hence,

Coding for Discrete Noiseless Channels

$$|\mathbf{D}_T| \le \sum_{t \le T} |\mathbf{F}_t| \le \frac{T}{T_1} |\mathbf{D}_T| \tag{65}$$

Taking logarithms gives

$$\frac{1}{T} \log |\mathbf{D}_T| \le \frac{1}{T} \log \left(\sum_{t \le T} |\mathbf{F}_t| \right)$$

$$\le \frac{1}{T} \log \frac{1}{T_1} + \frac{1}{T} \log T + \frac{1}{T} \log |\mathbf{D}_T| \tag{66}$$

Since

$$\lim_{T \to \infty} \frac{1}{T} \log T = 0$$

upper and lower bounds in Equation 66 converge as T becomes large. Hence, from Equations 63 and 66

$$C = \lim_{T \to \infty} \sup \frac{1}{T} \log \sum_{t \le T} |\mathbf{F}_t| \tag{67}$$

Notice that

$$|\mathbf{D}_T| \quad \text{and} \quad \sum_{t \le T} |\mathbf{F}_T|$$

are nowhere near the same magnitude, yet they yield the same result in the capacity computation because their exponential variation with T is the same. Thus e^{TC} (with C in nats) is generally only a very crude indicator of the actual number of messages that can be communicated in T seconds.

Example 2.6. Consider a binary noiseless channel with symbols α and β of duration 1 sec and $e \approx 2.718$ sec, respectively. Durations of sequences from this alphabet are illustrated in Figure 2.16 with branch length representing time. The tree will be complete for $T \le 6$ sec. For simplicity in drawing, some nodes are shown on top of each other, and hence all paths moving from left to right through the resulting lattice represent distinct code symbol sequences. Table 2.7 indicates the corresponding sequence sets.

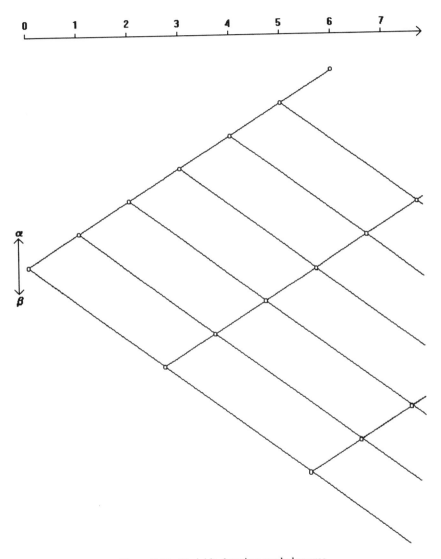

Figure 2.16. Variable duration symbol source.

Table 2.7. An Example of a Variable Duration Symbol Source

t	$\mathbf{F}_t$	$|\mathbf{F}_t|$	$\mathbf{D}_t$	$|\mathbf{D}_t|$	$\mathbf{R}_t$	$\dfrac{1}{t}\log_2\left(\sum_{t'\le t}|\mathbf{F}_{t'}|\right)$
0	—	1	$\mathbf{F}_0$	1	0	0
1	α	1	$\mathbf{F}_1 \cup \mathbf{F}_0$	2	1	1
2	$\alpha\alpha$	1	$\mathbf{F}_2 \cup \mathbf{F}_1 \cup \mathbf{F}_0$	3	0.79	0.79
e	β	1	$\mathbf{F}_e \cup \mathbf{F}_2 \cup \mathbf{F}_0$	3	0.58	0.74
3	$\alpha\alpha\alpha$	1	$\mathbf{F}_3 \cup \mathbf{F}_e \cup \mathbf{F}_1 \cup \mathbf{F}_0$	4	0.67	0.77
$1+e$	$\alpha\beta, \beta\alpha$	2	$\mathbf{F}_{1+e} \cup \mathbf{F}_3 \cup \mathbf{F}_e \cup \mathbf{F}_2$	5	0.62	0.76
4	$\alpha\alpha\alpha\alpha$	1	$\mathbf{F}_4 \cup \mathbf{F}_{1+e} \cup \mathbf{F}_3 \cup \mathbf{F}_e \cup \mathbf{F}_2$	6	0.65	0.75
$2+e$	$\alpha\alpha\beta, \alpha\beta\alpha, \beta\alpha\alpha$	3	$\mathbf{F}_{2+e} \cup \mathbf{F}_4 \cup \mathbf{F}_{1+e} \cup \mathbf{F}_3 \cup \mathbf{F}_e$	8	0.64	0.73
5	$\alpha\alpha\alpha\alpha\alpha$	1	$\mathbf{J}_2 \cup \mathbf{F}_{2+e} \cup \mathbf{F}_4 \cup \mathbf{F}_{1+e} \cup \mathbf{F}_3 \cup \mathbf{F}_e$	9	0.63	0.72
$2e$	$\beta\beta$	1	$\mathbf{F}_{2e} \cup \mathbf{F}_5 \cup \mathbf{F}_{2+e} \cup \mathbf{F}_4 \cup \mathbf{F}_{1+e} \cup \mathbf{F}_3$	9	0.58	0.68
$3+e$	$\alpha\alpha\alpha\beta, \alpha\alpha\beta\alpha, \alpha\beta\alpha\alpha, \beta\alpha\alpha\alpha$	4	$\mathbf{F}_{3+e} \cup \mathbf{F}_{2e} \cup \mathbf{F}_5 \cup \mathbf{F}_{2+e} \cup \mathbf{F}_4 \cup \mathbf{F}_{1+e}$	12	0.63	0.72
6	$\alpha\alpha\alpha\alpha\alpha\alpha$	1	$\mathbf{F}_6 \cup \mathbf{F}_{3+e} \cup \mathbf{F}_{2e} \cup \mathbf{F}_5 \cup \mathbf{F}_{2+e} \cup \mathbf{F}_4 \cup \mathbf{F}_{1+e}$	13	0.62	0.69

Equation 67 is valuable because it is relatively easy to evaluate $|\mathbf{F}_t|$ inductively. In evaluating C, we make use of the recursion

$$|\mathbf{F}_t| = |\mathbf{F}_{t-T_1}| + |\mathbf{F}_{t-T_2}| + \cdots |\mathbf{F}_{t-T_{|\mathbf{X}|}}| \qquad (68)$$

Equation 68 simply states that the collection $\mathbf{F}_t$ of sequences of symbols of duration t must be composed of sequences from $\mathbf{F}_{t-T_1}$ each followed by the symbol of duration T_1, the sequences from $\mathbf{F}_{t-T_2}$ followed by the symbol of duration T_2, etc. To make Equation 68 tally for all values of t, we must let $|\mathbf{F}_t| = 0$ for $t < 0$ and $|\mathbf{F}_0| = 1$, corresponding to the empty sequence. The fact that we artificially include a symbol sequence of duration zero in our calculations does not affect the convergence of Equation 67 to the desired value of capacity, since an increase by 1 does not affect the exponential growth of $\sum_{t \leq T} |\mathbf{F}_t|$.

We can now define a generating function that relates the parameters of importance: $|\mathbf{F}_t|$, T_i, $i = 1, 2, \ldots, |\mathbf{M}|$, and C. Let

$$g(s) = \sum_t |\mathbf{F}_t| e^{-ts} \qquad (69)$$

In mathematical terminology, $g(s)$ is known as a Dirichlet series. The theory of Dirichlet series states that a series with nonnegative coefficients converges for all values of s such that $\mathrm{Re}\{s\} > \sigma_0$ and diverges for all s where $\mathrm{Re}\{s\} < \sigma_0$. [See Apostol (1976), pp. 224–26.] The number σ_0 having this property is known as the *abscissa of convergence* of the series. Furthermore, if the abscissa of convergence is nonnegative ($\sigma_0 \geq 0$), then it is given by

$$\sigma_0 = \limsup_{T \to \infty} \frac{1}{T} \log_e \sum_{t \leq T} |\mathbf{F}_t| \qquad (70)$$

Comparing Equations 67 and 70 indicates that

$$\sigma_0 = C \qquad (71)$$

Thus, the abscissa of convergence of the generating function is the capacity of the channel. We now proceed to find the abscissa of convergence.

The generating function can be written in a relatively closed form by applying Equation 68 to Equation 69 as follows:

$$g(s) = \sum_{t>0} \left(\sum_{i=1}^{|\mathbf{X}|} |\mathbf{F}_{t-T_i}| \right) e^{-ts} + 1 \qquad (72)$$

Coding for Discrete Noiseless Channels

Here, the one term added corresponds to $N(0)e^{-0s}$ in the Dirichlet series expansion of $g(s)$; the expansion in Equation 68 is not valid for $t = 0$. Equation 72 can be rewritten as

$$g(s) = \sum_{i=1}^{|\mathbf{X}|} e^{-T_i s} \left[\sum_{t>0} |\mathbf{F}_{t-T_i}| e^{-(t-T_i)s} \right] + 1 \tag{73}$$

The expression in brackets is simply the generating function $g(s)$. Thus, the generating function must satisfy the simple relation

$$g(s) = \left[\sum_{i=1}^{|\mathbf{X}|} e^{-T_i s} \right] g(s) + 1 \tag{74}$$

By letting $x = e^{-s}$, then taking the Laplace transform of $g(s)$ to obtain $G(x)$, we can eliminate the exponential nature of Equation 74 to obtain

$$G(x) = \frac{1}{1 - \sum_{i=1}^{|\mathbf{X}|} x^{T_i}} \tag{75}$$

The abscissa of convergence C of $g(s)$ is related to the radius of convergence x_0 of $G(x)$ by the equation

$$C = -\log x_0 \tag{76}$$

Figure 2.17 indicates the transformation of the region of convergence under the mapping $x = e^{-s}$. By solving for the smallest real value x_0 of $|x|$ corresponding to an x that makes the denominator in Equation 75 go to zero, we can determine the capacity of the channel using Equation 76. Note that the zero in the denominator of Equation 75 must occur on the positive real axis between 0 and 1. We have proved the Theorem 2.4.

Theorem 2.4. The capacity of a discrete noiseless channel having symbol time durations $T_1, T_2, \ldots, T_{|\mathbf{X}|}$ is given by $C = -\log x_0$, where x_0 is the smallest positive root of the polynomial equation

$$1 - x^{T_1} - x^{T_2} - \cdots - x^{T_{|\mathbf{X}|}} = 0 \tag{77}$$

Example 2.7. We can check our work by using $|\mathbf{X}|$ signals of equal duration T_1. Equation 77 implies that $|\mathbf{X}| x_0^{T_1} = 1$ and hence,

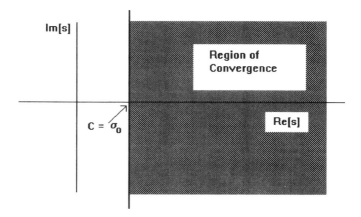

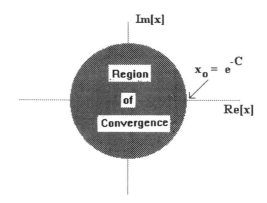

Figure 2.17. Convergence regions.

$$x_0 = |\mathbf{X}|^{-1/T_1} \tag{78}$$

It follows that the capacity of the channel is

$$C = -\log x_0 = \frac{1}{T_1} \log_2 |\mathbf{X}| \text{ bits/sec} \tag{79}$$

which agrees with previous results for the noiseless channel.

Example 2.8. With symbols of duration 1 sec. and e sec.

Coding for Discrete Noiseless Channels

$$G(x) = \frac{1}{1 - x^1 - x^e} = 1 + \sum_{k=1}^{\infty} (x + x^e)^k$$

$$= x^0 + x^1 + x^e + [x^2 + 2x^{(1+e)} + x^{2e}]$$
$$+ [x^3 + 3x^{e+2} + 3x^{2e+1} + x^{3e}] + \cdots$$

In the Dirichlet series expansion for $G(x)$, the exponents are the symbol sequence durations that occur, and the coefficients are the number of sequences in that duration. The root of the equation $1 - x - x^e$ nearest zero is given by $x_0 = 0.667168 \cdots$ so that

$$\text{capacity } C = -\log_2 0.667168 \cdots \sim 0.58338 \text{ bits/sec}$$

Now let us reconsider the definition of capacity C given in Equation 63. A seemingly more reasonable definition may be that the channel capacity C' is the least upper bound (lub) over the set of attainable information rates

$$C' = \operatorname*{lub}_{0 < T < \infty} R_T \tag{80}$$

We may be tempted to assume that C' and C (from Equation 63) are identical, which is not true. It is possible for C' to exceed C, as in the case of a binary channel with 1-sec and e-sec symbols, where $R_1 = 1$ bit/sec. To operate at the capacity C' as opposed to the capacity C, in addition to transmitting the known symbols, we must have the ability to transmit either fractions of channel symbols or *nothing*. That is, to communicate at rate R_T, we would divide time into units of T seconds and use the sequences of $\mathbf{D}_T$, some of which may be of duration less than T, for transmission during each interval.

On the other hand, C' is uninteresting, since it *is not* the capacity of the channel when *arbitrary* periods of no transmission are allowed. For example, with a 1-sec duration signal and arbitrary pauses, it is possible to transmit an infinite amount of information through a noiseless channel in slightly over 2 seconds as follows: Let a_i, $i = 1, 2, \ldots$, be a binary data sequence and let

$$\tau = \sum_{i=1}^{\infty} a_i 2^{-i} \tag{81}$$

Hence, τ is a number between 0 and 1. By first transmitting the 1-sec signal followed by a pause of duration τ and then the 1-sec signal again, a *perfect* measurement of the pause τ will recover all the data exactly. The capacity of a noiseless channel with such a signal set is infinite! Our point about the uninteresting nature of C' is now clearer. If we are not interested in periods of no transmission, we are interested in C rather than C'. If we are interested in functional periods of no transmission, we might as well have arbitrary periods of no transmission and (theoretically) infinite capacity. (Admittedly, perfect measurements are not physically possible.) In most cases, C' is a poor compromise.

We now relate C to the rate of transmission and earlier results on controllable rate codes.

When a variable symbol duration channel is used in conjunction with a fixed-rate source $\mathbf{M}$ having messages of duration T_S, then reliable communication using unique encoding of ϵ-typical sequences requires the number N_k of ϵ-typical sequences of k messages to be less than the number of distinct channel sequences available. This is guaranteed for large enough k when

$$N_k \leq e^{k[H(\mathbf{M}|\mathbf{M}^\infty)+\epsilon]} < |\mathbf{D}_{kT_s}| \tag{82}$$

Taking logarithms and dividing by kT_s gives

$$\frac{H(\mathbf{M}|\mathbf{M}^\infty)}{T_S} + \frac{\epsilon}{T_S} < R_{kT_s} \tag{83}$$

Since ϵ can be arbitrarily small and Equation 83 indicates that R_{kT_s} can approach C for large k, the necessary constraint for typical sequence coding is

$$\frac{H(\mathbf{M}|\mathbf{M}^\infty)}{T_S} < C \tag{84}$$

but since the sequences in $\mathbf{D}_{kT_s}$ do not all have the same duration, a buffering problem exists.

Controllable rate sources are easily adapted to variable symbol duration channels. Good block codes for a source $\mathbf{M}$ should have small values of average code-word duration $\bar{t}$

$$\bar{t} = \sum_{M \in \mathbf{M}} Pr(M)t(M) \tag{85}$$

where $t(M)$ is the duration of the code word assigned to M. Since $\bar{t}$ is measured in seconds/message, it could be used to convert entropy $H(\mathbf{M})$ information units/message into a time rate for comparison to C. Based on what capacity represents, we expect reliable communication using block codes to be established when

Coding for Discrete Noiseless Channels

$$\frac{H(\mathbf{M})}{\bar{t}} < C \tag{86}$$

and based on this conjecture, an appropriate definition of efficiency can be made. Notice that $H(\mathbf{M})/\bar{t}$ can be interpreted as

$$\frac{H(\mathbf{M})}{\bar{t}} = \sum_{M \in \mathbf{M}} Q(M) \left[\frac{1}{t(M)} \log \frac{1}{Pr(M)} \right] \tag{87}$$

where the $Q(M)$, given by

$$Q(M) = \frac{t(M) Pr(M)}{\bar{t}} \tag{88}$$

are nonnegative and sum to 1. Therefore, to determine the source probabilities that are well matched to the channel, i.e., $H(\mathbf{M})/\bar{t} \approx C$, we need only equate

$$\frac{1}{t(M)} \log \frac{1}{Pr(M)} = C \tag{89}$$

or

$$Pr(M) = e^{-t(M)C} \tag{90}$$

where C is measured in nats/sec. A specialization of Equation 90 to the case of fixed-symbol duration channels yields the equivalent result in Equation 54b.

Problems

1. Consider the noiseless channel using code symbols of duration 1 sec, 3 sec, and 4 sec.

    (a) What is the channel capacity in bits/second?

    (b) A memoryless information source output with the following message probabilities is transmitted over the channel:

$$P(s_1) = 0.30,\ P(s_2) = 0.20,\ P(s_3) = 0.15,\ P(s_4) = 0.12,$$

$$P(s_5) = 0.1,\ P(s_6) = 0.08,\ P(s_7) = 0.03,\ P(s_8) = 0.01,\ P(s_9) = 0.01$$

Find a good code for this source and calculate the information rate of the coded source in bits/second.

2. Consider the noiseless communication channel operating with three symbols of duration 2 sec and four symbols of duration 4 sec.
   (a) What is the channel capacity in bits/second?
   (b) A seven-message zero-memory source provides information to be transmitted. What set of source symbol probabilities would you use to maximize the information rate in bits/second?

2.10. Lempel–Ziv Coding Procedure

2.10.1. Agent 00111's Problem

The Huffman coding procedure and the extension to Markov sources helped Agent 00111 solve a large subset of his problems, namely, those involving a fairly clear assignment, a reasonably specific idea of what to report, and a set of *a priori* probabilities. For such missions, Agent 00111 constructed code books based on *a priori* probabilities, left a copy with headquarters, then departed. Precomputed code books did not work for some missions. For these, often the most lucrative given his pricing scheme, Agent 00111 either went on a mission with few preconceptions or found on arrival a completely different scenario than expected. In such cases, his prearranged code books were completely inadequate, and instead of saving transmission time or storage, they could cost him more than simply using straight text. What was worse, if communication was one way or at a premium, he could not reestablish a new code book for text compression with his headquarters by a separate channel. What Agent 00111 needed was a way of taking any sequence from an alphabet, transmitting it into a compressed form to his headquarters while instructing them *at the same time* how to decompress it. The Lempel–Ziv algorithm answers these needs.

2.10.2. The Algorithm

There are several ways of viewing the Lempel–Ziv algorithm. In its original form, it was used to measure the complexity of a sequence, and we introduce the concept from this perspective.

Coding for Discrete Noiseless Channels

If someone asked which numbers between 0 and 9 are random numbers, you would be justified in laughing. However, if someone asked which decimal sequences of length 100 were random, the question requires some thought. How does this question differ from the 1–9 question except in size? Human intuition tends to regard some sequences as less random than others.* To give the question some precision, we will attempt to define the complexity of a specific sequence.

We measure the complexity of a sequence in terms of a computer that has to be programmed to generate the sequence. We define the complexity of a sequence as the number of program lines in the shortest possible program that will on execution produce the sequence. (Of course, we have to observe minor caveats, making sure that program lines carry the same number of machine cycles and counting prestored data as program lines, etc.) Based on this definition, a complex sequence requires a lot of code. It follows that instead of sending a sequence, Agent 00111 could write the program and transmit the program to headquarters, presuming he were good at programming the computer in question.

Our definition depends on the machine that is to execute the programs. Are we allowed to design a machine suitable for the type of sequences expected, for example, a sequence of binary, ternary or Chinese symbols? Who declares the program optimal? We compensate in our definition by specifying the machine to be used and by making the machine extremely simple.

In the Lempel–Ziv algorithm, the machine is of the type shown in Figure 2.18. It contains a shift register of length n, a pointer to one of the register stages, and an input to the register. We refer to the rightmost stage as s_1 and the k-th stage from the right as s_k. We refer to the contents of s_k as x_{-k}. Instructions or program lines are either of the form (L, ρ, c), $L > 0$, or $(0, c)$, where

1. ρ (when present) instructs the machine where to place the pointer on the register, i.e., the pointer is placed on s_ρ. When the pointer is placed, the machine is wired as a feedback register. That is to say, for every clock pulse applied, $x_{-\rho}$ will be copied and placed into s_1. In addition, the contents of the register move one to the left

$$y = x_{-\rho}(old)$$

$$x_{-(k)}(new) = x_{-(k-1)}(old) \qquad \text{for } k \text{ decrementing from } n \text{ to } 2$$

$$x_{-1}(new) = y$$

---

* As an aside, human intuition is very bad at probability theory. Witness, for example, human suspicion and outrage if two people in the same family win two prizes in a lottery.

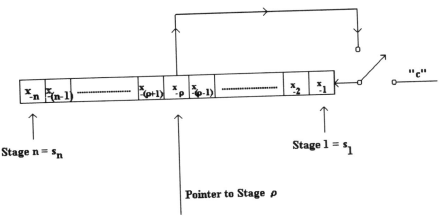

Figure 2.18. Schematic for the Lempel–Ziv decompression algorithm.

2. L indicates how many times the machine is to be clocked with the pointer in place. If $L = 0$, we can economize by not specifying the pointer.

3. c is an element of the alphabet. After the register is stepped L times with the pointer at s_p, the pointer is removed and the register is clocked once more moving the sequence to the left. The symbol c is entered into the register

$$y = c$$
$$x_{-(k)}(new) = x_{-(k-1)}(old) \quad \text{for } k \text{ decrementing from } n \text{ to } 2$$
$$x_{-1}(new) = y$$

Example 2.9. Suppose headquarters has a machine holding the following decompressed data in the register:

$$\_|\_|\_|\_|\_|\_|\_|\_|\_|\_|H|U|M|P|T|Y|\_|D$$

where | partitions the stages of the register and _ indicates a null character. Suppose Agent 00111 sends $(6, 7, s)$. As $p = 7$, the pointer is placed on the seventh stage from the right, i.e., the memory element holding U. Since $L = 6$, the register is clocked for 6 cycles, feeding $UMPTY\_$ into the right-hand side. Finally, the register is clocked once more to place s into the right-hand element; this gives

Coding for Discrete Noiseless Channels

$$\_|\_|\_|\_|H|U|M|P|T|Y|\_|D|U|M|P|T|Y|\_|s$$

This completes a description of the Lempel–Ziv algorithm if we wish to go from a stream of program lines to the original sequence. However, there are at least four major questions that remain to be addressed.

1. How does Agent 00111 send an arbitrarily large positive integer in an efficient format; i.e., how are p and L sent? The answer to this question is the subject of Chapter 4. In fact, the Lempel–Ziv algorithm motivates the question, although the algorithm did not precede the study of integer compression.

2. How does Agent 00111 generate the list of instructions? In essence, by virtue of the machine's simplicity, good quality code may be automatically generated. We discuss this later.

3. How does the machine behave vis-à-vis sequences generated from a stationary ergodic information source? It has been shown (Wyner and Ziv, 1989) that as n approaches infinity, the scheme achieves, with high-probability, the theoretical optimal economy of transmitting only H bits per source symbol, where H is, of course, the entropy of the information source. The results are discussed in Section 2.10.3.

4. Could a more complicated machine achieve more, in some sense of the term more? This has several possible interpretations. One interpretation is that for a stationary ergodic information source, a more complicated machine may need less memory to approximate the limit of H. Another interpretation is that a more complicated machine may be better at handling information sources that are not stationary nor ergodic. Note that this ties in with the discussion of language generation in Section 1.7.

In spite of the preceding questions, we have in principle solved Agent 00111's problem with a lack of *a priori* information about messages. He can now examine the sequence he wants to send, generate instructions on how to manufacture the sequence, and send instructions. He can even regenerate the sequence from his instructions to cross-check his programming. However, if Agent 00111 does make an error and transmits it, it could have a catastrophic effect on reconstructing the sequence at headquarters: Programs with bugs are not forgiving. This would be a major problem for Agent 00111 if instruction generation were not automated.

Concerning automatic code generation, it is important to distinguish:

- the complexity of the machine with which the transmitter and receiver are equipped,

- the redundancy of the instruction set transmitted by the machine
- the efficiency of the transmitter in generating optimal (or near-optimal) code for generating a sequence

It is possible to put greater functionality in the decompression machine than previously described. The transmitted program does not *necessarily* have to use this greater versatility any more than a programmer has to use complex multiplication in an integer problem. However, if the sequence happens to fall into a class where additional versatility is appropriate, further savings can be made. The penalty for additional versatility is that each transmitted instruction may have to be longer, i.e., to instruct the receiver when and when not to use an added feature.

Example 2.10. Suppose the receiving computer can be instructed to take a section of the buffer and add it to the buffer in reverse order, i.e., a palindromic facility. This may be of little or no use for some sequences. However, suppose the register of decompressed text held the following:

_|_|_|_|_|_|_|_|_|_|_|_|_|_|
_|_|_|_|_|_|A|B|L|E|_|W|A|S|_|I|_|E|R

and the instruction was

1. Place ABLE_WAS_L_E in a separate buffer (which requires only a pointer ρ and a length L instruction, as before).
2. Empty the buffer into the register starting from the right rather than the left (which requires one more bit of transmitted data per instruction).
3. Add . to the end.

The result is the rapid construction of

_|_|_|_|_|_|_|_|_|_|A|B|L|E|_|W|A|S|
_|I|_|E|R|E|_|I|_|S|A|W|_|E|L|B|A|.|

In this case, the palindromic upgrade was worth the overhead in instruction set size.

More general examples can obviously be considered. It follows that the decision about computer functionality at the point of decompression involves the compactness of the transmitted instruction set and expectations (if any)

Coding for Discrete Noiseless Channels

about sequence structure. If the latter is unknown (or varies across different sequences of interest), published results indicate it is better to keep it simple.

We use the Lempel–Ziv instruction sets (L, ρ, c) and $(0, c)$, $L > 0$ to develop an algorithm for automatically generating instructions from a sequence. In the following, we fix the source alphabet to be A, where $|A|$ is finite and nonzero.

Let the data sequence be $(x_i, -\infty \le i \le \infty)$. We write x_i^j for the sequence that begins with x_i and ends with x_j whenever $j \ge i$, e.g., $x_{-3}^2 = (x_{-3}, x_{-2}, x_{-1}, x_0, x_1, x_2)$. The same notation is extended to include $x_{-\infty}^j$, x_i^∞ and $x_{-\infty}^\infty$ in the obvious manner.

We assume without loss of generality that the text previous to x_0 has been compressed, i.e., $x_{-\infty}^{-1}$ has been processed. The logic of the next instruction generation is summarized in Figure 2.19. The algorithm begins by checking to see if x_0 has ever appeared in the previous history retained in memory, i.e., does x_0 appear as an element of x_{-n}^{-1}? If not, we send the instruction $(0, x_0)$ and proceed to process x_1^∞. If x_0 does appear, we go to the main algorithm. This algorithm finds the values of $L > 0$ and ρ: $0 < \rho \le n$ for which

1. $x_{-\rho}^{L-\rho-1} = x_0^{L-1}$.

2. L is the maximal possible length subject to (1) being true.

3. ρ is minimal subject to $0 < \rho \le n$ and (1) and (2) being true.

Let L_{max} and ρ_{min} be the values found by the algorithm. The instruction $(L_{max}, \rho_{min}, x_{L_{max}})$ is sent if L_{max} is finite. If L_{max} is infinite,[†] (∞, ρ_{min}, c) is sent where c is arbitrary.

In explanation, note that the existence of x_0 in x_{-n}^{-1} implies that there exists a pair (L, ρ) for which (1) is true: If ρ_1 is minimal subject to $x_{-\rho_1} = x_0$ and $0 < \rho_1 \le n$, set $L = 1$ and $\rho = \rho_1$. Hence, there exists (L_{max}, ρ_{min}) that satisfies (1), maximizes L, and then minimizes ρ. Moreover, if we assume that $L = L_{max}$ is finite and maximal, x_L is *not* equal to $x_{L-\rho}$, since equality implies that (L, ρ) can be replaced by $(L + 1, \rho)$. The aim of the main algorithm is to find values for (L_{max}, ρ_{min}). The starting point for the main algorithm is a pair (L, ρ) for which $x_{-\rho}^{L-\rho-1} = x_0^{L-1}$. In the first step of the main algorithm, the matching pattern is extended whenever possible. That is, if $x_{-\rho}^{L-\rho-1}$ equals x_0^{L-1} and $x_{L-\rho}$ equals x_L, we replace L by $L + 1$. If and when the process of extension terminates, $x_{L-\rho}$ does not equal x_L. Given a termination, we try and match x_0^L within the contents of x_{-n}^{-1}. If such a match does *not* exist we already have the maximal value of L and the desired minimal

---

[†] In practice, a maximum possible size for L_{max} is set so that the maximum possible value of L_{max} is always finite.

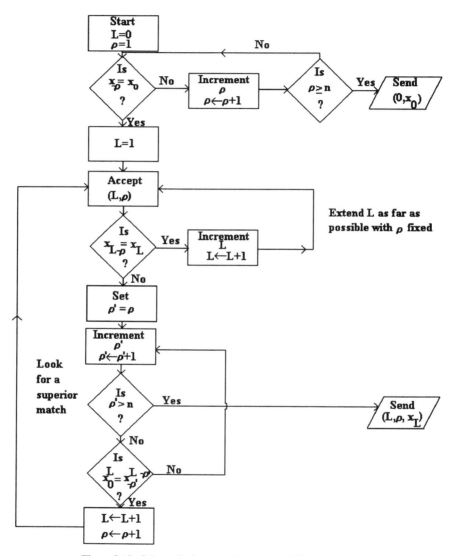

Figure 2.19. Schematic for generating Lempel–Ziv instructions.

Coding for Discrete Noiseless Channels

value of ρ. If such a match does exist, suppose that ρ' is minimal subject to $x_{-\rho'}^{L-\rho'} = x_0^L$ and $0 < \rho' \le n$. We replace L by $L + 1$, replace ρ by ρ', then return to the first step of the algorithm. (Note that $\rho' > \rho$ and we have only to check $x_{-\rho'}^{L-\rho'}$ against x_0^L for the pointers $\rho' > \rho$ for which $x_{-\rho'}^{L-\rho'-1} = x_0^{L-1}$. Such a check reduces to seeing if $x_{L-\rho'} = x_L$ or not.) It should be apparent that the algorithm closes with the desired maximal value of L and associated minimal pointer.

2.10.3. The Lempel–Ziv Algorithm and Entropy

Work by A. D. Wyner and J. Ziv (1989) established the connection between the Lempel–Ziv algorithm and the entropy of a stationary ergodic information source. The essence of their work is to apply Kac's lemma, a result on the average recurrence time between two identical observations obtained by sampling a stationary ergodic process, to the expected time between occurrences of a pattern generated by a stationary ergodic information source. To state the results, we need some notation. Let

$$M = (X_k)_{k=-\infty}^{\infty}$$

be a stationary ergodic information source, where X_k is a random variable that takes values in a finite alphabet A. Let X_i^j denote the sequence of random variables that starts with X_i and ends with X_j whenever $j \ge i$. (Conceptually, we might sample X_i^j to get values x_i^j.) For any values of $j \ge i$ for which $j - i = m$, let the mth order probability distribution for M be

$$P_M^{(m)}(x_i^j) = Pr\{X_i^j = x_i^j\}.$$

From Chapter 1, the entropy of the source is

$$H(M) = \lim_{m \to \infty} E\left[-\frac{1}{m} \log_2 P_M^{(m)}(X_0^{m-1})\right], \qquad (91)$$

where E denotes the expected value operator. For any specified buffer length n and any sequence $x_{-\infty}^{\infty}$ generated from M, the integers $L_{\max}$ and $\rho_{\min}$ have been defined in Section 2.10.2. We define random variables $L_n(M)$, $N_{l,n}(M)$, and $N_l(M)$ from the following definitions:

When $X_{-\infty}^{\infty} = x_{-\infty}^{\infty}$ and the shift register in Figure 2.17 is of length n, $L_n(M) = L_{\max}$.

When $\mathbf{X}_{-\infty}^{\infty} = x_{-\infty}^{\infty}$, the shift register in Figure 2.17 is of length n, and $L_{max} = l$, $N_{l,n}(M) = \rho_{min}$.

$N_l(M) = \lim_{n \to \infty} N_{l,n}(M)$.

Since a Lempel–Ziv instruction takes the form $(0, c)$ or (L_{max}, ρ_{min}, c), the asymptotic performance of the compression scheme depends on the behavior of these random variables.

Theorem 2.5. [*Wyner and Ziv (1989)*].

(a) As $n \to \infty$,

$$\frac{\log n}{L_n(M)} \to H(M) \tag{92}$$

in probability.

(b) As $l \to \infty$,

$$\frac{\log N_l(M)}{l} \to H(M), \tag{93}$$

in probability.

We must also know the number of bits necessary to transmit arbitrarily large positive integers, a subject discussed in Chapter 4. In Theorem 2.6, x^+ denotes the smallest integer for which $x^+ \geq x$ and all logarithms are base two and $O(z)$ denotes a function which is bounded as z goes to infinity.

Theorem 2.6. [*Elias (1975)*]. An integer N can be encoded with $(\log N)^+ + O(\log \log N)$ bits when N is large.

Consider a Lempel–Ziv instruction (L, ρ, c) derived from a register of length n. Since ρ lies in the range $[1, n]$, it takes at most $(\log n)^+$ to encode ρ. Similarly, it takes at most $(\log |A|)^+$ bits to encode c. [Actually, it takes at most $\log(|A| - 1)^+$ bits, since x_L is not equal to $x_{L-\rho}$]. Note that (L, ρ, c) encodes $L + 1$ source symbols. It follows that for a large enough value of L, the Lempel–Ziv instruction (L, ρ, c) uses a number of encoded bits per source symbol that equals

$$\frac{(\log L)^+}{L+1} + \frac{O(\log \log L)^+}{L+1} + \frac{(\log n)^+}{L+1} + \frac{(\log |A|)^+}{L+1} \tag{94}$$

Coding for Discrete Noiseless Channels

For large values of L and n,

$$\frac{(\log n)^+}{L+1}$$

is the only term in Equation 94 that does not *necessarily* go to zero.

As n tends to infinity, Equation 92 states that with high probability, L also tends to infinity. This has three corollaries:

1. The probability of $L = 0$, i.e., a $(0, c)$ instruction, tends to zero as n increases.

2. For a large enough value of n, the number of encoded bits per source symbol converges in probability to the quantity in Equation 94.

3. The number of encoded bits per source symbol is either zero or dominated by the third term in Equation 94. Applying Equation 92 to this term, the number of encoded bits per source symbol is with high probability close to the entropy $H(M)$ as n increases.

We have deduced the following corollary to Theorems 2.5 and 2.6.

Corollary 2.1. When applied to a sequence from a stationary ergodic information source M with a large enough amount of memory, the Lempel–Ziv algorithm will achieve an average number of encoded bits per source symbol that with high probability is close to the entropy $H(M)$.

The idea behind the proof may be heuristically demonstrated for a data compression scheme similar to the Lempel–Ziv algorithm. In this scheme, we assume the following:

Both the encoder and decoder know $x_{-\infty}^{-1}$, i.e., they both have an infinite memory of the stream prior to the symbol at time 0.

We compress blocks of m bits when m is a constant.

Let $b = x_0^{m-1}$. Using the same pattern-matching approach, we find the least value of ρ for which $x_{-\rho}^{m-\rho-1}$ is equal to b. If b never appears in the previous history of the stream, we send b in uncompressed form with a bit instructing the receiver about what is happening. We can view the process as sliding a correlator matched to b through the history of the sequence, stopping when we first obtain a match. In more detail, let y_b be a correlator matched to b. y_b is a function of an m-long vector of inputs $a = (a_0, a_1, \ldots, a_{m-1})$, $a_i \in A$ and has an output equal to

$$y(a, b) = \begin{cases} 0 & \text{if } a \neq b \\ 1 & \text{if } a = b \end{cases} \quad (95a)$$

We set $y_k(b)$ equal to the output of $y(a, b)$ when a is x_{-k}^{m-k-1}. Note that $y_0(b) = 1$, and we are interested in the minimal value of ρ for which $y_\rho(b) = 1$, $\rho > 0$. This minimal value is known as the first-recurrence time for the sequence $(y_k(b) | k \geq 0)$. The first-recurrence time depends on the precise values of both b and $x_{-\infty}^{-1}$. For a fixed b, we would like to estimate the *average* efficiency of compressing b. To define the average, we need some more notation.

We define:

$$Y(\mathbf{X}_{-\infty}^{m-1}, \mathbf{X}_0^{m-1}) = [y(\mathbf{X}_0^{m-1}, \mathbf{X}_0^{m-1}), y(\mathbf{X}_{-1}^{m-2}, \mathbf{X}_0^{m-1}),$$

$$\ldots, y(\mathbf{X}_{-k+1}^{m-k}, \mathbf{X}_0^{m-1}), \ldots] \quad (95b)$$

We let $Y(b) = Y(X_{-\infty}^{m-1}, X_0^{m-1} = b)$. Since M is a stationary ergodic process, it follows that $Y(b)$ is a stationary ergodic process taking values on the binary set $\{0, 1\}$.

We use this notation to find the average first-occurrence time for sequences obtained by sampling the stationary ergodic stochastic process $Y(b)$, a task now phrased completely in the language of stochastic processes. That is, the average is taken over all sequences that are possible histories for b; i.e., we average over all possible sequences $x_{-\infty}^{m-1}$ with $x_0^{m-1} = b$. We need the formal machinery because any individual $x_{-\infty}^{-m-1}$ with $x_0^{m-1} = b$ could have zero probability. Let the average first-recurrence time be $\mu(b)$. Set $Pr(b)$ equal to $P_M^m(b)$.

The Wyner and Ziv theorem relies on Theorem 2.7.

Theorem 2.7. [*Kac's lemma (1947)*]. When M is a stationary ergodic stochastic process

$$\mu(b) = \frac{1}{Pr(b)} \quad (96)$$

Note that $Pr(b)$ is the unconditioned probability that $\mathbf{X}_0^{m-1} = x_0^{m-1} = b$. We are assuming from definition that $Pr(b) > 0$.

With this result, we can indicate that our block compression scheme approaches capacity. The reasoning is as follows:

Coding for Discrete Noiseless Channels

1. By Theorem 2.6 for large values of m, we need on the order of

$$\log[\mu(b)] = \log\left[\frac{1}{Pr(b)}\right] \text{ bits}$$

to encode the pattern b.

2. Averaging the length of the preceding encoding over all possible patterns of b, we obtain the average length of an instruction as

$$H(\mathbf{X}_0^{m-1}) = \sum_{b \in \mathbf{X}_0^{m-1}} Pr(b) \log\left[\frac{1}{Pr(b)}\right] \tag{97}$$

(Note that since $0 \log(0)$ is zero, we can extend the summation to all patterns of b, whether or not they are possible.)

3. Dividing by m, we obtain the average number of encoded bits per source symbol. Letting m go to infinity, we see from Equation 91 that the average number of encoded bits per source symbol tends towards $H(M)$.

2.10.4. The Lempel–Ziv Approach and Sequence Complexity Results

We introduced the Lempel–Ziv algorithm from the perspective of sequence complexity. In Section 2.10.4, we discuss what Lempel and Ziv proved in their early papers.

The encoder in the original Lempel and Ziv variable-to-block compression scheme takes a variable amount of source data belonging to an alphabet A and sends it as a block of L_c symbols, where L_c is a preset design parameter. The encoder consists of:

A buffer of total length n

A division of the buffer divided into a front section of length $n - L_s$ and a back section of L_s symbols

The parameters are related by

$$L_c = 1 + \log(n - L_s)^+ + \log(L_s)^+$$

where the logs have a base equal to the alphabet size and logs in the equation are rounded up to the nearest integer. It follows that the two design parameters are L_c and the buffer size n.

Text that has already been processed is placed in the front section of the buffer, while text yet to be compressed is placed in the rear section. The encoder finds the maximum *reproducible extension* of the front buffer to the rear. That is, the encoder considers which position of the pointer in the front part of the buffer allows the machine to produce the maximum match with data in the rear. Let the maximum pointer value be ρ and the length of match be l. Finally, let the next symbol be c. Then, the Lempel-Ziv algorithm would send

$$(\rho - 1, l - 1, c)$$

where the numbers are expressed in radix $|A|$ arithmetic, where $|A|$ is the alphabet size. This notation and the (L, ρ, s) or $(0, s)$ instruction set in Section 2.10.2 are clearly related. Since $\rho \leq n - L_s$ and $l \leq L_s$, the length of the transmitted instruction in alphabet symbols is a constant as claimed, i.e., $L_c = 1 + \log(n - L_s)^+ + \log(L_s)^+$.

The performance of the algorithm varies from sequence to sequence, and since we do not know what sequence will be used with what probability, the idea of an average sequence is not computable. However, for a particular source, it is possible to compute the worst compression achieved on any infinite sequence. Once the buffer has been optimized for the source statistics, this compression ratio ρ_r was shown by Lempel and Ziv to be given by

$$\frac{L_c}{L_s} \leq \rho_r \leq \frac{L_c}{L_s - 1}$$

Moreover, they proved that for the same block length, no variable-to-block or block-to-variable compression scheme can asymptotically outperform the Lempel–Ziv algorithm when criteria for the worst compression ratio are adopted. These results are based on analysis assuming a knowledge of the source statistics. In fact, knowledge of the statistics is not critical; a lack of knowledge can be compensated for by increasing the buffer until it surpasses the (unknown) optimal length.

2.11. Notes

The algorithm for determining if a code is U_F or U_S was first published by Sardinas and Patterson (1953). One possible state diagram approach to testing for unique decipherability was given by Even (1963). Shannon (1948) contains many of the fundamental ideas of this chapter, including a heuristic statement of the asymptotic equipartition property (which is the basis for the coding theorem in Section 2.3), the controllable-rate source coding theorem in Section 2.6, and the capacity computation in Section 2.9. The lower bound on the probability of ambiguous decoding follows Fano (1961, chap. 4).

The existence theorem for binary trees was demonstrated by Kraft (1949) for instantaneous codes and later extended to all U_S and U_F codes by McMillan (1956). Schützenberger and Marcus (1959) and Jaynes (1959) employ generating function and partition function concepts to study constraints placed on the structure of code-word dictionaries by unique decodability requirements for possibly variable symbol duration situations. Norwood (1967) has enumerated the number of distinct code trees with a given number of terminal nodes. Huffman (1952) was the first to develop an algorithm that would locate a compact block code for any source probability distribution. Karp (1961) investigated the analogous problem for coding with symbols of unequal time duration. Gilbert (1971) considered the design of tree codes when source message probabilities are not known precisely. Kolmogorov (1965) was the first to propose judging the complexity of a sequence by the shortest binary program required to produce it. This method was extended by Martin-Löf (1966); the Lempel–Ziv paper on finite-sequence complexity (1976) is the one closest to this exposition. Other papers by Lempel–Ziv (1978, 1986) and Rodeh et al. (1981) have extended the idea closer to conventional data compression. Welch (1984) took the algorithm, exposed many implementational issues, and resolved several. It remains essential reading for those interested in implementation. In fact, it is noticeable that recent articles refer to the Lempel–Ziv–Welch (LZW) algorithm. Our exposition of why the Lempel–Ziv algorithm can approach capacity follows Wyner–Ziv (1989). We quote Kac's lemma and reference the proof in Kac (1947); Wyner–Ziv (1989) contains a proof of Kac's lemma that is both shorter and more general in scope. The proof is credited to Ozarow and Wyner.

For a thorough treatment of Dirichlet series, see Hardy and Riesz (1952). For a simpler introduction, see Apostol (1976).

Schemes for handling the problems of buffer overflow and exhaustion when encoding fixed-rate sources into variable-length codes have been developed by Jelinek (1968) and Jelinek and Schneider (1974).

References

T. M. Apostol. 1976. *Introduction to Analytic Number Theory.* Springer–Verlag, UTM.
S. Even. 1963. "Test for Unique Decipherability." *IEEE Trans. Inform. Theory* **IT-9**:109–12.
R. M. Fano. 1961. *Transmission of Information.* MIT Press and Wiley, New York.

E. Gilbert. 1971. "Codes Based on Inaccurate Source Probabilities." *IEEE Trans. Inform. Theory.* Vol 15-17:304-14.

S. W. Golomb. 1980. "Sources Which Maximize the Choice of a Huffman Coding Tree." *Inf. Contr.* **45**:263-72.

G. H. Hardy and M. Riesz. 1952. *The General Theory of Dirichlet's Series.* Cambridge, UK: Cambridge University Press.

D. A. Huffman. 1952. "A Method for the Construction of Minimum-Redundancy Codes." *Proc. IRE* **40**:1098-1101.

E. T. Jaynes. 1959. "Note on Unique Decipherability." *IRE Transactions on Information Theory* **IT-5**:98-102.

F. Jelinek. 1968. "Buffer Overflow in Variable-Length Coding of Fixed-Rate Sources." *IEEE Trans. Inform. Theory* **IT-4**:490-501.

F. Jelinek and K. S. Schneider. 1974. "Variable-Length Encoding of Fixed-Rate Markov Sources for Fixed-Rate Channels." *IEEE Trans. Inform. Theory* **IT-20**:750-55.

M. Kac. 1947. "On the Notion of Recurrence in Discrete Stochastic Processes." *Bull. Amer. Math. Soc.* **53**:1002-10, 1947.

R. M. Karp. 1961. "Minimum-Redundancy Coding for the Discrete Noiseless Channel." *IRE Trans. Inform. Theory* **IT-7**:27-38.

A. N. Kolmogorov. 1965. "Three Approaches to the Quantitative Definition of Information." *Probl. Inf. Transm.* **1**:1-7.

L. G. Kraft. 1949. "A Device for Quantizing, Grouping, and Coding Amplitude Modulated Pulses." M.S. thesis, MIT.

A. Lempel and J. Ziv. 1976. "On the Complexity of Finite Sequences." *IEEE Trans. Inform. Theory* **IT-22**:75-81.

———. 1978. "Compression of Individual Sequences via Variable-Rate Coding." *IEEE Trans. Inform. Theory* **IT-24**:530-36.

———. 1986. "Compression of Two-Dimensional Data." *IEEE Trans. Inform. Theory* **IT-32**: pp. 2-8.

B. McMillan. 1956. "Two Inequalities Implied by Unique Decipherability." *IRE Trans. Inform. Theory* **2**:115-16.

P. Martin-Löf. 1966. "The Definition of Random Sequences." *Inf. Contr.* **9**:602-19.

E. Norwood. 1967. "The Number of Different Possible Compact Codes." *IEEE Trans. Inform. Theory* **IT-3**:613-16.

M. Rodeh, V. R. Pratt, and S. Even. 1981. "Linear Algorithm for Data Compression via String Matching." *J. Ass. Comput. Mach.* **28**:16-24.

A. A. Sardinas and G. W. Patterson. 1953. "A Necessary and Sufficient Condition for the Unique Decomposition of Coded Messages." *1953 IRE Convention Record,* Part 8, 106-08.

M. P. Schützenberger and R. S. Marcus. 1959. "Full Decodable Code-Word Sets." *IRE Transactions on Information Theory,* vol 5, 12-5.

C. E. Shannon. 1948. "A Mathematical Theory of Communication." *Bell Sys. Tech. J.* **27**:379-423, 623-56.

T. A. Welch. 1984. "A Technique for High-Performance Data Compression." *IEEE Comput.* **C-17**:8-19.

A. D. Wyner and J. Ziv. 1989. "Some Asymptotic Properties of the Entropy of a Stationary Ergodic Data Source with Applications to Data Compression." *IEEE Trans. Inform. Theory* **IT-35**:1250-58.

3

Synchronizable Codes

3.1. An Untimely Adventure

Agent 00111 continued to muse over his success as a secret agent. His compatriots, adversaries, and contacts had all declared him to be without parallel in the history of espionage, and who was he to disagree? However, he knew full well that their praise was to some degree self-serving. After all, it was better to be outwitted by someone brilliant than to confess to one's own stupid mistakes. Obviously, it was not in his interest to dispute his own high standing, but, in truth, he had ran into some very stupid people in the course of his career. Sinking back into his overstuffed chair in front of the glowing fireplace, sipping a large brandy, he started to reminisce. He could not help but chuckle as his thoughts ran to a particularly strange sequence of events.

It was a long time ago, during a period of great activity between his present agency and Country X. Agent 00111 was on espionage and terrorist duty and had an extremely urgent assignment to destroy one of X's two remaining clandestine munitions stores hidden in the capital of a neutral country. He had already located one of the stores and was planning furiously to effect its destruction. He knew full well that his organization had been compromised and it was only a matter of time before the unknown double agent told Country X which store had been located. In fact, the double agent had agreed to transmit the identity of the store at precisely midnight that night. The code the double agent had agreed on with his agency involved repeating $S10$ or $S01$ continuously for 1 hour depending on whether Store A or Store B had been located. Fortunately for Agent 00111, the code clerk was a bumbling amateur with as little common sense as, it transpired, sense of preservation. In the interest of data compression, this clerk suggested dropping the letter S from the messages. The simple truth was that the double agent and code clerk did not have synchronized watches or accurately timed transmissions. By the time the clerk had picked up the transmission, he and the assembled party bigwigs heard "... 010101010101 ...." The party reaction to the benefits of this compression at the expense of any useful intelligence was not recorded; neither was the fate of the code clerk.

Agent 00111 laughed out loud; the munitions store had been destroyed, and his reputation was given a boost. In reality, he had been very lucky. The code clerk had not known how to select a code that allowed time synchronization to be unambiguously determined. The solution—using a U_I dictionary—is the subject of Chapter 3.

3.2. Identifying U_I Dictionaries

Let us first consider the hypothetical situation where Agent 00111 expects to receive a doubly infinite sequence of symbols from a field agent but to receive no additional information concerning the location of word synchronization points (i.e., the location of commas used to separate words). Agent 00111 has to determine what conditions to specify on the code-word dictionary to guarantee unique decodability of the doubly infinite sequence. If a code-word dictionary allows a decoder to guarantee unique decodability of a doubly infinite sequence, recall from Chapter 2 that the dictionary is called U_I or infinitely uniquely decodable.

As in the case of the Sardinas–Patterson algorithm, we must investigate the type of ambiguities that can occur. Let $\{x_I\}$ be a doubly infinite sequence of symbols from alphabet X representing a doubly infinite sequence of code words. The four typical forms of unresolvable ambiguities are shown in Figure 3.1 (with ⌞⎯⎯⎯⌟ indicating a code word):

(a)

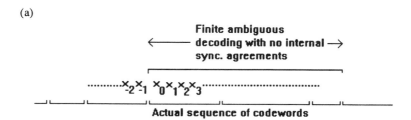

Figure 3.1. Four types of unresolvable ambiguity.

Synchronizable Codes

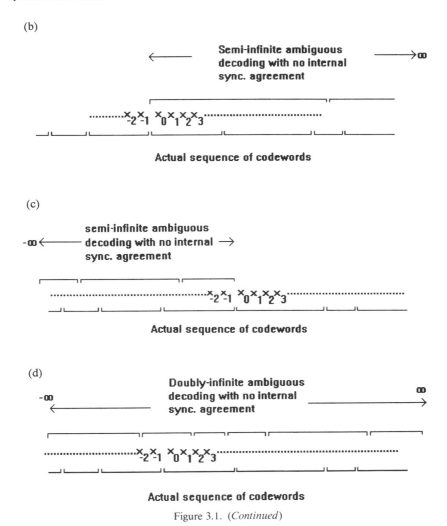

Figure 3.1. (*Continued*)

Case (a) shows a finite period of ambiguity from a fixed starting point in time. Case (b) shows a semi-infinite period of ambiguity from a fixed starting point in time. Case (c) shows a semi-infinite period of ambiguity ending at a fixed point in time; note that Case (c) can be regarded as a time reversal of Case (b). Case (d) shows a doubly infinite period of ambiguity. None of the cases permits word synchronization internal to the period of ambiguity. Notice that ambiguous decodings of the types in Cases (a) and (b) are impossible if and only if the dictionary is U_F and U_S, respectively. Thus, a U_I dictionary must be both U_F and U_S.

Interestingly, even though a code is U_S, a counterexample of the form in Case (c) can exist. As an example, consider the U_S dictionary composed of 0, 10, and 11, from which the sequence $\cdots$ 11111110, $\cdots$ can be constructed. Obviously, the interpretations $\cdots$, 11, 11, 11, 10, $\cdots$ and $\cdots$, 11, 11, 11, 11, 0, $\cdots$ are both possible when all digits to the left of 0 are 1. Thus, if each word in a U_S dictionary is reversed, the resultant dictionary is not necessarily U_S.

If any one of the four forms of unresolvable ambiguities exists for a given dictionary, then an ambiguously decodable sequence of code words exists having the property that *the number of times that a code word in one decoding of the sequence overlaps a code word in another decoding of the sequence is countably infinite.* In a Case (b), (c), and (d) ambiguous sequence, this follows from the definition, since a countably infinite number of words is decoded ambiguously with no agreements on word synchronization anywhere in the ambiguous portion of the decoding. If a Case (a) finite ambiguous decoding exists, the finite segment can be concatenated with itself a countably infinite number of times to yield a code-word sequence whose decoding contains a countably infinite number of overlapping code words, proving the preceding statement.

Note that the number of distinct overlaps of two code words is bounded by the number of code-word tails, which is finite for any finite dictionary. Thus, at least one of the code-word tails, which is an overlapping of two code words, must appear a countably infinite number of times in the ambiguously decodable word sequence. Obviously, whenever a finite segment of a sequence of code words can be found that begins and ends with the same overlap, then a periodic ambiguously decodable sequence of code words exists, since the sequence can be constructed from an overlapping concatenation of the segment with itself an infinite number of times. We have proved that an ambiguously decodable doubly infinite sequence of code words exists if and only if a periodic ambiguously decodable sequence exists.

Example 3.1. Consider the code {00001, 01100, 10011, 10010, 11011, 01101}. Since all words have length 5, the code is automatically U_S. However, the code is not U_I. Notice that beginning with the overlap 01 (the length 2 tail of 00001), a sequence of overlapping code words can be constructed that end in the overlap 01 (the length 2 prefix of 01100; see Figure 3.2a).

(a)

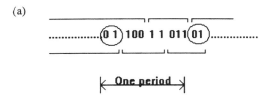

(b)

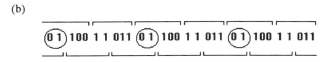

Figure 3.2. A code example.

But the sequence of digits 1001101101 is simply one period in a periodic ambiguously decodable sequence (see Figure 3.2b).

A systematic check for periodic ambiguous sequences can now be conducted. The method used, due to Levenshtein, is a generalization of the Sardinas–Patterson algorithm. Let Seg 0 contain all the code words from a U_S dictionary and let Seg 1 be the collection of all possible code-word suffixes excluding the code words themselves. Note that a code word cannot be a suffix of another code word due to the U_S condition. Compute the remainder of the Seg table as in the Sardinas–Patterson algorithm. If the Seg table becomes periodic, the code is not a U_I dictionary, since a doubly infinite ambiguous sequence exists. If the Seg table ends in an empty Seg, the code is a U_I dictionary, since no periodic ambiguous sequence exists. However, computations do not have to be carried this far. Insert arrows to indicate the elements of Seg K produced by each element of Seg $K-1$, for $K = 2, 3, \ldots$. If two elements connected by a sequence of arrows are identical, a period of an ambiguous sequence has been located. Computation time may be saved by calculating across rather than down in the Seg table. This allows suffixes in higher Segs that stem from one suffix in Seg 1 to be dropped from Seg 1 if they did not lead to a periodic ambiguous sequence. In fact, if any fruitless suffixes appear in later calculations, no further calculations based on them need be made.

Example 3.2. Consider the code defined in Figure 3.3.

Code: 000001, 011111, 000011, 001111, 000111,

001011, 001101, 000101, 010111

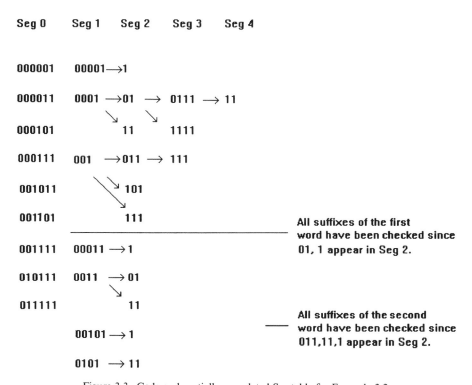

Figure 3.3. Code and partially completed Seg table for Example 3.2.

Fill in the remainder of the table shown in Figure 3.3 to show that the code in the example is a U_I dictionary.

Example 3.3. Consider a similar dictionary defined in Figure 3.4.

Synchronizable Codes 137

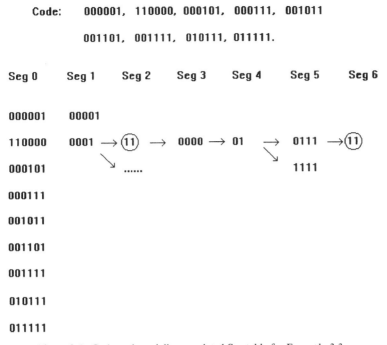

Figure 3.4. Code and partially completed Seg table for Example 3.3.

Without completing the table, it is apparent that we can construct a doubly infinite periodic sequence that is ambiguous, namely, the one shown in Figure 3.5.

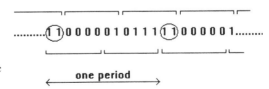

Figure 3.5. A doubly infinite periodic ambiguous sequence.

In Examples 3.1–3.3, the Seg tables beginning with Seg 1 look very much like the trellis diagram of a Markov source. Since the algorithm for constructing Seg K, $K = 2, 3, \ldots$, does not change as K changes, it seems that the equivalent state diagram would be a more compact way of describing the table. Loops in the diagram would correspond to periodic ambiguously decodable sequences.

Example 3.4. The complete equivalent state table for Example 3.2 is shown in Figure 3.6.

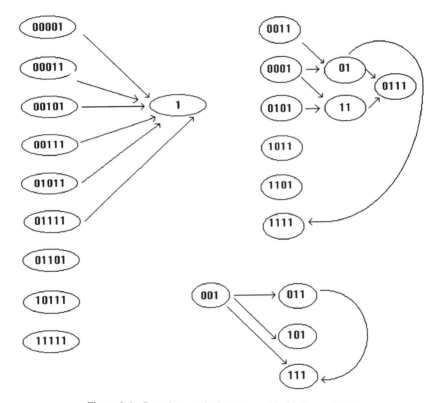

Figure 3.6. Complete equivalent state table for Example 3.2.

Since no loops are present the code is U_I.

Example 3.5. The complete equivalent state table for Example 3.3 is shown in Figure 3.7.

Synchronizable Codes

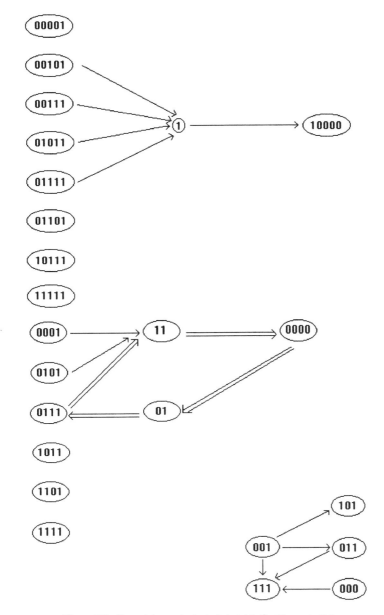

Figure 3.7. Complete equivalent state table for Example 3.3.

The loop (shown with double arrows) indicates that the code is not U_1.

In summary, we can modify the Sardinas–Patterson algorithm to give a compact test of whether a dictionary is U_I or not.

Exercise

1. Are the following codes U_I?

Code A	Code B
100	101
101	1000
1000	1100
1001	1011
1011	10000
10000	10001
10001	10011
10011	10111
10111	10010
10010	
10110	

3.3. The Hierarchy of Synchronizable Codes

The property of being or not being U_I does not exhaust all of a dictionary's possible synchronization capabilities. Many dictionaries are *statistically synchronizable*, but they are not U_I dictionaries. This weak constraint requires that as the length of the coded message sequence examined tends to infinity, the probability of being able to specify correct word synchronization tends to 1. To satisfy such a weak constraint with a dictionary of n-tuples over an $|\mathbf{X}|$ symbol alphabet, it is not surprising that little of the potential dictionary size of $|\mathbf{X}|^n$ words has to be sacrificed. For example, if only one word is dropped from the dictionary and the remaining $|\mathbf{X}|^n - 1$ words occur independently and randomly in messages, then with probability 1, the dropped word will ultimately show up in all the out-of-phase positions, leaving only the true phase as a contender for word synchronization.

Synchronizable Codes

It is even possible to use all $|X|^n$ words but to have one of them occur with a lower probability than the others. In this case, the in-phase position is the one where the abnormal word occurs least often, and an arbitrarily high confidence level can be established by examining longer and longer message sequences, provided that the information source is ergodic. (This method also permits banned word schemes to function in the presence of small numbers of symbol errors.) In another variation, we can use a fixed symbol that is used periodically without banning the symbol from further use within these periods. This allows statistical synchronization, on the premise that the information scheme is unlikely to generate long *periodic* repeats of the fixed symbol. Note that synchronization applies only to the number of code words between repetitions.

At the opposite end of the synchronization spectrum, one of the $|X|$ code symbols can be used exclusively as the last symbol for each code word. This reserved symbol is then an explicit flag for word boundaries. We refer to this synchronization-indicating symbol as a *comma* and to the resultant code as a *comma code*. The major drawback of this coding system is the associated reduction in the information-bearing capacity of the system, since only $|X| - 1$ symbols of the $|X|$ symbol alphabet are available for communicating information. Moreover, capacity may be wasted transmitting the commas on a once-per-word basis.*

These extremes in synchronizable codes have different appeals to Agent 00111. Suppose a comma code has a maximum word length of L. Then Agent 00111 need only inspect, at worst, $L - 1$ symbols to locate a synchronization point; that is, if the $L - 1$ symbols contain a comma, the synchronization point has been determined. If it does not, the next symbol must conclude the code word, so the synchronization point has still been determined.

On the other hand, statistical synchronizable codes allow Agent 00111 more data per transmitted symbol but with the unpleasant possibility (however remote) of having to wait an extremely long time before being able to read the transmission. In several scenarios (e.g., short messages), this is unacceptable to Agent 00111.

Between the extremes of comma codes and statistically synchronizable codes, there are many intermediate cases. Clearly, some criteria are necessary for comparing the advantages and disadvantages of synchronization schemes. There are three types of synchronization cost that a code can incur in the absence of noise:

Synchronization acquisition delay: This is the number of consecutive code symbols we must observe to determine a synchronization point. In Agent

---

* Both criticisms can be reduced if $|X|$ is large and the length of the comma is short, or if the code is variable word length.

00111's terms, this is the time he must spend observing an incoming transmission before he can begin reading it.

Decoding delay: Given the location of the first symbol of a code word, the decoding delay is the number of symbols, after the last symbol in the same code word, that must be observed before the end of the code word can be located definitely.

Synchronization loss: If we determine a word synchronization point in a long finite sequence of code words, it may be possible to decode forward and backward from that point over the complete sequence of symbols. Synchronization loss equals the total number of symbols in the observed sequence that cannot be uniquely decoded into code words. This loss is caused by the existence of an observation boundary.

All three of these synchronization costs are random variables depending on the transmitted sequence. In the case of statistically synchronizable codes, it is possible to find examples of individual code-word sequences with infinite synchronization acquisition delay and infinite synchronization loss, even though the delay and loss averaged across all sequences is not only finite but quite small. Again at the opposite end of the synchronization spectrum, a comma code has no decoding delay and both its synchronization acquisition delay and synchronization loss are upper bounded by $L - 1$, where L is the maximum word length of the code.

In the remaining sections of Chapter 3, we study U_I dictionaries. In fact, U_I dictionaries always have bounded synchronization acquisition delay. This can be verified by observing that the maximum synchronization acquisition delay that can occur corresponds to the length of the longest ambiguously interpretable segment of a sequence of code words, which in turn corresponds to the longest construction in the Seg table for Levenshtein's algorithm. This construction can easily be bounded as follows: Consider the longest sequence of code-word tails connected by arrows appearing in the Seg table. No two code-word tails can appear twice in this sequence, since the dictionary is U_I. In the worst possible scenario, every code-word tail appears once in the longest ambiguously interpretable sequence. Hence, the maximum synchronization acquisition delay of a code is upper bounded by the sum of the lengths of all code-word tails (including the words themselves). A single word of length l can contribute at most $1 + 2 + 3 + \cdots + l = l(l + 1)/2$ symbols to the synchronization delay. Thus, a U_I code with $|\mathbf{M}|$ words and maximum word length L clearly has synchronization delay bounded by $|\mathbf{M}| L(L + 1)/2$.

Synchronizable Codes

Exercises

1. Consider the full tree code 0, 10, 110, 111 corresponding to messages from a memoryless source with probabilities P_1, P_2, P_3, and P_4, respectively.

   (a) Let $P_1 = 1/2$, $P_2 = 1/4$, $P_3 = P_4 = 1/8$. Find $P_S(n)$, the probability of finding synchronization after observing exactly n consecutive symbols from a doubly infinite sequence of code words.

   (b) What is the average synchronization delay in (a)?

   (c) Repeat (a) when P_1, P_2, P_3, and P_4 are arbitrary.

2. Consider a noiseless channel with input alphabet $|\mathbf{X}|$ that transmits one symbol per second.

   (a) If a comma code with all words of length n is used, what is the maximum transmissible information rate?

   (b) If a variable word length is allowed, what is the maximum (over all comma codes and all source probability distributions) transmissible information rate?

3. Find the maximum synchronization delay of the following code:

   100, 101, 1000, 1001, 1011, 10000, 10001, 10011, 10111, 10010, 10110

4. If all words in a U_I dictionary of $|\mathbf{M}|$ words have the same length L, determine an upper bound on the maximum synchronization acquisition delay of the code. (Improve on the general bound given in the text.)

5. Explain how the maximum decoding delay can be determined from the Sardinas–Patterson algorithm. Illustrate your technique on the code 100, 110, 1001, 11000, 0111.

6. Show that U_I dictionaries have bounded synchronization loss and bounded decoding delay.

7. The code consisting of all 3-tuples except 000 is statistically synchronizable. Lowerbound the probability of attaining synchronization after observing at most L binary symbols. Assume that the seven code words are equally likely, and the code words in the sequence are independently selected.

3.4. A Bound on U_I Dictionary Size

We begin our examination of U_I dictionaries by considering the maximum possible size of such dictionaries. Bounds on dictionary size are not

Figure 3.8. An ambiguously decodable sequence.

difficult to obtain. The key observation is that certain n-tuples from an alphabet cannot be members of a U_I dictionary regardless of what other n-tuples are in the dictionary. A simple example of a nonadmissible n-tuple is 00, which when repeated many times, gives the ambiguously decodable sequence shown in Figure 3.8. In general, let d be a divisor of n. An n-tuple $(a_0 a_1 \cdots a_{n-1})$ is said to have *period d* if d is the smallest number such that

$$a_i = a_{i \oplus d} \qquad \text{for } 0 \leq i < n \tag{1}$$

where $\oplus$ denotes modulo n arithmetic. For example, if $n = 6$, then

000000 has period 1

010101 has period 2

001001 has period 3

001100 has period 6

Obviously, the period d of an n-tuple must be a divisor of n, and we denote this fact by writing $d\,|\,n$. Those n-tuples having period n are called *nonperiodic;* n-tuples having period $d < n$ are called *periodic. Periodic n-tuples cannot be words in a U_I dictionary* because their starting points are ambiguous in the context of a doubly infinite sequence. Consequently, we have an immediate restriction on the size of a U_I dictionary. We may extend this argument. The presence of an n-tuple in a U_I dictionary also implies that $n - 1$ other n-tuples cannot be in the dictionary. For example, if 001 is in the dictionary, the possibility of ambiguous decodings of the form shown in Figure 3.9 can

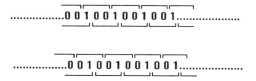

Figure 3.9. Another ambiguously decodable sequence.

Synchronizable Codes

be prevented only if 100 and 010 are *not* in U_I. In general, we say that two n-tuples $(a_0 a_1 \cdots a_{n-1})$ and $(b_0 b_1 \cdots b_{n-1})$ are *cyclically equivalent* if

$$a_i = b_{i \oplus k} \qquad \text{for } 0 \leq i < n \qquad (2)$$

The n cyclically equivalent n-tuples $(a_0 a_1 \cdots a_{n-1}), (a_1 a_2 \cdots a_{n-1} a_0), \ldots, (a_{n-1} a_0 a_1 \cdots a_{n-2})$ are said to form a *cyclic equivalence class*. A U_I dictionary can contain at most one n-tuple from each cyclic equivalence class. In Table 3.1, the 2^6 binary 6-tuples are divided into cyclic equivalence classes.† There are nine cyclic equivalence classes of nonperiodic 6-tuples, and thus, a U_I binary dictionary can have at most nine code words of length 6 symbols.

The following facts should be evident:

1. All n-tuples in a cyclic equivalence class have the same period.

2. The number of distinct n-tuples in a cyclic equivalence class is equal to the period of the n-tuples in the class.

3. If d is a divisor of n, then there is a one-to-one correspondence between the cyclic equivalence classes of nonperiodic d-tuples (nonperiodic cyclic-equivalence [NPCE] classes of d-tuples) and the cyclic equivalence classes of n-tuples having period d.

To illustrate Fact 3, the binary 6-tuple example must have word periods 1, 2, 3, and 6 corresponding to all divisors of 6. The binary NPCE classes of binary 1, 2, and 3 tuples are

$$
\begin{array}{ccc}
d = 1 & d = 2 & d = 3 \\
\overbrace{[0], [1]} & \overbrace{\begin{bmatrix} 1 & 0 \\ 0 & 1 \end{bmatrix}} & \overbrace{\begin{bmatrix} 1 & 1 & 0 \\ 0 & 1 & 1 \\ 1 & 0 & 1 \end{bmatrix} \begin{bmatrix} 1 & 0 & 0 \\ 0 & 1 & 0 \\ 0 & 0 & 1 \end{bmatrix}}
\end{array}
$$

Comparing these to Table 3.1 indicates that the preceding d-tuples are the length d prefixes (1 period) of binary 6-tuples having period d, $d = 1, 2, 3$.

---

† This classification will be useful in Sections 3.8 and 3.11.

Table 3.1. Cyclic Equivalence Classes of Binary 6-Tuples and Their Decimal Representations

A. Primitive NPCE classes (period 6)

1	100000	5	101000	11	110100
2	010000	10	010100	22	011010
4	001000	20	001010	44	001101
8	000100	40	000101	25	100110
16	000010	17	100010	50	010011
32	000001	34	010001	37	101001
13	101100	23	111010	31	111110
26	010110	46	011101	62	011111
52	001011	29	101110	61	101111
41	100101	58	010111	59	110111
19	110010	53	101011	55	111011
38	011001	43	110101	47	111101

B. Nonprimitive NPCE classes (period 6)

3	110000	7	111000	15	111100
6	011000	14	011100	30	011110
12	001100	28	001110	60	001111
24	000110	56	000111	57	100111
48	000011	49	100011	51	110011
33	100001	35	110001	39	111001

C. Nonprimitive periodic classes

Period 3:		9	100100	27	110110
		18	010010	54	011011
		36	001001	45	101101
Period 2:		21	101010		
		42	010101		
Period 1:		0	000000		

D. Degenerate class

| Period 1: | | 63 | 111111 | | |

The immediate task is to enumerate the number of cyclic equivalence classes of n-tuples for all n. The number $B(|\mathbf{X}|, n)$ of NPCE classes of n-tuples over an alphabet with $|\mathbf{X}|$ symbols can be determined by noting that

$$|\mathbf{X}|^n = \sum_{d|n} dB(|\mathbf{X}|, d) \qquad n = 1, 2, \cdots \qquad (3)$$

Synchronizable Codes

Both sides of Equation 3 count the number of n-tuples with symbols in $\mathbf{X}$. Using Fact 3, the right side of Equation 3, counts the number of elements $dB(|\mathbf{X}|, d)$ in the $B(|\mathbf{X}|, d)$ cyclic equivalence classes of period d and sums over all possible periods of n-tuples. Equation 3 can be solved for $B(|\mathbf{X}|, n)$ using Möbius inversion (see Appendix 3A) to give

$$B(|\mathbf{X}|, n) = \frac{1}{n} \sum_{d|n} \mu(d) |\mathbf{X}|^{n/d} \qquad (4)$$

$$= \frac{1}{n} \sum_{d|n} \mu(n/d) |\mathbf{X}|^{d} \qquad (5)$$

It should be realized that $B(|\mathbf{X}|, n)$ is an *upper bound* on the number of n-tuples in a U_I dictionary and the dictionary may not have *any* representatives of some NPCE classes.

Example 3.4.

$$\mathbf{X} = \{0, 1\} \qquad n = 6$$

$$B(\{0, 1\}, 6) = \frac{1}{6} [\mu(1)2^6 + \mu(2)2^3 + \mu(3)2^2 + \mu(6)2]$$

$$= \frac{1}{6} (64 - 8 - 4 + 2) = 9$$

Not every U_I dictionary has fixed word length n. In the next section, we examine fixed word-length U_I dictionaries in greater detail. Sections 3.8–3.10 examine the more general case.

Exercises

1. How many NPCE classes of 18-tuples over the alphabet $\{0, 1, 2\}$ exist?
2. How large an alphabet $\mathbf{X}$ is required to generate 10^6 NPCE classes of 10-tuples?

3.5. Fixed-Word-Length U_I Dictionaries

The obvious challenge in Section 3.4 is to find U_I codes with exactly $B(|\mathbf{X}|, n)$ words of length n but no words of other lengths. Such a dictionary is called a *maximal U_I code*. In Section 3.5, we catalog various U_I dictionary designs and determine some of their major parameters, e.g., size, synchronization delay, etc.; in particular, we consider the following cases:

Maximal comma-free codes

Prefixed comma-free codes

Path-invariant comma-free codes

Lexicographic U_I codes

A collection **C** of n-tuples is said to be *comma free* if for every choice of the n-tuples $(a_1, a_2, \ldots, a_n)$ and $(b_1, b_2, \ldots, b_n)$ from **C**, the n-tuple $(a_j, a_{j+1}, \ldots, a_n, b_1, b_2, \ldots, b_{j-1})$ is not in **C** for every j in the range $2 \le j \le n$. Equivalently, **C** is a comma-free code if no overlap of two code words is a code word. Since all code words have length n, a comma-free code **C** is obviously U_S, and the no-overlap property implies that code **C** must also be U_I.

Paradoxically, the one example of a comma-free code we have already seen is the comma code in Example 3.2, where $n = 6$. However, this comma code contains fewer than $B(|\mathbf{X}|, n)$ code words and hence is not a maximal comma-free code.

3.5.1. Maximal Comma-Free Codes

Maximal comma-free codes exist for at least some values, as shown in the following example.

Example 3.5. Assume that the elements of the alphabet **X** have been ordered ($<$) and define a code **C** with the property that a three-tuple (x_1, x_2, x_3) is in **C** if and only if

$$x_1 < x_2 \ge x_3 \qquad (6)$$

C is comma free, since the overlaps of two words

$$(x_1, x_2, x_3)(x_4, x_5, x_6)$$

take the forms

$$(x_2, x_3, x_4) \qquad \text{or} \qquad (x_3, x_4, x_5) \qquad (7)$$

Since $x_2 \ge x_3$ and $x_4 < x_5$, neither of these forms is allowed in the dictionary **C** as described in Equation 6. Note that the dictionary **C** is maximal, since

Synchronizable Codes

every nonperiodic 3-tuple can be cyclically shifted to yield a 3-tuple satisfying Equation 6, and hence every NPCE class of 3-tuples contains a code word. In more detail, the 3-tuple is cyclically permuted until the largest element is in the middle position. If two elements qualify as the largest, the 3-tuple is cycled to put the smallest element at the left. If these three elements all qualify as largest, the triple is periodic and ineligible for inclusion.

A general construction algorithm for maximal comma-free dictionaries of any odd word length n is discussed in Section 3.10. Maximal comma-free dictionaries for even values of n do not necessarily exist, as shown by Theorem 3.1.

Theorem 3.1 (Golomb, Gordon, and Welch). If n is an even integer, then maximal comma-free dictionaries do not exist when

$$|\mathbf{X}| > 3^{n/2} \tag{8}$$

Proof: Let $n = 2j$, and let $\mathbf{C}$ be a comma-free dictionary. We define S_1 to be the set of all j-tuples $(a_1, a_2 \cdots a_j)$ that form the first-half of some code word in $\mathbf{C}$ and S_2 to be the set of j-tuples $(a_{j+1}, a_{j+2}, \cdots, a_n)$ that form the second-half of some word in $\mathbf{C}$. Relative to the set $\mathbf{J}$ of all j-tuples, we set

$$A = S_1 \cap S_2' \quad B = S_1 \cap S_2 \quad C = S_1' \cap S_2 \quad D = S_1' \cap S_2'$$

where the prime denotes complementation in $\mathbf{J}$. The four sets A, B, C, and D are mutually exclusive and mutually exhaustive, so that any j-tuple is in one and only one set. Hence, every n-tuple may be associated with a pair (AA), (AB), $\cdots$, or (DD), depending on which set contains the n-tuple's j-symbol prefix and which set contains the n-tuple's j symbol suffix. Words from the comma-free dictionary $\mathbf{C}$ can have only the forms (AB), (AC), and (BC), since (BB) is excluded by the comma-free constraint applied to the overlap of j bits.

We now demonstrate that when Equation 8 is true, there is a nonperiodic n-tuple whose cyclic permutations are not of the form (AB), (AC), or (BC). It follows that there exists an NPCE class that is not represented by a code word in $\mathbf{C}$, and $\mathbf{C}$ is not a maximal comma-free code. Consider the following blocks of length j:

$$(\underbrace{1, 1, 1, \ldots, 1}_{j \text{ symbols}}, m) \quad 1 \leq m \leq |\mathbf{X}|$$

Let T_i be the cyclic permutation that shifts each letter i units to the left. Define

$$F_m(i) = \begin{cases} 1 \text{ if } T_i(1, 1, \ldots, m) \in A \cup D \\ 2 \text{ if } T_i(1, 1, \ldots, m) \in B \\ 3 \text{ if } (1, 1, \ldots, m) \in C \end{cases}$$

For each m in $\mathbf{X}$, $F_m(i)$ is a function with a domain of j elements and a range of three elements. There can be at most 3^j such functions that are distinct. Since the number of functions F_m is $|\mathbf{X}|$ where

$$|\mathbf{X}| > 3^{n/2} = 3^j$$

there exist two distinct integers p and m such that $F_p \equiv F_m$ for all i. We claim that no cyclic permutation of the nonperiodic n-tuple

$$w = (1, 1, \ldots, 1, p, 1, 1, \ldots, 1, m)$$

has the form (AB), (AC), or (BC). Any cyclic permutation of w consists of a cyclic permutation of $(1, 1, \ldots, p)$ followed by the same cyclic permutation of $(1, 1, \ldots, m)$, or vice versa. Since $F_p \equiv F_m$, these cyclic permutations come from the same set, i.e., $A \cup D$, B, or C. Therefore, w has one of the forms (AA), (AD), (DA), (BB), or (CC), and the theorem is proved.

The inequality in Theorem 3.1 is not tight; for example, when $n = 4$, the Theorem 3.1 states that maximal dictionaries do not exist for $|\mathbf{X}| > 9$. It can be shown that in the $n = 4$ case, maximal dictionaries do not exist for $|\mathbf{X}| \geq 4$.

Theorem 3.1 gives no clue to the existence of maximal binary dictionaries for even n. However, it is known that maximal comma-free binary dictionaries exist for $n = 2, 4, 6, 8$, and 10, and maximal ternary dictionaries are known for 2 and $2k$, $k > 1$.

Proof that no maximal comma-free dictionary of even word length $n = 2j$ exists when $|\mathbf{X}| > 3^j$ has been refined twice. In Jiggs (1963), nonexistence is proved for $|\mathbf{X}| > 2^j + j$, and in Tang et al. (1987), this is further improved for whenever $|\mathbf{X}| > j^{\log_2 j} + j$ [for a simpler proof of the latter result, see van Lint (1985)]. All of these refinements are still based on the problem of representing only the

$$\binom{|\mathbf{X}|}{2}$$

Synchronizable Codes

cyclic equivalence classes of the form $\{(1, 1, \ldots, 1, p, 1, 1, \ldots, 1, m)\}$ in a comma-free dictionary; and if only these classes are considered, no better general result seems possible than "whenever $|\mathbf{X}| > j^{3/2} + j$." However, when *all* cyclic equivalence classes are considered, empirical data suggest that "whenever $|\mathbf{X}| > 3$" is a sufficient condition for comma-free dictionaries of even word length not attaining the upper bound corresponding to all non-periodic classes being represented.

Note that after observing $2n - 2$ symbols from an unsynchronized sequence of comma-free code words, we are guaranteed either to observe (1) a code word (which cannot be an overlap of two code words) or (2) the $n - 1$ length suffix of a code word followed by the $n - 1$ length prefix of another code word. Hence, the maximum synchronization delay of a comma-free code word is $2n - 2$ symbols.

3.5.2. Prefixed Comma-Free Codes

The only objection to a comma code as a U_I dictionary is its dictionary size. As an extreme example, a binary fixed-word-length comma code contains exactly one word regardless of the choice of word length n; i.e., if we use 1 for a comma, we have only 0 left for a signalling set! This situation can be alleviated to some degree by using a sequence $a_1, a_2 \cdots a_p$ rather than a single symbol as a marker. A dictionary **C** is *comma-free with prefix* $a_1, a_2 \cdots a_p$, $2p \leq n + 1$, if every n-tuple **n** in the dictionary starts with $a_1 \cdots a_p$ and if **n** equals $(a_1 \, a_2 \cdots a_n)$, **n** has the property that

$$a_1 a_2 \cdots a_p \neq a_i a_{i+1} \cdots a_{i+p-1}$$

for all choices of i in the range $1 < i \leq n + 1 - p$. Equivalently, in a prefixed comma-free code, the prefix $a_1, a_2 \cdots a_p$ cannot occur anywhere in a code word except at the beginning sequence of p symbols.

To see that the code is comma free, consider the overlaps of two code words from **C**, as shown in Figure 3.10. Since $n + 1 \geq 2p$, the overlap must include p symbols from either the first or the second code word. If the overlap includes p or more symbols from the first code word, the overlap cannot be a code word, since the overlap cannot begin with $a_1 \, a_2 \cdots a_p$. Similarly, if

Figure 3.10. The overlaps of two code words from C.

the overlap contains p symbols from the second code word, the overlap cannot be a code word, since it contains $a_1 a_2 \cdots a_p$ as an interior sequence.

To count the code words in a comma-free code $\mathbf{C}$ with prefix $a_1 a_2 \cdots a_p$, draw the state diagram of a pth-order Markov source having alphabet $\mathbf{X}$. Every code word $\mathbf{C}$ corresponds to a sequence of $n - p$ transitions beginning at state $a_1 a_2 \cdots a_p$ and never returning to that state. By removing all transitions going *to* the state $a_1 a_2 \cdots a_p$, every $n - p$ length path through the state diagram beginning at $a_1 a_2 \cdots a_p$ denotes a code word. These paths may be counted as follows. Denote $a_1 a_2 \cdots a_p$ as State 1 and the other states as State i, $1 < i \le |\mathbf{X}|^p$. For $i > 1$, let $n_i(t)$ be the number of distinct paths that begin at State 1 and without returning to state 1, reach State i after t transitions. Regarding time $t = 0$ as the point of departure from State 1, we set $n_1(0) = 1$ and $n_i(0) = 0$ for $i > 1$. Since we are never allowed back to State 1, we have $n_1(t) = 0$, $t > 0$. For $i > 1$, let $\mathbf{S}_i$ be the set of states having transitions going to State i. Then,

$$n_i(t) = \sum_{k \in \mathbf{S}_i} n_k(t - 1) \tag{9}$$

or in vector notation

$$N(t) = TN(t - 1) \tag{10}$$

where

$$N(s) = \begin{bmatrix} n_1(s) \\ n_2(s) \\ \vdots \\ n_{|\mathbf{X}|^p}(s) \end{bmatrix} \tag{11}$$

and T is a matrix whose $(i, j)^{th}$ entry is 1 when $i > 1$ and there is a transition from State j to State i or zero otherwise. Using the fact that $N(0)$ is a vector with a single 1 representing the initial state and the remainder zeros, the dictionary size for words of length n is given by

$$|\mathbf{C}| = j^t N(n - p) = j^t T^{n-p} N(0) \tag{12}$$

where j is the all 1 column vector. $\mathbf{C}$ is not maximal, i.e., $|\mathbf{C}|$ is less than $B(|\mathbf{X}|, n)$ because there generally exist NPCE classes whose elements either do not contain the subsequence $a_1 a_2 \cdots a_p$ or do contain the subsequence twice in the same n-tuple.

3.5.3. Path-Invariant Comma-Free Codes

Suppose one specifies (e.g., by an incidence matrix) which of the $|X|$ symbols may occur in which of the n positions in a code word. That is, we form a $|X|$ by n matrix with rows indexed by the symbols and columns indexed by the code-word positions. We set the $(s, i)^{\text{th}}$ entry at 1 if symbol

Symbol	Position		
	1	2	3
1	1	0	1
2	1	1	1
3	0	1	1

121
122
131
132
133
231
232
233

**Comma-free code,
8 words, does not
use all combinations**

Symbol	Position		
	1	2	3
1	1	0	1
2	0	1	1
3	0	1	1

121
122
123
131
132
133

**Path Invariant
comma-free code,
6 words**

Symbol	Position		
	1	2	3
1	1	1	0
2	1	1	0
3	0	0	1

113
123
213
223

Comma code, 4 words

Figure 3.11. Dictionaries with $n = |X| = 3$.

s occurs in position *i* of a code word. If the dictionary formed by constructing all combinations of symbols in permitted positions happens to be comma free, it is called a *path-invariant* comma-free dictionary. The comma code is the simplest example of a path-invariant comma-free code.

To make the discussion more concrete, we give examples in Figure 3.11 of maximum-sized dictionaries for the case $n = |\mathbf{X}| = 3$ that show both the incidence matrix and the code. The 8-word comma-free code is maximal in size, but not all paths through its incidence matrix represent code words; e.g., 222 is not in the code. The 6-word comma-free code on the other hand is path invariant comma free. To demonstrate this, write two incidence matrices side by side, representing all possible sequences of two words from the code in Table 3.2.

Notice that the symbols in positions 2*a*, 3*a*, 1*b* cannot form a code word, since column 2*a* of the incidence matrix has no 1s (allowed symbols) in common with the initial symbols permitted for code words, as shown in columns 1*a* or 1*b*. Similarly, the symbols in positions 3*a*, 1*b*, and 2*b* cannot form a code word, since column 1*b* is orthogonal to the second column in the code-word incidence matrix.

The judicious use of pairs of orthogonal columns can be generalized to give a class of incidence matrices yielding path-invariant comma-free codes of length *n* over an $|\mathbf{X}|$ symbol alphabet. A typical incidence matrix is shown in Figure 3.12.

Notice that the number *L* of 1s in the first column of the incidence matrix has not been specified. Thus, *L* can be chosen to maximize the dictionary size $D(L)$, where $D(L)$ is equal to the number of distinct paths through the incidence matrix, i.e.,

$$D(L) = L(|\mathbf{X}| - L)^{\lfloor n/2 \rfloor} |\mathbf{X}|^{\lfloor (n-1)/2 \rfloor} \tag{13}$$

It is easily verified that the derivative of *D* with respect to *L* changes sign only once, and therefore, if *L* were a continuous variable, there would be only one maximum-attaining value for *L*. Thus since *L* is discrete, we need look only for the unique *L* for which

Table 3.2. Position of Symbols

Symbol	Position					
	1*a*	2*a*	3*a*	1*b*	2*b*	3*b*
1	1	0	1	1	0	1
2	0	1	1	0	1	1
3	0	1	1	0	1	1

Synchronizable Codes

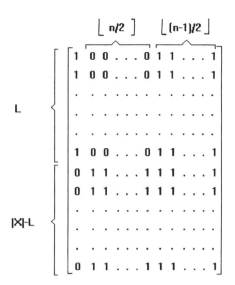

Figure 3.12. An incidence matrix.

$$D(L-1) \leq D(L) > D(L+1) \tag{14}$$

The special case when $L = 1$ is optimal occurs if and only if $D(1) \geq D(2)$ or equivalently when

$$\left[\frac{|\mathbf{X}| - 1}{|\mathbf{X}| - 2}\right]^{\lfloor n/2 \rfloor} \geq 2 \tag{15}$$

Obviously, this is true for all $\lfloor n/2 \rfloor$ greater than zero when the alphabet size $|\mathbf{X}|$ is 2 or 3. In fact, this inequality is satisfied for any fixed $|\mathbf{X}|$ with a large enough n.

3.5.4. Lexicographic U_I Codes

The following algorithm shows a simple way of selecting one code word from each of $B(|\mathbf{X}|, n)$ nonperiodic cyclic equivalence classes to form a U_I code.

Lexicographic Code Algorithm.

Order the symbols in $\mathbf{X}$.

Form the NPCE classes of n-tuples over the alphabet $\mathbf{X}$.

Put the n-tuples of each NPCE class in lexicographic order, i.e., in dictionary form as specified by the ordering in Step 1.

The code is formed by taking the first n-tuple from each lexicographically ordered NPCE class.

The algorithm is illustrated in Example 3.6.

Example 3.6. $X = \{0, 1, 2\}$, $n = 3$. Assume $0 < 1 < 2$. Then the lexicographically ordered NPCE classes are

$$\begin{bmatrix} 001 \\ 010 \\ 100 \end{bmatrix} \begin{bmatrix} 002 \\ 020 \\ 200 \end{bmatrix} \begin{bmatrix} 011 \\ 101 \\ 110 \end{bmatrix} \begin{bmatrix} 012 \\ 120 \\ 201 \end{bmatrix}$$

$$\begin{bmatrix} 021 \\ 102 \\ 210 \end{bmatrix} \begin{bmatrix} 022 \\ 202 \\ 220 \end{bmatrix} \begin{bmatrix} 112 \\ 121 \\ 211 \end{bmatrix} \begin{bmatrix} 122 \\ 212 \\ 221 \end{bmatrix}$$

and the lexicographic code is

$$C = \{001, 002, 011, 012, 021, 022, 112, 122\}$$

This code is U_I, as can be seen from Levenshtein's test (see Figure 3.13).

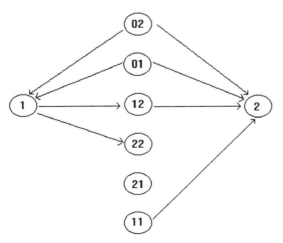

Figure 3.13. Levenshtein's test applied to Example 3.6.

Synchronizable Codes

Since there are no loops in the segment state diagram, the code is U_I.

All words in a lexicographic code satisfy the property that any k symbol suffix of a code word is *strictly* preceded by the k symbol prefix of the same code. To verify this result, consider a code word of the form pms, where p is the prefix, s is the suffix, and m is the (possibly nonexistent) center. Assume that p and s have the same length, either k or $n - k$. Lexicographically earliest selection within an NPCE class implies that

$$pms < msp \qquad pms < spm \qquad (16)$$

from which we can conclude that

$$pm \leq ms \qquad p \leq s \qquad (17)$$

Both inequalities in Equation 17 must in fact be strict for the following reasons:

1. If $p = s$, then Equation 16 implies that

$$pm < ms \qquad ms < pm$$

which is clearly impossible. Therefore,

$$p < s \qquad (18)$$

2. If $pm = ms$, then Equation 16 implies that

$$s < p \qquad p < s$$

which is also impossible. Hence,

$$pm < ms \qquad (19)$$

Note that the left-side inequality in Equation 17 is a comparison of an overlapping prefix and suffix of the word pms. The right-side inequality in Equation 17 is a comparison of a nonoverlapping prefix and suffix of the same word. With these results, the synchronizability of lexicographic codes can be verified.

Let p_{ik} and s_{ik} be the k symbol prefix and suffix of code word c_i. Suppose that code words c_i and c_j can overlap as shown in Figure 3.14a.

Using Equation 18 if $k \leq n/2$ and Equation 19 otherwise, it follows that

$$p_{ik} < s_{ik} = p_{jk} \qquad (20)$$

Hence, the *code words* must be ordered

$$c_i < c_j \qquad (21)$$

Thus, in a finite-length ambiguously decodable sequence of the form shown in Figure 3.14b, it follows that

$$c_1 < c_2 < c_3 < c_4 < c_5 \cdots \qquad (22)$$

Since the dictionary is *finite*, Equation 22 can involve at most $B(|\mathbf{X}|, n) - 1$ inequalities, so the indicated overlapping construction must terminate. Seg $B(|\mathbf{X}|, n)$ in Levenshtein's algorithm must be empty, and lexicographic codes are U_I dictionaries.

In an ambiguous construction, such as the one in Equation 22, the initial sequence of successive words $c_1, c_3, c_5, \ldots$, must be a strictly ordered sequence of k symbols. Therefore, the sequence $c_1, c_3, c_5, \cdots$ can contain at most $|\mathbf{X}|^k$ words. Similarly, the overlapping sequence $c_2, c_4, c_6 \cdots$ can contain at most $|\mathbf{X}|^{n-k}$ words, and therefore, the maximum number of words that can occur in the ambiguous encoding is $|\mathbf{X}|^{\min(k, n-k)}$. Choosing the worst possible value for k gives a (loose) upper bound to the synchronization delay of at most $n|\mathbf{X}|^{\lfloor n/2 \rfloor}$ symbols.

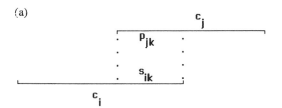

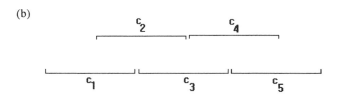

Figure 3.14. Overlaps in a lexicographic U_I code.

Synchronizable Codes

Exercises

1. If the test in Section 3.2 is applied to a comma-free code, for what value of m is Seg m guaranteed to be zero? Is Example 3.2 comma-free? Why?

2. Design a maximal comma-free code (a) with words of length 4 over a three-symbol alphabet; (b) with words of length 5 over a binary alphabet.

3. Consider the generating function for a prefixed comma-free code defined by

$$G(z) = \sum_{n=0}^{\infty} |\mathbf{C}_n| z^n$$

   where $|\mathbf{C}_n|$ is the size of $\mathbf{C}_n$, a prefixed comma-free dictionary of word length n.

   (a) How is the quantity R,

$$R = \lim_{n \to \infty} \frac{1}{n} \log |\mathbf{C}_n|$$

   related to $G(z)$? (Similar questions were discussed in Section 2.9.)

   (b) Develop a method for evaluating $G(z)$.

   (c) Evaluate R for the code using 001 as a prefix.

4. Show that the incidence matrix

$$\begin{bmatrix} 1 & 0 & 1 \\ 0 & 1 & 1 \\ 0 & 1 & 1 \end{bmatrix}$$

   yields the largest possible path-invariant comma-free code for length 3 words over a three-symbol alphabet.

5. What is the maximum dictionary size for a path-invariant comma-free code with (a) $|\mathbf{X}| = 10$ and $n = 7$; (b) $|\mathbf{X}| = 12$ and $n = 8$?

6. Determine the maximum synchronization acquisition delay for a path-invariant comma-free code.

7. Find the largest possible binary prefixed comma-free code with word length 7. Compare this to the dictionary size of a similar maximal comma-free code and a path-invariant comma-free code.

8. Design a word length 8 binary lexicographic U_I code with bounded synchronization acquisition delay. Is it comma free? What is its maximum synchronization acquisition delay?
9. Show that the maximum synchronization acquisition delay for a lexicographic code with (a) $n = 2$ is $|X| - 1$; (b) $n = 3$ is 3 when $|X| = 2$; (c) $n = 3$ is 6 when $|X| = 3$.

3.6. Comparing Fixed-Word-Length Synchronizable Codes

So far we have considered five basic U_I dictionaries: (1) the comma code, (2) the maximal comma-free code, (3) the prefixed comma-free code, (4) the path-invariant comma-free code, and (5) the lexicographic code. Since these are all fixed-word-length codes, their synchronization losses are all bounded by their word length n, and their decoding delays are all zero. The fundamental trade-off occurs between synchronization acquisition delay and dictionary size. These two properties are tabulated in Table 3.3.

The fourth column in Table 3.3 indicates the asymptotic transmission rate or effective capacity of the code described. This was computed from the dictionary $\mathbf{C}_n$ in the following manner:

$$R = \lim_{n \to \infty} \frac{1}{n} \log |\mathbf{C}_n| \qquad (23)$$

We know that the effective capacity of the synchronized noiseless channel is $\log|X|$ and from R in Table 3.3, the capacity of the nonsynchronized channel

Table 3.3. Properties of Codes

Code C_n	Maximum synchronization acquisition delay	Dictionary size $|C_n|$	Transmission rate R
Comma code	$n - 1$	$(|X| - 1)^{n-1}$	$\log(|X| - 1)$
Prefixed comma-free code	n	$<B(|X|, n)$	$<\log|X|$
Path-invariant comma-free code	n	$L(|X| - L)^{(n/2)[(n-1)/2]}$	$\frac{1}{2}[\log(|X| - 1) + \log|X|]$
Comma-free code	$2(n - 1)$	$\leq B(|X|, n)$	$\log|X|$
Lexicographically ordered, bounded delay code	$\leq n|X|^{(n/2)}$	$B(|X|, n)$	$\log|X|$

Synchronizable Codes

is identical to that of the synchronized channel. This statement is incontrovertible once synchronization has been acquired, but notice that in computing effective capacity, we have allowed the synchronization acquisition delay to go to infinity. A proper interpretation of the effective capacity computation in Equation 23 is that a doubly infinite, nonsynchronized sequence of code words can be communicated at any information rate strictly less than the capacity $\log|\mathbf{X}|$ of the noiseless channel (there is no synchronization acquisition delay present in this case).

The computation of effective capacity or asymptotic transmission rate for the comma code and the path-invariant comma-free code (with $L = 1$) is straightforward. Notice from Table 3.3 that neither code uses the capacity of the channel effectively, and the capacity loss becomes smaller as $|\mathbf{X}|$ increases.

The effective capacity computation for the remaining cases in Table 3.3, i.e., the comma-free codes, uses the fact that $B(|\mathbf{X}|, n)$ words can be found for all odd n. With this information, we can write the following set of inequalities:

$$|\mathbf{X}|^n > B(|\mathbf{X}|, n) = \frac{|\mathbf{X}|^n}{n} \sum_{d|n} \mu(n/d) |\mathbf{X}|^{d-n}$$

$$\geq \frac{|\mathbf{X}|^n}{n} \left(1 - |\mathbf{X}|^{-n} \sum_{d=1}^{[n/2]} |\mathbf{X}|^d \right)$$

$$= \frac{|\mathbf{X}|^n}{n} \left\{ 1 - |\mathbf{X}|^{-n} \left[\frac{|\mathbf{X}|^{(n/2)+1} - 1}{|\mathbf{X}| - 1} \right] \right\}$$

$$> \frac{|\mathbf{X}|^n}{n} [1 - 2|\mathbf{X}|^{-(n/2)}]. \tag{24}$$

The first inequality is due to the fact that $B(|\mathbf{X}|, n)$ is the number of disjoint classes in a collection of $|\mathbf{X}|^n$ sequences. The second inequality follows from the fact that aside from n, only numbers from 1 to $n/2$ can divide n. To obtain an inequality, we assume that (1) all these numbers are actually divisors of n (which is true for $n = 1, 2, 3, 4, 6$) and (2) in each case, $\mu(n/d) = -1$. The last inequality results from removing -1 from the numerator of the innermost brackets and replacing $|\mathbf{X}| - 1$ by $|\mathbf{X}|/2$. Since the log function is a monotonically increasing function of its argument, we can take logarithms in Equation 24 and divide by n to obtain

$$\log|\mathbf{X}| > \frac{\log B(|\mathbf{X}|, n)}{n} > \log|\mathbf{X}| - \frac{\log n}{n} + \frac{\log(1 - 2|\mathbf{X}|^{-n/2})}{n} \tag{25}$$

As n goes to infinity

$$\frac{\log n}{n} \quad \text{and} \quad \frac{\log(1 - 2|\mathbf{X}|^{-n/2})}{n}$$

both go to zero, and the effective capacity results in Table 3.3 have been demonstrated for the remaining cases, i.e., maximal for comma-free and lexicographically ordered codes.

In Equation 25, the term $(1/n)\log(1 - 2|\mathbf{X}|^{-n/2})$ behaves like

$$(2/n)|\mathbf{X}|^{-n/2}$$

for large n; i.e., it decreases rapidly as n increases. This term is a bound on the cost of eliminating periodic words from the code book. A more expensive cost is the $(\log n)/n$ term in Equation 25. This term decreases slowly with n, and it is open to the following information-theoretic interpretation: When each code word is required to carry complete synchronization information (as is the case in a comma-free code), it must contain $\log n$ bits in its n symbols to indicate the proper one of n possible synchronization positions, expending $(\log n)/n$ bits/symbol of the channel capacity for synchronization information.

All U_I codes with fixed-word-length n can be synchronized with a device of the form shown in Figure 3.15. As drawn, both the punctuation and data shift registers are n units long. The punctuation (1 = comma, 0 = no comma) stored in unit i in the punctuation register indicates the appropriate possible punctuation between coded data stored in units i and $i + 1$. When the synchronizer is turned on, the punctuation shift register is loaded with 1s, indicating that synchronization could be anywhere within the data stored in the data register. The code-word recognizer outputs a 1 if the data register contains a code word, and 0 otherwise. As n-tuples occur in the data register that are not code words, 1s in the punctuation register are erased. When a single 1 remains in the punctuation register, there is only one candidate for the synchronization position in the data register. The synchronization indicator then indicates when a code word is present in the data register.

Exercises

1. Using the concept of a typical sequence and fixed rate source encoding, derive coding theorems for

    (a) A comma-free coding system for the $|\mathbf{X}|$-ary noiseless channel

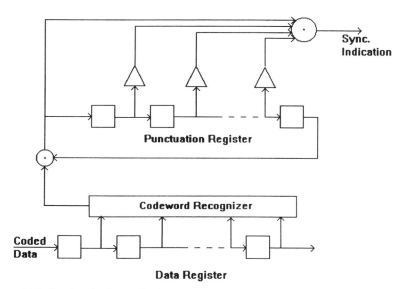

Figure 3.15. Synchronization device for U_l codes. △, inverter; □, storage device; ⊙, multiplier.

(b) A lexicographic coding system for the $|\mathbf{X}|$-ary noiseless channel

What constraints are required in (a) that are not required in (b)?

2. Notice the certain types of codes may automatically be classified as more general codes. Consider the following implications for fixed $|\mathbf{X}|$ and n:

Comma Code $\Rightarrow$

Path-Invariant Comma-Free Code $\Rightarrow$

Code with Maximum Synchronization Acquisition Delay $n \Rightarrow$

Comma-Free Code $\Rightarrow$

Code with Maximum Synchronization Acquisition Delay $2(n-1)$

Supply counterexamples showing that none of the preceding implications can be reversed.

3. Show how to mechanize synchronizers for the following binary, word length 7 codes:

   (a) A path-invariant code

   (b) A comma code

   (c) A prefixed comma-free code using prefix 000

   Simplify your design as much as possible.

3.7. Variable-Word-Length Synchronizable Codes

A code **C** is called *synchronizable* if any finite sequence of code words can be entered into a machine called a synchronizer and in a fixed amount of time, the words come out of the synchronizer suitably punctuated with commas between words.

After studying U_I dictionaries of fixed word length, we are in a good position to study the *iterative* construction of synchronizable codes with variable word length. The resulting codes occupy the same position in the theory of U_I dictionaries that tree codes have in the theory of U_S dictionaries; i.e., if it is possible to construct a U_I code with n_i words of length i, $i = 1, 2, 3, \ldots$, then it can be done using the iterative procedure that follows.

We construct codes $\mathbf{C}^{(0)}, \mathbf{C}^{(1)}, \ldots, \mathbf{C}^{(j)}$ in an iterative fashion. The starting point is the code alphabet, i.e., $\mathbf{C}^{(0)} = \mathbf{X}$. It is always possible to consider the code alphabet $\mathbf{C}^{(0)}$ to be synchronizable; i.e., the decoder inserts commas after every symbol. Given a starting point, let $\mathbf{C}^{(j)}$ represent the jth synchronizable code in an iterative construction procedure. The next synchronizable code $\mathbf{C}^{(j+1)}$ can be constructed by selecting a word $w^{(j+1)}$ from $\mathbf{C}^{(j)}$ and using it as a suffix k times on all of the remaining words in $\mathbf{C}^{(j)}$, where k can take any nonnegative integer value. First, select a $w^{(j+1)}$ from $\mathbf{C}^{(j)}$. Then,

Synchronizable Codes

$$\mathbf{C}^{(j+1)} = \{c(w^{(j+1)})^k \mid c \in \mathbf{C}^{(j)}, c \neq w^{(j+1)}, k \in \mathbf{Z}, k \geq 0\}$$

where $ab^k = abb \cdots b$, and b appears k times.

The synchronizer for $\mathbf{C}^{(j+1)}$ consists of the synchronizer for $\mathbf{C}^{(j)}$ followed by a device that erases the first comma in ",$w^{(j+1)}$," every time it appears in the output stream. This obviously synchronizes the stream of $\mathbf{C}^{(j+1)}$, since $w^{(j+1)}$ is used *only* as a suffix in $\mathbf{C}^{(j+1)}$.

Example 3.7. Let $n = 2$ and $\mathbf{X} = \mathbf{C}^{(0)} = \{0, 1\}$.

We could select either 1 or 0, then construct the next code in the sequence. Let $w^{(1)} = 0$. Obviously, $\mathbf{C}^{(1)}$ consists of all words that start with 1 and are followed by a finite number of zeros. To save space, we continue with the subset of words of length at most 5:

$$\mathbf{C}^{(1)} = 1 \quad 10 \quad 100 \quad 1000 \quad 10000$$

Continuing, select $w^{(2)} \in \mathbf{C}^{(1)}$ as $w^{(2)} = 1$.

$$\begin{array}{llll} \mathbf{C}^{(2)} = & 10 & 100 & 1000 & 10000 \\ & & 101 & 1001 & 10001 \\ & & & 1011 & 10011 \\ & & & & 10111 \end{array}$$

Our choice of $w^{(j+1)} \in \mathbf{C}^{(j)}$ in this procedure is arbitrary. Suppose we do two more constructions, selecting words of length 3 and removing all code words of length exceeding 5:

$$w^{(3)} = 100$$

$$\begin{array}{lllll} \mathbf{C}^{(3)} = & 10 & 101 & 1000 & 10000 \\ & & & 1001 & 10001 \\ & & & 1011 & 10011 \\ & & & & 10111 \\ & & & & 10100 \end{array}$$

$$w^{(4)} = 101$$

$$\mathbf{C}^{(4)} = \begin{array}{lll} 10 & 1000 & 10000 \\ & 1001 & 10001 \\ & 1011 & 10011 \\ & & 10111 \\ & & 10100 \\ & & 10101 \end{array}$$

Note that we have completely removed every word of length 3 in $\mathbf{C}^{(4)}$.

There are several points immediately worth noting. Obviously, using a word $w^{(j+1)} \in \mathbf{C}^{(j)}$ as a prefix instead of a suffix to construct $\mathbf{C}^{(j+1)}$ will also work. The synchronizer in this case is the synchronizer for $\mathbf{C}^{(j)}$ followed by a device that erases the comma following $w^{(j+1)}$ in ",$w^{(j+1)}$," every time it appears. To some degree, we can use our selection to adjust the number of low-weight code words. For example, suppose we always select a minimum-length code word in $\mathbf{C}^{(j)}$ for $w^{(j+1)}$. This particular selection procedure shows that for a preset value k, after a finite number of constructions, we can ensure that $\mathbf{C}^{(j)}$ has no code words of length less than k.

A variety of codes can be constructed depending on the choice of words to be eliminated in the construction, the order in which they are eliminated, and the choice of prefix or suffix construction for each step. The code structure used in Example 3.7 is completely specified by the following:

Alphabet = $\{1, 0\}$

Maximum word length = 5

$$\begin{array}{ll} w^{(1)} = 0 & \text{suffix} \\ w^{(2)} = 1 & \text{suffix} \\ w^{(3)} = 100 & \text{suffix} \\ w^{(4)} = 101 & \text{suffix} \end{array}$$

The resultant code $\mathbf{C}^{(4)}$ in Example 3.7 is uniquely specified, and in addition, if all the words are used, the specification is unique.

The synchronizer can be constructed in a modular fashion, where the jth module is a device that erases the first comma in ",$w^{(j)}$," when $w^{(j)}$ is used as a suffix, and the second comma when $w^{(j)}$ is used as a prefix. As an example,

Synchronizable Codes

assume that $w^{(j)} = 100$, the code under consideration is binary, and we are deleting the first comma; i.e., 100 is a suffix. A shift register mechanization of the jth module is shown in Figure 3.16. The upper register stores the punctuation information (i.e., 1 denotes comma, 0 denotes no comma), which appears in the lower register data. Notice that the output of the multiplier is 1 if and only if the register's contents are $x^1 0^0 0^0 1^1$, indicating that 100 is in the lower register preceded and followed by a comma in the upper register. In this case, the initial comma is erased by binary addition of the multiplier output and the first comma. Since the input data and punctuation are represented by a sequence of words from a U_I dictionary, when 100 is preceded and followed by commas, no internal commas in 100 can be present and there is no need to check for their absence in the data register.

When the complete synchronizer is constructed by connecting a sequence of modules, the output of the complete synchronizer is a Boolean function of the contents of the modules' data registers. This output can be significantly simplified by logical design techniques. The synchronization delay is simply the length of the complete synchronizer's data register, which is approximately

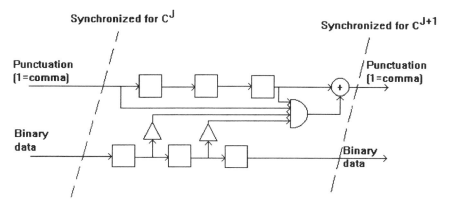

Figure 3.16. Synchronizer for $w^{(j)} = 100$ (suffix). D, "and" gate, *mod* 2 multiplication; □, binary storage device; △, inverter; ⊕, "exclusive or" gate, *mod* 2 addition.

the sum of the lengths of suffixes and prefixes $w^{(j)}, j = 1, 2, \ldots$, used in the iterative construction algorithm.

The construction procedure has the pleasant property that if several suffix words of different lengths are used in the construction process, the number of words of each length in the resultant dictionary does not depend on the order in which words are used in the modification process. Furthermore, this property remains true if we restrict the code to the subset of words having a code-word length that is less than some finite constant. To show this, let the unmodified dictionary word lengths be represented by the polynomial

$$P_j(x) = \sum_{l=1}^{\infty} N_l^{(j)} x^l \tag{26}$$

where $N_l^{(j)}$ denotes the number of words of length l in the jth dictionary and x is an arbitrary number with magnitude less than unity. If the construction procedure is performed using a word $w^{(j+1)}$ of length α_{j+1}, then the new dictionary's set of word lengths is specified by the polynomial

$$P_{j+1}(x) = \sum_{k=0}^{\infty} (x^{\alpha_{j+1}})^k [P_j(x) - x^{\alpha_{j+1}}] \tag{27}$$

This representation follows from the construction procedure:

1. $P(x) - x^{\alpha_{j+1}}$ is a polynomial representing the original dictionary word lengths with a word of length α_{j+1} removed.
2. $(x^{\alpha_{j+1}})^k [P(x) - x^{\alpha_{j+1}}]$ represents word lengths added to the dictionary when a word of length α_{j+1} is repeated k times and appended to all the remaining words of the dictionary.

Equation 27 can be rewritten as

$$1 - P_{j+1}(x) = \frac{1 - P_j(x)}{1 - x^{\alpha_{j+1}}} \tag{28}$$

In this form, it is obvious that another suffix modification using a word of length α_{j+2} will result in the polynomial representation

$$1 - P_{j+2}(x) = \frac{1 - P_j(x)}{(1 - x^{\alpha_{j+1}})(1 - x^{\alpha_{j+2}})} \tag{29}$$

Synchronizable Codes

and furthermore, the order in which the modifications were made does not affect the structure of the resulting dictionary's word length. Note that restricting code words by a maximal length N amounts to truncating the expansion of $P_j(x)$ to exclude terms x^k, $k > N$. It follows that the order of the modifications cannot affect this latter directory's word length structure.

Example 3.8. In Example 3.7, the original binary elementary dictionary is represented by the polynomial $P(x) = 2x$. After removing two words of length 1 and two words of length 3, the modified dictionary word lengths for the complete code can be found as follows:

$$P_4(x) = 1 - \frac{1 - 2x}{(1 - x)^2(1 - x^3)^2} = x^2 + 3x^4 + 6x^5 + 6x^6 + \cdots \quad (30)$$

Limiting ourselves to words of length 5 or less, the resultant dictionary $\mathbf{C}^{(4)}$ contains 1 word of length 2, three words of length 4, and six words of length 5, as confirmed by the explicit construction used in Example 3.7.

Exercises

1. (a) Construct a synchronizable binary code with words of length 6 or less by using all words of length two or less as suffixes.

   (b) Draw the modular synchronizer for your code. Simplify your result until the output function is simply a function of the contents of the data register.

2. A bounded synchronization acquisition delay code is being designed for a binary noiseless channel. Design a minimum average word length code for the memoryless source with message probabilities 0.5, 0.2, 0.1, 0.05, 0.05, 0.05, 0.05.

3. (a) Show that the following algorithm achieves the largest iteratively constructed dictionary of words of length less than or equal to L: Systematically choose $w^{(j+1)}$ to be the shortest word in $\mathbf{C}^{(j)}$ and continue until the maximum dictionary size has been achieved.

   (b) What is the maximum dictionary size when $\mathbf{X} = \{0, 1, 2\}$ and $L = 7$? $L = 8$?

3.8. Necessary Conditions for the Existence of U_I Dictionaries

The key to U_I dictionary existence comes from Levenshtein's algorithm, which states that a code **C** is not U_I if and only if a periodic ambiguously decodable sequence exists.

Example 3.9
Consider the code

$$\mathbf{C} = \{c_1 = 001, c_2 = 101, c_3 = 01, c_4 = 1010, c_5 = 0110\}.$$

By confining the test to periodic sequences, the following simplification occurs. An ambiguous interpretation of a periodic sequence, e.g., the one shown in Figure 3.17a, indicates that the cyclic equivalence class containing one period of the ambiguous decoding actually contains at least two cyclically distinct sequences of code words. For example, if we cycle the periodic sequence 00110101, we obtain both $c_1 c_2 c_3$ and $c_4 c_5$, as well as the code-word cycles $c_3 c_1 c_2$, $c_2 c_3 c_1$ and $c_5 c_4$, as shown in Figure 3.17b.

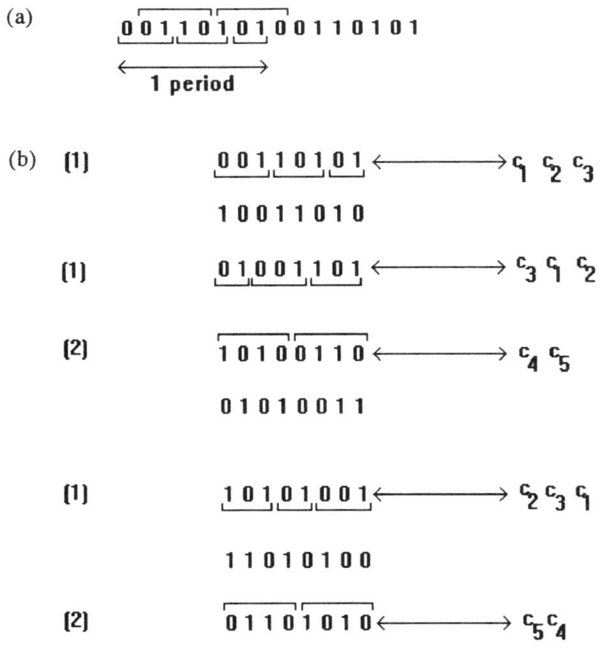

(a) 0 0 1 1 0 1 0 1 0 0 1 1 0 1 0 1
    ←⎯ 1 period ⎯→

(b) [1] 0 0 1 1 0 1 0 1 ⟷ $c_1\ c_2\ c_3$
         1 0 0 1 1 0 1 0

    [1] 0 1 0 0 1 1 0 1 ⟷ $c_3\ c_1\ c_2$

    [2] 1 0 1 0 0 1 1 0 ⟷ $c_4\ c_5$
         0 1 0 1 0 0 1 1

    [1] 1 0 1 0 1 0 0 1 ⟷ $c_2\ c_3\ c_1$
         1 1 0 1 0 1 0 0

    [2] 0 1 1 0 1 0 1 0 ⟷ $c_5\ c_4$

Figure 3.17. A code example.

Synchronizable Codes

We conclude from Example 3.9 that if a cyclic equivalence class contains a code-word sequence $c_1 c_2 c_3$, it also contains $c_3 c_1 c_2$ and $c_2 c_3 c_1$. This can be generalized.

Let $\mathbf{W}$ be a finite set of words. Two word sequences, $w_{i_0}, w_{i_1}, \ldots, w_{i_{n-1}}$ and $w_{j_0}, w_{j_1}, \ldots, w_{j_{m-1}}$ are *cyclically equivalent as word sequences* if:

They have the same number of words, i.e., $d = n = m$, and there exists a k: $0 \le k < d$ for which $j_l = i_l + k$ for all l: $0 \le l < d$, where the addition is modulo d.

Less formally, the subscripts of the first sequence can be cycled to yield the subscripts of the second sequence. Two word sequences are *cyclically distinct* word sequences if they are not cyclically equivalent. Note that we have split the set of word sequences into equivalence classes under the heading of cyclically equivalent as word sequences. In the following discussion, we refer to a word sequence as a *code-word sequence* if the words in the sequence are code words from an agreed code. Example 3.9 demonstrates that a periodically ambiguous sequence contains two cyclically distinct code-word sequences $c_1 c_2 c_3$ and $c_4 c_5$. The generalization follows in Theorem 3.2

Theorem 3.2. A necessary condition for a code $\mathbf{C}$ to be U_I is, for all choices of n, the total number of cyclically distinct, nonperiodic, code-word sequences containing n symbols from $\mathbf{X}$ must be less than or equal to the number $B(|\mathbf{X}|, n)$ of available NPCE classes.

Proof: If this condition were not satisfied, two cyclically distinct code-word sequences would be present in the same NPCE class, so the code would not be U_I.

We proceed to count the number of cyclically distinct nonperiodic code-word sequences of length n symbols from $\mathbf{X}$.

Let N_l, $l = 1, 2, \cdots$ denote the number of words of length l symbols in the code $\mathbf{C}$. The formation of compound words of length n from the words in $\mathbf{C}$ is evaluated by taking semiordered partitions of the positive integer n. For example, if $n = 6$, a semiordered partition of 6 is given by

6
51
411
3111
21111
111111
42
312
321
2211
2121
33
222

Partitioning 6 indicates that a compound word of length 6 can be constructed from a code word of length 5 followed by a code word of length 1, or by a code word of length 4 followed by two code words of length 1, etc. The trivial partition of 6, namely, 6 itself, indicates that a single code word of length 6 is a compound word of length 6. The number of words formed by adding a length 1 word as a suffix to a length 5 word is $N_5 N_1$ words. Since the code must be uniquely decodable, the $N_5 N_1$ words formed in this manner cannot be periodic nor cyclic shifts of one another; therefore, they occupy $N_5 N_1 = B(N_5 N_1, 1)$ distinct NPCE classes of length 6 words. Notice that the cyclic permutation of 51, namely, 15, is not included in the semiordered partition of 6, since the compound words formed by adding a length 5 word as a suffix to a length 1 word are cyclically equivalent to the words that 51 represents. This type of analysis can be used to evaluate the number of NPCE classes occupied by word sequences obtained from each of the partitions of n. The last three partitions of 6 present a more difficult counting problem than the 51 partition, since the partitions are periodic. For example, among the $N_2^2 N_1^2$ compound words represented by the 2121 partition, several are cyclically equivalent word sequences or periodic word sequences. To count the number of cyclically distinct nonperiodic word sequences properly, consider the basic alphabet of symbol sequences of length 3 formed by adding a length 1 word as a suffix to a length 2 word. There are $N_2 N_1$ compound symbols in this new alphabet, and the number of ways that two of these compound symbols can be combined, avoiding cyclically equivalent and periodic compound words, is by definition $B(N_2 N_1, 2)$ words.

The number of cyclically distinct nonperiodic code-word sequences containing n symbols is given in general by

Synchronizable Codes

$$\sum_{\bar{p}\in \mathbf{P}(n)} B[N_{p_1}N_{p_2}\cdots N_{p_q}, r(\bar{p})]$$

where $\mathbf{P}(n)$ is the collection of semiordered partitions of n; $p_1 p_2 \cdots p_q$ is one period of the semiordered partition $\bar{p}$; and $r(\bar{p})$ is the number of periods in the partition $\bar{p}$. Substituting this counting result into the existence condition for U_I codes yields the result.

Theorem 3.3 (Golomb and Gordon). A necessary condition for the existence of a U_I code over an alphabet $\mathbf{X}$ with N_l words of length l, $l = 1, 2, \ldots$, is

$$\sum_{\bar{p}\in \mathbf{P}(n)} B(N_{p_1}N_{p_2}\cdots N_{p_q}, r(\bar{p})) \leq B(|\mathbf{X}|, n) \qquad (31)$$

for all choices of n.

Example 3.10. Can a ternary U_I dictionary with one word of length 1, two words of length 2, six words of length 3, and nine words of length 4 exist? Here,

$$N_1 = 1 \qquad N_2 = 2 \qquad N_3 = 6 \qquad N_4 = 9$$

We must check Equation 31 for $n = 1, 2, 3, 4, \ldots$.

For $n = 1$, the only partition of 1 is 1; hence,

$$1 = B(N_1, 1) \leq B(3, 1) = 3$$

and the inequality is satisfied.

For $n = 2$, the semiordered partitions of 2 are

$\bar{p}$	$r(\bar{p})$
2	1
11	2

$$2 = B(N_2, 1) + B(N_1, 2) \leq B(3, 2) = 3.$$

Hence, the $n = 2$ inequality is satisfied.

For $n = 3$,

$$\begin{array}{cc} \bar{p} & r(\bar{p}) \\ 3 & 1 \\ 21 & 1 \\ 111 & 3 \end{array}$$

$$8 = B(N_3, 1) + B(N_2 N_1, 1) + B(N_1, 3) \le B(3, 3) = 8.$$

Thus, the $n = 3$ inequality is satisfied.
For $n = 4$.

$$\begin{array}{cc} \bar{p} & r(\bar{p}) \\ 4 & 1 \\ 31 & 1 \\ 22 & 2 \\ 211 & 1 \\ 1111 & 4 \end{array}$$

$$18 = B(N_4, 1) + B(N_3 N_1, 1) + B(N_2, 2) + B(N_2 N_1^2, 1)$$
$$+ B(N_1, 4) \le B(3, 4) = 18.$$

Thus, Equation 31 is satisfied for $n = 1, 2, 3, 4$, and it may be possible to construct the desired code.

3.9. Cyclic Equivalence Class Occupancy and the Sufficiency of Iteratively Constructed Codes

It is often desirable to determine quickly whether or not a sequence of symbols is a code word. After the jth step in an iterative code construction procedure, a code word can be considered as a sequence of words from the jth synchronizable code.

Example 3.11. Consider the word $c = 10100$ in Example 3.7. We denote by $c_{(i)}$ the code word c interpreted as a sequence of words from $\mathbf{C}^{(i)}$. Thus,

$$c_{(0)} = ,1,0,1,0,0,$$

The interpretation $c_{(0)}$ is unique and always possible provided all symbols in c are in the alphabet and the word length is admissible. Recall that in con-

Synchronizable Codes

structing $\mathbf{C}^{(1)}$, $w^{(1)} = 0$ (suffix), and the representation of c as a sequence of words from $\mathbf{C}^{(1)}$ can be found by removing the comma before $w^{(1)}$ everywhere that ,$w^{(1)}$, appears in $c_{(0)}$; i.e., we change ,0, to 0, everywhere ,0, appears. In this case,

$$c_{(1)} = ,10,100,$$

In general, to construct $c_{(j+1)}$ from $c_{(j)}$ for any code word, remove the comma before or after $w^{(j+1)}$, depending on whether $w^{(j+1)}$ is used as a suffix or prefix, respectively, everywhere that ,$w^{(j+1)}$, appears in $c_{(j)}$. Continuing Example 3.11 gives

$$w^{(2)} = 1 \text{ (suffix)} \qquad c_{(2)} = ,10,100,$$
$$w^{(3)} = 100 \text{(suffix)} \qquad c_{(3)} = ,10100,$$
$$w^{(4)} = 101 \text{(suffix)} \qquad c_{(4)} = ,10100,$$

At every step in the preceding method for constructing $c_{(j)}, j = 1, 2, \ldots$, the construction from the previous step is unique, with no other choices being possible. Since the initial representation $c_{(0)}$ is also unique, each of the preceding representative sequences is unique. We have proved Theorem 3.4.

Theorem 3.4. If the initial sequence c is a code word, $c_{(i)}$ is a sequence of code words from $\mathbf{C}^{(i)}$, with each word preceded and followed by a comma. Conversely, if we start with an arbitrary sequence c and after n constructions find that $c_{(n)}$ consists of a concatenation of alphabet symbols preceded and followed by a comma, then c must be a code word in $\mathbf{C}^{(n)}$.

Proof: The preceding method describes how to construct the sequence $c_{(i)}$.

The Theorem 3.4 always identifies a code word if c is an iteratively constructed code word. Given a sequence of symbols c, we can ask a more general question, "Which (if any) of the words cyclically equivalent to c is a code word?" This is answered by using the method in Theorem 3.4 *simultaneously* on all elements of the cyclic equivalence class $\{c\}$ containing c. This can be approached conceptually by placing the sequence clockwise on a circle, putting commas between symbols to obtain $c_{(0)}$, then proceeding to delete commas in a clockwise manner.

Example 3.12. Let $c = 10110$. We want to see if c cycles to yield a valid code word in $\mathbf{C}^{(4)}$. We consider the pattern constructed by writing c clockwise on a circle, as shown in Figure 3.18a.

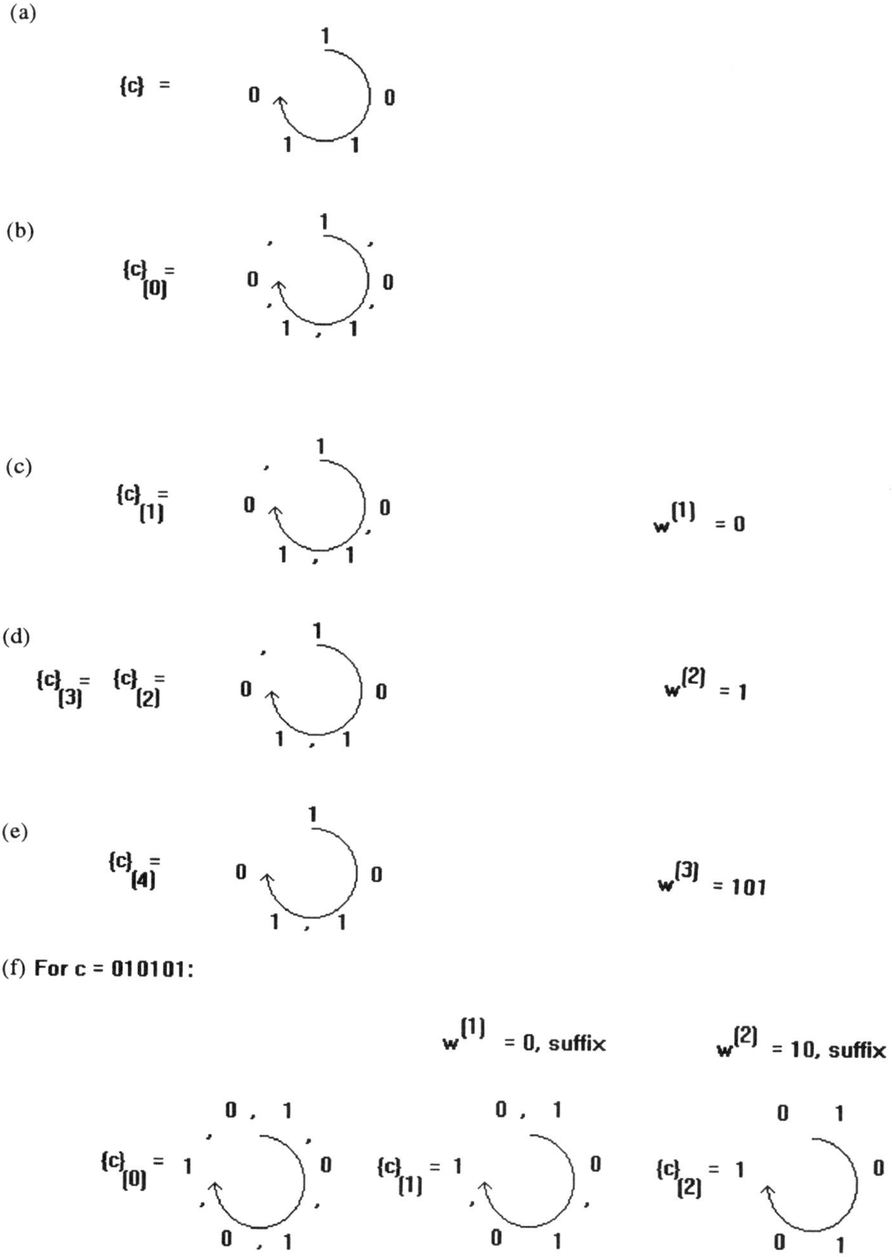

176   Chapter 3

(a)

{c} =   0 ↶ 1) 0
              1   1

(b)

{c}_{[0]} = 0 ↶ 1) 0
              1   1

(c)

{c}_{[1]} = 0 ↶ 1) 0 $w^{(1)} = 0$
              1   1

(d)

{c}_{[3]} = {c}_{[2]} = 0 ↶ 1) 0 $w^{(2)} = 1$
                          1   1

(e)

{c}_{[4]} = 0 ↶ 1) 0 $w^{(3)} = 101$
              1   1

(f) **For c = 010101:**

              $w^{(1)} = 0$, suffix $w^{(2)} = 10$, suffix

{c}_{[0]} = 1 ↶ 0,1) 0   {c}_{[1]} = 1 ↶ 0,1) 0   {c}_{[2]} = 1 ↶ 0 1) 0
               0,1   0 1   0   1

Figure 3.18. Examples of cycling patterns.

All words in the cyclic equivalence can be found by reading clockwise around the circle, beginning with each of the five different positions. We can insert commas to construct $\{c\}_{(0)}$, as shown in Figure 3.18b.

Each word in the class can be interpreted as a sequence of words from $\mathbf{C}^{(0)}$. Beginning with $w^{(1)} = 0$ (suffix), erase the comma preceding $w^{(1)}$ everywhere that it occurs, reading the sequence in a clockwise direction. We refer to the final pattern as $\{c\}_{(1)}$, as shown in Figure 3.18c. Reading clockwise, members of the equivalence class that are words or sequences of words from $\mathbf{C}^{(1)}$ begin immediately after the comma in $\{c\}_{(1)}$.

We repeat this procedure with $w^{(2)} = 1$ (suffix) to obtain $\{c\}_{(2)}$, with $w^{(3)} = 100$ (suffix) to obtain $\{c\}_{(3)}$, and with $w^{(4)} = 101$ (suffix) to obtain $\{c\}_{(4)}$, as shown in Figures 3.18d and 3.18e.

We conclude that c does cycle to give a code word in $\mathbf{C}^{(4)}$, and reading clockwise from the comma, the code word is 10101.

Notice that there are no alternatives in the preceding identification procedure; hence, each cyclic equivalence class of punctuated symbols is occupied, if at all, by a sequence of code words unique to within cyclic permutations of the words (e.g., $\{c\}_{(2)}$ is occupied by ,101, 10, and ,10, 101,). This unique-occupancy characteristic is another verification that the iterative construction in Section 3.7 produces directories that satisfy Levenshtein's test, i.e., the construction produces U_I codes. This follows, since the unique cyclic equivalence class occupancy characteristic is identical to the absence of *periodic* counterexamples in Levenshtein's test.

What happens when the cyclic procedure fails, i.e., when c does not cycle to give a sequence of code words? A cyclic equivalence class is not occupied by a sequence of code words from $\mathbf{C}^{(j)}$ if and only if one of the construction words $w^{(1)}, w^{(2)}, \ldots, w^{(j)}$ is identical to one period of the cyclic equivalence class. For example, if $c = 010101$, we obtain the patterns shown in Figure 3.18f. In the special case of c, a nonperiodic cyclic n-tuple, the preceding observation can be sharpened to give a condition for whether or not a NPCE class is occupied.

Lemma 3.1. An NPCE class $\{c\}$ is unoccupied by a sequence of code words from $\mathbf{C}^{(j)}$ if and only if $\{c\}$ is occupied by a word used as a suffix or prefix in constructing $\mathbf{C}^{(i)}$, $i \leq j$.

Equivalence class occupancy and Lemma 3.1 are fundamental to proving Theorem 3.5.

Theorem 3.5. (Scholtz). If the numbers N_l, $l = 1, 2, \ldots, L$ satisfy the inequalities

$$\sum_{\bar{p} \in P(n)} B[N_{p_1} N_{p_2} \cdots N_{p_q}, r(\bar{p})] \leq B(|\mathbf{X}|, n) \tag{32}$$

for $n = 1, 2, \ldots, L$, then there exists an iteratively constructed U_l dictionary with N_l code words of length l, $l = 1, 2, \ldots, L$.

The proof of this result uses the following:
Construction: Let $\mathbf{C}^{(j)}$ represent the jth dictionary in the construction procedure and let $N_l^{(j)}$ be the number of words of length l in $\mathbf{C}^{(j)}$. To construct $\mathbf{C}^{(j+1)}$ from $C^{(j)}$,

1. Find the minimum word length n for which

$$N_n^{(j)} > N_n$$

2. Select a word of length n from $C^{(j)}$ to serve as a suffix (or prefix) $w^{(j+1)}$.

3. Construct $\mathbf{C}^{(j+1)}$ using the selected prefix.

Repeat Steps 1–3, increasing j each time by 1 until enough words exist in code $\mathbf{C}^{(j)}$ to compose the desired code.

Before proving Theorem 3.4, let us consider Example 3.13.

Example 3.13. Using the construction process in Theorem 3.4, we construct a ternary code with word lengths

$$N_1 = 1 \quad N_2 = 2 \quad N_3 = 6 \quad N_4 = 9$$

In Example 3.10, it was shown that these numbers satisfied Golomb's and Gordon's inequalities (Equation 31). The initial step in the construction procedure is

$$\mathbf{C}^{(0)} = 0, 1, 2$$

Using both $w^{(1)} = 0$ and $w^{(2)} = 1$ as suffixes, we have

$\mathbf{C}^{(2)} = 2, 10, 20, 21, 100, 200, 101, 201, 211,$

$\qquad\qquad 1000, 2000, 1001, 2001, 1011, 2011, 2111$

Hence,

$$N_1^{(2)} = 1 \quad N_2^{(2)} = 3 \quad N_3^{(2)} = 5$$

At this point, $N_1^{(2)} = N_1$, and we cannot remove any more length 1 words. Now $N_2^{(2)} > N_2$, so we choose $w^{(3)} = 10$ as a prefix:

Synchronizable Codes 179

$\mathbf{C}^{(3)}$ = 2, 20, 21, 100, 200, 101, 201, 211, 102,
1000, 2000, 1001, 2001, 1011, 2011, 2111, 1020, 1021

At this point, $N_i = N_i^{(3)}$, $i = 1, 2, 3, 4$, and $\mathbf{C}^{(3)}$ is the desired dictionary.

Proof: The proof is based on establishing the following properties of the constructed sequence of U_l dictionaries. Let n_j be the length of the suffix $w^{(j)}$ used in constructing $\mathbf{C}^{(j)}$. Then,

1. $n_j \geq n_{j-1}$

2. All code words of length less than n_j have been constructed, i.e.,

$$N_l^{(j)} = N_l \text{ for } l < n_j$$

3. All NPCE classes of sequences longer than n_j are occupied by code-word sequences from $\mathbf{C}^{(j)}$, i.e.,

$$\sum_{\bar{p} \in \mathbf{P}(l)} B(N_{p_1}^{(j)} \cdots N_{p_q}^{(j)}, r(\bar{p})) = B(|\mathbf{X}|, l) \quad \text{for } l > n_j$$

4. There are always enough words in $\mathbf{C}^{(j)}$ to continue the construction if necessary, i.e., there exists an integer $m_j \geq n_j$ for which

$$N_l^{(j)} = N_l, \text{ for } l < m_j, \quad \text{and } N_{m_j}^{(j)} > N_{m_j}$$

We prove that these properties are preserved by the construction procedure and hence the procedure may be repeated until for large enough j, $N_l^{(j)} = N_l$ for all $l \leq L$.

First we must show that (1)–(4) are true for $j = 0$. Evaluating Equation 32 for $n = 1$ gives

$$N_1 \leq |\mathbf{X}| = N_1^{(0)} \tag{33}$$

If the equality in Equation 32 holds, every symbol in the alphabet must be used as a code word, and thus, the construction is complete. If strict inequality holds, then choose $n_0 = 1$, and the conditions for (1)–(4) are satisfied.

We must now use induction. Assume that 1–4 are true for dictionary $\mathbf{C}^{(j)}$; we then show that they remain true for $\mathbf{C}^{(j+1)}$. Under the construction procedure, we have to use a word of length m_j as $w^{(j)}$. Hence, from (2) and (4):

$$n_{j+1} = m_j \geq n_j, \quad \text{and}$$

(1) is verified for $j + 1$.

From 4, since the construction cannot create nor eliminate words of length less than n_{j+1},

$$N_l^{(j+1)} = N_l \quad \text{for } l < n_{j+1}, \quad \text{and} \tag{34}$$

(2) is verified for $j + 1$.

Since no words of length greater than n_{j+1} have been used as suffixes or prefixes in the construction procedure, all NPCE classes of length greater than n_{j+1} are occupied by sequences of words from $\mathbf{C}^{(j+1)}$.

$$\sum_{\bar{p} \in \mathbf{P}(l)} B(N_{p_1} \cdots N_{p_q}, r(\bar{p})) = B(|\mathbf{X}|, l) \quad \text{for } l > n_{j+1} \tag{35}$$

(3) is verified for $j + 1$.

If $N_{n_j}^{(j)} > N_{n_j} + 1$, then

$$n_{j+1} = n_j \quad N_{n_{j+1}}^{(j+1)} > N_{n_{j+1}} \quad \text{and} \quad m_{j+1} = n_{j+1} \tag{36}$$

If $N_{n_j}^{(j)} = N_{n_j} + 1$, then

$$n_{j+1} = n_j \quad N_{n_{j+1}}^{(j+1)} = N_{n_{j+1}}, \quad \text{and} \tag{37}$$

(4) is verified.

Since $\{N_l, l = 1, 2, \cdots\}$ satisfies Golomb's and Gordon's inequalities, Equations 31, 32, and 35 imply that

$$\sum_{\bar{p} \in \mathbf{P}(n_{j+1}+1)} B(N_{p_1} \cdots N_{p_q}, r(\bar{p})) \leq B(|\mathbf{X}|, n_{j+1} + 1)$$

$$= \sum_{\bar{p} \in \mathbf{P}(n_{j+1}+1)} B(N_{p_1}^{(j+1)} \cdots N_{p_q}^{(j+1)}, r(\bar{p})) \tag{38}$$

Equations 34 and 37 indicate that N_l and $N_l^{(j+1)}$ are identical for l less than $n_{j+1} + 1$ and the only partition $\bar{p}$ of $n_{j+1} + 1$ that contains $n_{j+1} + 1$ is the trivial one-element partition of $n_{j+1} + 1$ by itself. Canceling all identical terms in Equation 38 yields

$$B(N_{n_{j+1}+1}, 1) \leq B(N_{n_{j+1}+1}^{(j+1)}, 1) \tag{39}$$

which implies

$$N_{n_{j+1}+1} \leq N_{n_{j+1}+1}^{(j+1)} \tag{40}$$

Synchronizable Codes

If strict inequality holds in Equation 40, the proof of (4) is complete. If equality holds in Equation 40, the preceding technique can be iterated to show that

$$N_{n_{j+1}+2} \leq N_{n_{j+1}+2}^{(j+1)} \qquad (41)$$

We repeat this iterative technique and note that a strict inequality *must* eventually appear. When a strict inequality occurs, the recursive proof of (4) is complete, and (4) is verified. It follows that (1)–(4) are true for all j and eventually the construction process produces a dictionary with the desired word-length structure, proving Theorem 3.5.

In this construction, we required only Equation 32 to be satisfied for $n = 1, 2, \ldots, L$, where L is the maximum word length of the desired code. Hence, the necessary conditions in Section 3.8 for the existence of a U_I code require us only to check Golomb's and Gordon's inequality for $n \leq L$. Since satisfying Golomb's and Gordon's inequalities is equivalent to being able to carry out the preceding iterative construction, either one is an acceptable test for the existence of a U_I dictionary. Since enumerating semiordered partitions of n becomes very tedious for large n, the existence of a U_I dictionary for specified N_l, $l = 1, 2, \cdots$ may be easier to check by direct construction. Note that existence checking can be facilitated by using the generating functions defined in Equations 26–28.

It is worthwhile noting some analogies between synchronous results in Chapter 2 and asynchronous results in Chapter 3; in particular,

- The Kraft–MacMillan inequality (Equation 25) in Chapter 2, is comparable to the Golomb–Gordon inequality.

- The prefix code construction, when the Kraft–MacMillan inequality is satisfied, is comparable to the Scholtz construction when the Golomb–Gordon inequality is satisfied.

Exercise

1. (a) What is the semiordered partition of 7?

    (b) Using Golomb's and Gordon's inequalities, determine whether or not it is possible to construct a binary code with $N_0 = N_1 = N_2 = 0$, $N_3 = 1$, $N_4 = 3$, $N_5 = 6$, $N_6 = 9$, $N_7 = 16$.

    (c) Investigate the problem in (b) via the alternative approach of attempting construction.

(d) Using your results from (c), indicate the correspondence between code words or compound words and the terms in Golomb's and Gordon's inequalities that count them.

3.10. Constructing Maximal Comma-Free Codes of Odd Word Length

We can now show a simple procedure for constructing comma-free codes of odd word length n which have one word in every NPCE class of length n.

1. Specify odd word length n and alphabet size $|\mathbf{X}|$.

2. Construct a synchronizable code using the method in Section 3.9. Choose $w^{(j+1)}$ as the shortest odd-length word in $\mathbf{C}^{(j)}$ (if there is more than one, any of the shortest words will do) and use only suffix constructions. Continue this procedure until no odd-length word of length less than n remains in the dictionary. Let $\mathbf{C}_f$ denote this dictionary.

3. The words of length n in the synchronizable code $\mathbf{C}_f$ form a comma-free dictionary $\mathbf{C}$.

We will refer to this construction as the shortest odd-length suffix (SOLS) construction. We first verify the maximality of the number of code words in $\mathbf{C}$ generated by this construction and then verify the comma-free property.

Maximality of the code-word dictionary results from a property (described in Section 3.9) of the iterative construction method, namely, that every NPCE class of m-tuples not composed of a construction word $w^{(j)}$ of length m and its cyclic shifts contains an m-tuple that can be identified as a code word or a sequence of code words from the iteratively constructed code. We apply this observation to n and argue as follows. Since none of the $w^{(j)}$ in the SOLS construction are of length n, every NPCE class of length n must be identifiable as containing an n-tuple equivalent to a word or sequence of words from the code constructed in Step 2. Furthermore, the parameter n in the SOLS construction is odd, and the code-word sequence cannot consist entirely of even-length code words. Hence, each identified code-word sequence corresponding to an NPCE class of length n must contain at least one odd-length code word. However, in the SOLS construction process, all odd-length code words with length less than n are used as construction words; i.e., they are $w^{(j)}$'s in Step 2, and the only odd-length code words of length $\leq n$ are of length n. It follows that every NPCE class of length n contains at least one code word of length n. Since the construction produces a code $\mathbf{C}_f$ that is U_I, every NPCE class of

Synchronizable Codes

length n contains exactly one code word of length n, and the maximality of **C** is verified.

Proof of the comma-free property for SOLS-constructed codes is based on the fact that in $\mathbf{C}_f$, an odd-length code word can be a proper overlap of two code words only if the even-length suffix of one of the code words involved is the even-length prefix of the other code word involved. Therefore, to demonstrate comma freedom in such an odd-length code construction, it is sufficient to prove Theorem 3.6.

Theorem 3.6. Let **C** be a code with word length n over an r-symbol alphabet, generated by the SOLS construction and let s be any even integer, $0 < s < n$. Then an s-symbol suffix of a code word c cannot be the s-length prefix of a code word c', for all $c, c' \in \mathbf{C}$.

Before proving this theorem, let us illustrate the kind of calculations used in the proof.

Example 3.14. Let $n = 17$. The following 17-bit code word c belongs to a code generated by a SOLS construction that always chooses the lexicographically least of the qualifying words as the next construction word. A construction word is always used as a suffix.

$$c = 10111001011001100$$

Can the 17-tuple

$$g = 10110011000100101$$

which begins with the last 10 bits of c be in the same code? The answer is obtained by testing g to see if it is the code word selected from its cyclic equivalence class of 17-tuples by Theorem 3.4. We modify the presentation here by placing the bits of g, separated by commas, in a straight line rather than around a circle, recalling that the comma preceding the last bit is the same as the comma following the first bit. We are actually representing a cyclic equivalence class. We apply the selection and identification processes described in Theorem 3.4 to both c and g, then display both results so the reader can compare what happens within the shared sequence of 10 bits. We indicate the construction word $w^{(j)}$ for each iteration, subscript c and g, with the test iteration count (j), and mark key commas for later reference. Results are shown in Figure 3.19.

$c_{[0]}$ = $\overset{[2]}{,}1,0,1,1,1,0,0\overset{[3]}{,}1,0,1,1,0,0,1,1,0\overset{[1]}{,}\overset{[4]}{0},$

$g_{[0]}$ = $\overset{[3]}{,}1,0,1,1,0,0,1,1,0\overset{[1]}{,}\overset{[4]}{0},0,1,0,0,1,0,1\overset{[6]}{,}$

$w^{[1]}$ = 0

$c_{[1]}$ = $\overset{[2]}{,}1\;0,1,1,1\;0\;0\overset{[3]}{,}1\;0,1,1\;0\;0,1\overset{[1]}{,}1\;0\;0\overset{[4]}{,}$

$g_{[1]}$ = $\overset{[3]}{,}1\;0,1,1\;0\;0,1\overset{[1]}{,}1\;0\;0\;0\overset{[5]}{,}1\;0\;0,1\;0,1\overset{[6]}{,}$

$w^{[2]}$ = 1

$c_{[2]}$ = $\overset{[2]}{,}1\;0\;1\;1,1\;0\;0\overset{[3]}{,}1\;0\;1,1\;0\;0\;1\overset{[1]}{,}1\;0\;0\overset{[4]}{,}$

$g_{[2]}$ = $\overset{[3]}{,}1\;0\;1,1\;0\;0\;1\overset{[1]}{,}1\;0\;0\;0\overset{[5]}{,}1\;0\;0,1\;0\;1\overset{[6]}{,}$

$w^{[3]}$ = 1 0 0

$c_{[3]}$ = $\overset{[2]}{,}1\;0\;1\;1\;1\;0\;0\overset{[3]}{,}1\;0\;1\overset{[1]}{,}1\;0\;0\;1\;1\;0\;0\overset{[4]}{,}$

$g_{[3]}$ = $\overset{[3]}{,}1\;0\;1\overset{[1]}{,}1\;0\;0\;1\overset{[7]}{,}1\;0\;0\;0\;1\;0\;0\overset{[5]}{,}1\;0\;1\overset{[6]}{,}$

$w^{[4]}$ = 1 0 1

$c_{[4]}$ = $\overset{[2]}{,}1\;0\;1\;1\;1\;0\;0\;1\;0\;1\overset{[1]}{,}1\;0\;0\;1\;\;1\;0\;0\overset{[4]}{,}$

$g_{[4]}$ = $1\;0\;1\overset{[1]}{,}1\;0\;0\;1\overset{[7]}{,}1\;0\;0\;0\;1\;0\;0\;1\;0\;1$

Figure 3.19. Testing a sequence **g** to determine if it is in the same code as **c**.

If continued, this procedure eventually erases the comma (1) in $c_{(4)}$; c is a code word. Continuing the identification procedure on $g_{(4)}$ erases the unmarked comma, leaving comma (1) preceding the first symbol of the cyclic shift of g that is a code word. Notice that at each step before comma (3) was erased, the punctuation between commas (3) and (1) is identical, eventually forcing comma (3) to be erased at the same iteration in both $c_{(j)}$ and $g_{(j)}$ (on the fourth iteration in this example). This effect is the key to the proof of Theorem 3.6. However, to conclude Example 3.14, g is not in the same code as c. A cyclic shift of g, namely, 10011000100101101, is in the same code as c.

Proof: We will prove Theorem 3.6 by simultaneously applying the identification procedure in the previous section to a code word c and any sequence g that begins with an even-length suffix of c.

We denote the interpretations of c and g (if possible) as sequences of words from the jth iteratively constructed code $\mathbf{C}^{(j)}$ by $c_{(j)}$ and $g_{(j)}$, respectively. In describing what happens in this identification process, we denote an odd-length code word from the current dictionary by O and an even-length code word by E. When the length of a word does not matter, we represent such a word by A. Whenever E or A appears, we allow the possibility of no code word at all. We subscript A or E with a numeral m to indicate a sequence of m arbitrary or even-length words preceded and followed by commas. Thus,

$$,A_3, = ,A,A,A,$$

$$,E_4, = ,E,E,E,E,$$

We also subscript commas for later reference. For example, note that $c_{(j)} = ,A_{m_j},^{(1)}A$, for some $m_j \geq 0$. In fact, since $c_{(j)}$ will eventually become a single code word, it follows that $m_j = 0$ for a large enough value of j. Furthermore, when $m_j \neq 0$, comma (1) can be removed from this identification procedure only if the last word A in $c_{(j)}$ is of odd length $\leq n$, i.e., $c_{(j)}$ ends with a construction word $w^{(k)}$ for some $k > j$. It follows that a more refined representation of $c_{(j)}$ is

$$c_{(j)} = ,A_{m_j},^{(1)}O^{(j)},$$

for some $m_j \geq 0$ and for an odd-length word $O^{(j)}$ from $\mathbf{C}^{(j)}$.

As will be proved, the following form is general enough to remain invariant to the simultaneous application of the identification procedure to c and g as long as comma (3) is present. In terms of words from $\mathbf{C}^{(j)}$,

$$c_{(j)} = \overset{(2)}{,}A_{n_j}\overset{(3)}{,}A_{k_j}\overset{(1)}{,}O^{(j)}\overset{(4)}{,}$$
$$g_{(j)} = \overset{(3)}{,}A_{k_j}\overset{(1)}{,}E_{m_j},A^{(j)}\overset{(5)}{,}A_{p_j}\overset{(6)}{,} \tag{42}$$

where:

- s symbols between comma (3) and comma (4) denote the even-length suffix of c, and these are identical to the even-length prefix of g.
- Identically labeled segments $,^{(3)}A_{k_j},^{(1)}$ of $c_{(j)}$ and $g_{(j)}$ are identical in both symbols and punctuation; in other words, comma (1) indicates where the two sequences of code words, one for $c_{(j)}$ one for $g_{(j)}$, that start at comma (3) and are taken from the code $\mathbf{C}^{(j)}$, start to diverge.
- Comma (5) is defined to be the leftmost comma to the right of the first s symbols in $g_{(j)}$; in other words, read s symbols from comma (1). In general, the sequence ends at either the end or in the middle of a code word from $\mathbf{C}^{(j)}$. This code word is defined to be $A^{(j)}$, and we place comma (5) at the end of this code word.
- k_j, m_j, and p_j are nonnegative integers.

Denoting the distance between marked commas (i) and (j) by d_{ij}, Theorem 3.6 assumes that

$$d_{24} = n \text{ (odd integer)} \quad d_{34} = s \text{ (even integer)},$$

Since d_{14} is odd, it follows that

$$d_{31} = d_{34} - d_{14} = s - d \text{ (odd integer)} \Rightarrow k_j > 0$$

In other words, there have to exist code words in $\mathbf{C}^{(j)}$ between commas (3) and (1). E_{m_j} represents the longest sequence of even-length code words that occur in $g_{(j)}$, starting at comma (1). It remains to be proved that the $\mathbf{C}^{(j)}$ code words from comma (1) to $A^{(j)}$ in $g_{(j)}$ are all even length. m_j can be zero. Note that, $d_{35} \geq s = d_{34}$, i.e.,

$$d_{14} \leq d_{15}$$

We now demonstrate that the intermediate structure in Equation 42 exists for all j such that comma (3) preceding the length s suffix is present and the step in the SOLS construction that erases comma (3) in $c_{(j)}$ must also erase comma (3) in $g_{(j)}$.

Synchronizable Codes

The identification procedure always begins with every symbol in c and g separated by commas. Therefore, $c_{(0)}$ and $g_{(0)}$ can be described in the notation of Equation 42 as

$$n_0 = p_0 = n - s \quad k_0 = s - 1 \quad m_0 = 0 \quad A^{(0)} = O^{(0)} \quad d_{14} = d_{15} = 1$$

Assume that $c_{(j)}$ and $g_{(j)}$ satisfy the structure in Equation 42 and the SOLS procedure is used to construct code words in $\mathbf{C}^{(j+1)}$ from code words in $\mathbf{C}^{(j)}$ using $w^{(j+1)} \in \mathbf{C}^{(j)}$ as the suffix construction word. We consider what happens when commas preceding $w^{(j+1)}$ in the structure in Equation 42 of $c_{(j)}$ and $g_{(j)}$ are erased to produce $c_{(j+1)}$ and $g_{(j+1)}$.

1. Erase commas immediately preceding occurrences of the word $w^{(j+1)}$ in A_{n_j} and A_{k_j}.

   a. If the first word in A_{k_j} is $w^{(j+1)}$, then comma (3) is not present in either $c_{(j+1)}$ or $g_{(j+1)}$, g cannot be a code word, and the proof is complete.

   b. If the first word in A_{k_j} is not $w^{(j+1)}$, then comma (3) remains in the structure of $c_{(j+1)}$ and $g_{(j+1)}$ in Equation 42. Furthermore, this step produces $n_{j+1} \leq n_j$ and $1 \leq k_{j+1} \leq k_j$. In this case, the identification procedure continues.

2. Erase commas immediately preceding occurrences of the word $w^{(j+1)}$ that follow comma (1) in $c_{(j)}$ or lie between comma (1) and comma (5) in $g_{(j)}$.

   a. Suppose that $O^{(j)} \neq w^{(j+1)}$ and $A^{(j)} \neq w^{(j+1)}$. Then no further commas are erased in this step and the structure in Equation 42 recurs in $c_{(j+1)}$ and $g_{(j+1)}$.

   b. Suppose that $O^{(j)} \neq w^{(j+1)}$ and $A^{(j)} = w^{(j+1)}$. For this situation to arise in a shortest odd-length suffix construction, the lengths of $O^{(j)}$ and $A^{(j)}$ must be equal, and since this case specifies $O^{(j)} \neq A^{(j)}$, it follows that $m_j > 0$. The location of comma (1) is not changed in this iteration, $m_{j+1} = m_j - 1$, and $A^{(j)}$ is a proper suffix of $A^{(j+1)}$. In this case, the structure in Equation 42 recurs in $c_{(j+1)}$ and $g_{(j+1)}$.

   c. Suppose $O^{(j)} = w^{(j+1)}$ and $A^{(j)} \neq w^{(j+1)}$. Then $O^{(j)}$ is a proper suffix of $O^{(j+1)}$ and the comma designated by (1) in $c^{(j+1)}$ is the comma immediately preceding $O^{(j+1)}$. We define the sequence $A_{k_{j+1}}$ of words from $\mathbf{C}^{(j+1)}$ to lie between commas (3) and (1) in $c_{(j+1)}$. This sequence has an identical image in $g_{(j+1)}$ because the larger structure A_{k_j} containing it was similarly replicated in $c_{(j)}$ and $g_{(j)}$. No commas in $g_{(j)}$ were erased in this step, but the effect moving the comma (1) designation in this iteration leftward makes $m_{j+1} = m_j + 1$. In this case, the structure in Equation 42 recurs in $c_{(j+1)}$ and $g_{(j+1)}$.

d. Suppose $O^{(j)} = w^{(j+1)}$ and $A^{(j)} = w^{(j+1)}$. Then $O^{(j+1)}$, the comma (1) location in $c_{(j+1)}$ and $g_{(j+1)}$, and $A_{k_{j+1}}$ are all determined as in Step 2c. However, in this case, the comma preceding $A^{(j)}$ in $g_{(j)}$ also is erased. It follows that $m_{j+1} = m_j$ because E_{m_j} first has an even-length word added on the left by redesignating comma (1) and then has its rightmost word absorbed into $A^{(j+1)}$ by erasing the comma prior to $A^{(j)}$. Again, the structure in Equation 42 recurs in $c_{(j+1)}$ and $g_{(j+1)}$.

3. Erase any commas immediately preceding occurrences of $w^{(j+1)}$ that follow comma (5) in $g_{(j)}$. This step either leaves comma (5) in its original position or moves it to the right. This step preserves the structure in Equation 42 in $c_{(j+1)}$ and $g_{(j+1)}$.

Since eventually comma (3) must be erased in the process of identifying c as a code word and the preceding analysis indicates that when this erasure occurs in $c_{(j)}$, it also occurs in $g_{(j)}$, it follows that g has been identified as not being a code word, proving Theorem 3.6.

As pointed out, Theorem 3.6 proves that we obtain a comma-free code from SOLS, which is Corollary 3.1.

Corollary 3.1. The SOLS construction produces a maximal comma-free code of fixed word length n for any odd integer n.

Exercises

1. (a) Construct a word length 7 binary comma-free code.

   (b) How many more words could you add to this code if you only required the code to be a U_I dictionary with maximum word length 7?

2. Is there a variation of the comma-free code construction stated in Chapter 3 that will result in exactly the same comma-free code?

3. A generalization of the comma-free code concept is possible given the following definition: A code is said to be *comma-free* if no word is an overlap of two or more code words. With this definition, variable-word-length codes are allowed and one code word can be entirely within another code word. Show that the maximum synchronization acquisition delay is $2(n_{\max} - 1)$, where $n_{\max}$ is the length of the longest word in the code.

4. Using Corollary 3.1 show that the following construction yields a variable-word-length comma-free code as defined in Problem 3.

   (a) Do parts 1 and 2 of the comma-free code construction specified in this section, for a code of length n.
   (b) Choose an odd n_{max}, $n < n_{max}$.
   (c) As code words select all odd-length words of length l, $n \leq l \leq n_{max}$, that were generated in (a).

3.11. Automating Binary Bounded Synchronization Delay Codes

All of the work described thus far approaches the problem of code design from the viewpoint of selecting code words from NPCE classes and demonstrates their synchronizability at the decoder. This is all that is necessary when it is feasible for the encoder and decoder to use table look-up procedures for encoding and decoding after synchronization has been established. However, even for moderate word lengths n, the number of binary code words $B(2, n)$ can be quite large, with $B(2, n)$ behaving like 2^n for large n (see Table 3.4). This is one of the major obstacles to making the use of synchronizable codes practical.

We indicate a systematic procedure for mapping data sequences into primitive NPCE classes (the first step in encoding) and for performing the inverse mapping (the last step in decoding). By restricting our attention to primitive NPCE classes, the subject can be rephrased algebraically in terms of U_m, the group of units of the integers modulo $m = 2^n - 1$. Special attention is given to the problem of design, so that the data sequence can be chosen from the collection of elements in a direct product of cyclic groups. In this case, each coordinate of the data sequence can be arbitrarily chosen from a finite set of consecutive integers. Cases when the direct product representation can be achieved for maximal dictionaries are based on certain properties of the factorization of $2^n - 1$, a problem that has received much attention in number theory [Reisel (1968), Brillhart *et al.* (1983), and Birkhoff and Vandiver (1904)]. We have deliberately limited our attention to the case of $2^n - 1$ being the product of distinct primes, i.e., p^2 never divides $2^n - 1$ for a prime p. More general results are readily available in Scholtz and Welch (1970).

3.11.1. Cyclic Equivalence Class Representations

In general, a binary n-tuple $b = (b_0, b_1, \ldots, b_{n-1})$ can be viewed as representing the binary number b, $0 \leq b \leq 2^n - 1$

$$b = \sum_{i=0}^{n-1} b_i 2^i \qquad (43)$$

For the remainder of Section 3.11.1, we exclude the all-1 n-tuple from consideration; in other words, we study B'_n, where

$$B'_n = \{b \mid b < 2^n - 1\}$$

We define Z_m to be the integers modulo m. It follows that we have identified the elements of B'_n with the elements of Z_m, where $m = 2^n - 1$. Note that Z_m has operations of addition and multiplication modulo m that induce corresponding actions on B'_n. The multiplication operation in Z_m has two special cases of interest to n-tuples:

- Multiplication of $b \in Z_m$ by 2 (modulo m) corresponds to a cyclic shift of b provided $b \in B'_n$. More generally, cyclic shifts of the n-tuple are generated from b by multiplication by 2^i and reduction modulo $2^n - 1$. Hence, numbers in the same cyclic equivalence class with b are

$$b^{(i)} \equiv 2^i b \mod(2^n - 1) \qquad i = 1, \ldots, n-1 \qquad (44)$$

Note that $b \in B'_n$ is in a periodic cyclic class of period $j < n$ if and only if

$$(2^j - 1)b \equiv 0 \mod(2^n - 1) \qquad (45)$$

- Multiplication of b by -1 in Z_m corresponds to taking the binary complement of every entry in b.

The first property is the one on which we focus. In the language of groups, the cyclic group generated by 2 in Z_{2^n-1} (where the group operation is multiplication mod $(2^n - 1)$), written $\langle 2 \rangle$, acts on B'_n. The action splits B'_n into orbits, whereby two n-tuples are in the same orbit if and only if they are in the same cyclic equivalence class. An orbit has size dividing n (the size of $\langle 2 \rangle$), and the orbits of size equal to n are precisely the NPCE classes. Note that we have to exclude the all-1 vector to identify a cyclic shift with multiplication mod $(2^n - 1)$; the all-1 n-tuple is equal to $2^n - 1$. The equivalence between n-tuple and number, and the effect of cyclic shifts on an n-tuple are illustrated in Table 3.1 for $n = 6$.

The collection of integers modulo m, namely, Z_m, is never a group under the operation of multiplication mod m (0 has no inverse). However, Z_m contains a multiplicative group U_m, defined later, that is of interest. There are several equivalent definitions of U_m. The simplest is probably the following:

Synchronizable Codes

The collection of integers less than m and relatively prime to m form a group U_m under multiplication modulo m. If $x \in U_m$, then

$$(x, m) = 1 \tag{46}$$

An equivalent definition states that if $x \in U_m$, there has to exist an integer y for which

$$xy = 1 \text{ modulo } m \tag{47}$$

Another definition states that if $x \in U_m$, then for every nonzero element z in Z_m,

$$xz \neq 0 \text{ modulo } m. \tag{48}$$

Exercise

Show that these definitions are equivalent.

In Appendix 3.A, it is shown that U_m has $\phi(m)$ elements, where $\phi(m)$ is the Euler totient function. If $m = p_1^{e_1} p_2^{e_2} \cdots p_k^{e_k}$, then

$$\phi(m) = m \prod_{i=1}^{k} \left(1 - \frac{1}{p_i}\right) = \prod_{i=1}^{k} p_i^{e_i - 1}(p_i - 1) \tag{49}$$

Let $m = 2^n - 1$. We note that 2 is relatively prime (no common factor) to $2^n - 1$, implying that $2 \in U_m$. This implies that $\langle 2 \rangle$ is a subgroup of U_m and $\langle 2 \rangle$ has n elements. Note that if b is a *periodic* n-tuple, b does *not* belong to U_m. This follows, since if b is periodic, there exists $j < n$ for which $2^j b = b$, i.e., $(2^j - 1)b = 0$, and Equation 48 excludes b from U_m. This states that if $b \in U_m$, b is in an NPCE class. Not every element in an NPCE class is labeled by an element in U_m. Elements in B'_n that are labeled with elements in U_m are called *primitive* elements; the others are called imprimitive elements. It is not difficult to show that if one element in a cyclic class is primitive, all the elements in that class are primitive.

Example 3.15. If $n = 4$, $m = 15 = 5 \times 3$. It follows that there are $\phi(15) = 4 \times 2 = 8$ elements in U_{15}. However, there are 12 elements in NPCE classes, namely, the three classes generated by 1000, 1110, and 1100. Only the first two classes are primitive. In the integers modulo 15,

$$U_{15} = \{1, 2, 4, 8\} \cup \{7, 14, 13, 11\} \qquad \langle 2 \rangle = \{1, 2, 4, 8\}$$

It follows that the factor group, written as,

$$\frac{U_{15}}{\langle 2 \rangle}$$

is a group with two elements, namely, $\langle 2 \rangle$ and $7 \times \langle 2 \rangle$. These two elements are in 1-to-1 relation with the two primitive NPCE classes.

We now look at mechanizations that label the *primitive* NPCE classes, i.e., mechanizations labeling elements in

$$\frac{U_m}{\langle 2 \rangle}$$

Toward this end, we must analyze the group structures of U_m and $U_m/\langle 2 \rangle$ and factor them into simpler objects. There are many different approaches; one involves projecting elements in U_m onto the prime power factors of $2^n - 1$. Let

$$2^n - 1 = \prod_{i=1}^{k} p_i^{e_i} \qquad (50)$$

denote the decomposition of $2^n - 1$ into the product of distinct powers of primes $p_1, \ldots, p_k$. The sequence of projections of a number b onto the prime power factors $p_i^{e_i}, i = 1, \ldots, k$, is simply the k-tuple $(x_1, x_2, \ldots, x_k)$, where

$$b = x_i \bmod p_i^{e_i} \qquad i = 1, 2, \ldots, k. \qquad (51)$$

This projection may be thought of as a map from Z_m to

$$\prod_{i=1}^{k} Z_{p_i^{e_i}}$$

The Chinese remainder theorem ensures that b can be reconstructed from its set of projections $(x_1, \ldots, x_k)$.

Theorem 3.7. Given primes $p_1, \ldots, p_k$ and integers $x_1, \ldots, x_k$, the simultaneous congruences $b = x_i \bmod p_i^{e_i}$, $i = 1, 2, \ldots, k$, have a unique solution modulo $\prod_{i=1}^{k} p_i^{e_i}$.

Synchronizable Codes

The proof of this theorem [see Scholtz and Welch (1970) or Berlekamp (1968)] indicates further that x is reconstructed by a computation of the form

$$b = \sum_{i=1}^{k} \gamma_i x_i \bmod (2^n - 1) \tag{52}$$

where γ_i, $i = 1, \ldots, k$, are integer constants.

The map that sends b to $(x_i, \ldots, x_k)$ sends U_m onto a subgroup of

$$\prod_{i=1}^{k} Z_{p_i^{e_i}}$$

In fact, each of the components x_i must be a member of the appropriate ring of units, and we have

$$U_m \equiv \prod_{i=1}^{k} U_{p_i^{e_i}} \tag{53}$$

Exercise

Prove this result. (Hint: use Equations 46 or 48 to show that one side of Equation 53 is contained in the other, then prove the sets are equal in number.)

Example 3.16. Let $n = 4$. We have $p_1 = 5$, $p_2 = 3$, and $e_1 = e_2 = 1$. $U_5 = \{1, 2, 3, 4\}$ and $U_3 = \{1, 2\}$. b goes to $(b \bmod 5, b \bmod 3)$. Under this mapping, the elements of U_m become

$1 = (1, 1)$	$2 = (2, 2)$	$4 = (4, 1)$	$8 = (3, 2)$
$7 = (2, 1)$	$14 = (4, 2)$	$13 = (3, 1)$	$11 = (1, 2)$

Example 3.17. Let $n = 23$.

$$2^{23} - 1 = 47 \times 178481$$

$$p_1 = 47 \qquad p_2 = 178481$$

$$\gamma_1 = 2677215 \qquad \gamma_2 = 5711393$$

If $b = 50$, then the projections of b are

$$(x_1, x_2) = (3, 50)$$

Reconstruction of b follows from

$$b = 2677215 \times 3 + 2711393 \times 50$$
$$= 50 \bmod(2^{23} - 1)$$

We note that this is not the only algebraic factoring of U_m that can be performed; an alternative approach uses the fact that U_m is a commutative *cyclic* group to factor the group into a product of cyclic Sylow q_i subgroups, where q_i are the prime factors of $\phi(m)$. Note: A Sylow p subgroup of a group G is a subgroup of order p^j, where p^j is the largest power of p to divide $|G|$.

We now turn our attention to the special case where n has no square prime divisors, i.e., $e_i = 1$ for all i: $1 \leq i \leq k$. The more general case can be found in Scholtz and Welch (1970). In this case, $\phi(p_i^{e_i}) = \phi(p_i) = p_i - 1$, and there exists a one-to-one correspondence between the elements in U_m and the $\prod_i (p_i - 1)$ distinct k-tuples $(x_1, \ldots, x_k)$, with $1 \leq x_i \leq p_i - 1$ for all i. We now consider the structure of $U_m/\langle 2 \rangle$ under the mapping to the projections. We note that 2 has order n in U_m, i.e., the powers of 2 have n distinct values. The order of 2 in U_{p_i} divides n but may also be a proper divisor for some of the primes p_i; for example, in the $n = 4$ case, 2 has order 2 in U_3. In Birkhoff and Vandiver (1904), it is shown that there exists at least one prime p_1 for which the order of 2 in U_{p_1} remains equal to n, i.e., p_1 can be chosen so that 2^j, $j = 0, 1, \ldots, n - 1$, are all distinct modulo p_1, for $n > 1$ and $n \neq 6$. Under these conditions, it can be shown that

$$\frac{U_m}{\langle 2 \rangle} = \left(\frac{U_{p_1}}{\langle 2 \rangle}\right)\left(\prod_{i=2}^{k} U_{p_i}\right) \tag{54}$$

Since U_{p_1} is a cyclic group, the factor group $U_{p_1}/\langle 2 \rangle$ is also cyclic. In fact, there exists a generator g_1 for U_{p_1} for which

g_1^j, $j = 0, 1, \ldots, p_1 - 2$, are all distinct modulo p_1, i.e., $\langle g_1 \rangle = U_{p_1}$.

$g_1^{p-1/n} = 2$, i.e., $\langle g_1^{p-1/n} \rangle = \langle 2 \rangle$

The cosets $g_1^k \langle 2 \rangle$, $0 \leq k < \dfrac{p-1}{n}$, are distinct and map to the elements of $\dfrac{U_{p_1}}{\langle 2 \rangle}$.

Synchronizable Codes

It follows that the set of k-tuples of the form

$$(g_1^{d_1}, x_2, x_3, \ldots, x_k) \tag{55}$$

with

$$0 \le d_1 < \frac{p_1 - 1}{n} \tag{56}$$

and

$$1 \le x_i \le p_i - 1 \quad i = 2, \ldots, k \tag{57}$$

contains exactly one k-tuple representing a binary n-tuple from each primitive NPCE class. [Scholtz and Welch (1970) also show that certain x_i coordinates can have their ranges increased by 1 to include some nonprimitive NPCE classes in the mapping.] We are now in a position to consider the mechanization of encoding and decoding binary data into primitive NPCE classes.

Binary data being fed into the encoder must fit into the ranges of d_1 and $x_2, \ldots, x_k$. Since these ranges are not generally powers of 2, a fraction of a bit in efficiency may be lost for each factor of $2^n - 1$. Remember that the cost of fixed word length n and bounded synchronization delay is at least $\log_2 n$ symbols from each word (see Section 3.7).

Example 3.18. Let $n = 23$. In this case, either the range of x_1 or the range of x_2 can be reduced to produce the desired effect; however, decoding is much simpler if the shorter range is reduced. Hence, a representation for one element of each primitive NPCE class can be achieved with the set of all pairs (d_1, x_2) where $0 \le d_1 < 2$ and $0 < x_2 < 178481$. The actual element b can be reconstructed from (d_1, x_2) by the equation

$$b = 2677215 \cdot 5^{d_1} + 5711393 \cdot x_2 \mod(2^{23} - 1)$$

since 5 is a primitive element of the group under multiplication modulo 47.

For $n = 23$, the integer upper bound on $\log_2 23$ is 5, leaving at best $23 - 5 = 18$ binary information symbols fully available in each word. Since the range of d_1 is 2, one binary data symbol specifies d_1. Since $2^{17} = 131{,}072$, 17 binary data symbols can be used to specify x_2, and hence this binary data to primitive NPCE class mapping is the most efficient one possible. A block diagram of the mechanization is shown in Figure 3.20.

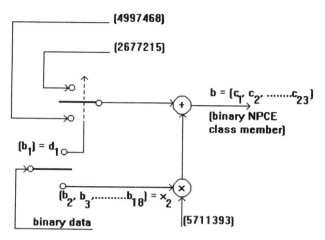

Figure 3.20. Encoding—mapping length 18 binary data vectors into length 23 binary primitive NPCE classes. $\oplus$ and $\otimes$ are mod $(2^{23}-1)$. In actual mechanization, decimal constants would be supplied in their binary form.

3.11.2. Encoding and Decoding

The encoding process to be used in conjunction with a comma-free code (see Section 3.6.1) or a lexicographically chosen code (see Section 3.6.4) is now simple:

1. Binary data is mapped into an element b representing an NPCE class via the algorithm described in 3.11.1.

2. The code word selection recognition algorithm described in 3.9 and 3.10 is applied to the particular code to select the n-tuple c (a cyclic shift of the binary vector represented by b in the NPCE). c is the code word to be transmitted.

Decoding involves three basic operations on a received symbol string:

Synchronizable Codes

1. Synchronization, i.e., segmenting the string into codewords. This process may also use the codeword selection/recognition algorithm described in 3.9 and 3.10.

2. Cyclically shifting code word c to a form b that represents the NPCE class in the algebraic structure of the data mapping

3. Inverse mapping from x to the data vector.

The problem of determining the cyclic shift of the code word c that equals the data representative b in its NPCE class is solved by noting that b is the only element in its NPCE class that has a projection $g_1^{d_1} = b \bmod p_1$ for $0 \le d_1 < (p_1 - 1)/n$ (this is the reason that the mapping was one-to-one from data to NPCE classes). Hence, the binary numbers x, corresponding to cyclic shifts of c, must be reduced $\bmod p_1$ to x_1', and it must be determined whether x_1' is equal to some $g_1^{d_1}$ with d_1 in the allowed range. Our discussion of receiver mechanization is completed.

Example 3.19. Let $n = 23$. Then x is equal to the NPCE class representative b if and only if

$$x = 1 \text{ modulo } 47 \text{ or } x = 5 \bmod 47$$

If $x = \alpha$ modulo 47, where α is not 1 or 5, it is possible to determine whether the jth cyclic shift of x is 1 or 5 by using the fact that

$$2^j x = 2^j \alpha \text{ modulo } 47$$

where, of course, the left side of the preceding equation is the jth cyclic shift of x written as an n-bit binary number. A diagram of the mechanization of this step is shown in Figure 3.21. In this mechanization, the vector x is read (most significant bit first) into the two shift registers that perform the indicated modulo 47 and 178481 reductions. When all the bits of x have been read into the registers, the process continues with zeros at the input. In this latter phase, the output is enabled when the binary representation of either 1 or 5 appears in the mod 47 register.

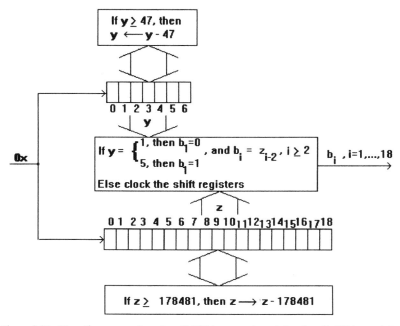

Figure 3.21. Decoding—mapping a length 23 binary vector x into a length 18 binary data vector. The vector x is read (most significant bit first) into the two shift registers which perform the indicated modulo 47 and 178481 reductions. When all the bits of x have been read into the registers, the process continues with zeros at the input. In this latter phase, the output circuit is enabled when the binary representation of either 1 or 5 appears in the mod 47 register.

The complexity of the decoder hinges on the range of values that d_1 can assume. For $e_1 = 1$, this is $(p_1 - 1)/n$. In Example 3.19, this range is only 2, namely, d_1 equals 0 or 1, corresponding to $x_1 = 1$ or 5. Certainly x_1 can in general be easily recovered, but the decision whether the corresponding d_1 is in the allowable range may be difficult to mechanize if the range is large.

In Table 3.4, we have tabulated the theoretical maximum binary dictionary size $B(2, n)$, the maximum size $C(n)$ for which the data mapping can be mechanized, the prime factor p_1 of $2^n - 1$ corresponding to the truncated range, the number of information bits k per word when the source is binary, and the size of the range of d_1.

Table 3.4. Code Mechanization Data

n	p_1	$B(2, n)$	$C(n)$	k	$\dfrac{(p_1 - 1)}{n}$
7	127	18	18	4	18
9	73	56	56	5	8
11	23	186	184	7	2
13	8191	630	630	9	630
15	151	2182	2170	9	10
17	131071	7710	7710	12	7710
19	524287	27594	27594	14	27594
21	337	99858	99568	15	16
23	47	364722	364720	18	2
25	601	1342176	1342176	18	24
27	262657	4971008	4971008	21	9728
29	233	18512790	18503928	24	8

3.12. Notes

For a thorough look at the various synchronization problems occurring in digital communication systems, see Stiffler (1971). Levenshtein (1961) originally stated the algorithm for identifying U_I codes.

The early work on comma-free codes was performed by Golomb, Gordon, and Welch (1958) and Jiggs (1963). Kendall and Reed (1962) suggested the path-invariant approach to comma-free codes. Gilbert (1960) developed the prefixed comma-free code concept. Niho (1973) found the binary code-word length 10 maximal comma-free code.

Lexicographically chosen U_I dictionaries were described in Golomb and Gordon (1965) and by Eastman and Even (1964). In the same work, Golomb and Gordon indicated the inequalities that word lengths of a U_I dictionary must satisfy. The sufficiency of these inequalities and the iterative construction procedure for synchronizable codes were demonstrated by Scholtz (1966).

Eastman (1965) first demonstrated the existence of maximal comma-free codes for all odd word lengths. The proof and construction algorithm given in 3.11 follows Scholtz (1969).

For studies of synchronizable codes that do not necessarily have bounded synchronization delay, see Scholtz and Storwick (1970). For a perspective on statistically synchronizable codes, see Wei and Scholtz (1980).

Devitt (1983) showed that the constructions of Eastman and Scholtz are distinct.

Appendix: The Möbius Inversion Formula

The following mathematical relation is basic to many applications, including synchronizable code construction. This relation involves the following equation:

$$g(n) = \sum_{d|n} f(d) \qquad n = 1, 2, \cdots \tag{A1}$$

where the sum is over all divisors of n including 1 and n. Since there are occasions when we know $g(n)$, $n = 1, 2, \ldots$, but not $f(d)$, a solution for $f(n)$ in terms of $g(d)$ is desired.

Solutions are attainable by hand for small values of n; for example,

$$g(1) = f(1)$$
$$g(2) = f(1) + f(2)$$
$$g(3) = f(1) + f(3)$$
$$g(6) = f(1) + f(2) + f(3) + f(6) \tag{A2}$$

implies

$$f(6) = g(6) - g(3) - g(2) + g(1) \tag{A3}$$

Notice that Equation (A3) is the solution for $f(6)$ regardless of the type of functions f and g represent. The form of Equation (A3) depends only on the prime decomposition of the number 6.

General solutions for f in terms of g can be derived in several ways. The general solution that we give here is based on properties of a function $\mu(d)$ known as the Möbius function.

$$\mu(d) = \begin{cases} 1 & \text{if } d = 1 \\ (-1)^r & \text{if } d \text{ is the product of } r \text{ distinct primes} \\ 0 & \text{if } d \text{ has a repeated prime factor} \end{cases} \tag{A4}$$

Hence, the first few values of $\mu(d)$ are

d	1	2	3	4	5	6	7	8	9	10	11	12	13
$\mu(d)$	1	-1	-1	0	-1	1	-1	0	0	1	-1	0	-1

Synchronizable Codes

We will need the following property of the Möbius function from Theorem A.1.

Theorem A.1. Let $\mu(d)$ denote the Möbius function. Then

$$\sum_{d|n} \mu(d) = \begin{cases} 1 & \text{if } n = 1 \\ 0 & \text{if } n \neq 1 \end{cases} \quad (A5)$$

Proof: Let us denote the prime decomposition of n by

$$n = p_1^{c_1} p_2^{c_2} \cdots p_k^{c_k} \quad c_i \geq 1 \quad i = 1, \ldots, k \quad (A6)$$

Then

$$\sum_{d|n} \mu(d) = \sum_{i_1=0}^{c_1} \cdots \sum_{i_k=0}^{c_k} \mu(p_1^{i_1} p_2^{i_2} \cdots p_k^{i_k}) \quad (A7)$$

Using the fact that the Möbius function is zero when its argument has repeated primes and substituting Equation (A4) into Equation (A7) gives

$$\sum_{d|n} \mu(d) = \sum_{i_1=0}^{1} \cdots \sum_{i_k=0}^{1} \mu(p_1^{i_1} \cdots p_k^{i_k})$$

$$= \sum_{i_1=0}^{1} \cdots \sum_{i_k=0}^{1} (-1)^{\sum_{j=1}^{k} i_j} = \prod_{j=1}^{k} \left(\sum_{i_j=0}^{1} (-1)^{i_j} \right) \quad (A8)$$

Each factor of the product in Equation (A8) is zero. Since every positive integer greater than 1 has a decomposition of the form given in Equation (A6), we have demonstrated Equation (A5) for $n \neq 1$. By definition, $\mu(1) = 1$, and hence Equation (A5) is proven.

We proceed to demonstrate the Möbius Inversion Formula, Theorem A.2, which provides a systematic solution to Equation (A1).

Theorem A.2. Given an integer m and a function $g(\cdot)$, which for all integers n that divide m, is defined by

$$g(n) = \sum_{d|n} f(d) \quad (A9)$$

then $f(m)$ is uniquely given by

$$f(m) = \sum_{n|m} g(n)\mu\left(\frac{m}{n}\right) \qquad (A10)$$

where $\mu(\cdot)$ is the Möbius function. Note that 1 and n are considered to be legitimate divisors of n.

Proof: Substituting Equation (A1) into the right side of Equation (A9) gives

$$\sum_{d|n} f(d) = \sum_{d|n} \sum_{k|d} g(k)\mu\left(\frac{d}{k}\right) \qquad (A11)$$

A change of variables in this sum can be made by noting the following relation:

$$\{d, k : k | d, d | n\} = \left\{d, k : k | n, \quad d = \alpha k, \quad \alpha \bigg| \frac{n}{k}\right\} \qquad (A12)$$

The set on the left side of Equation (A12) is the collection of values that d and k take in Equation A11. In the set on the right side of Equation (A12), d is replaced by αk, and hence Equation (A11) is reduced to a sum on k and α, with α the variable in the inner sum.

$$\sum_{d|n} f(d) = \sum_{k|n} \left[\sum_{\alpha | \frac{n}{k}} \mu(\alpha)\right] g(k) \qquad (A13)$$

Using Theorem A.1, the inner sum is zero unless $n/k = 1$, in which case the sum is 1. Hence, Equation (A13) reduces to Equation (A9), and we have demonstrated that Equation (A10) is one solution to Equation (A9).

Uniqueness of the solution in Equation (A10) is demonstrated by contradiction. Suppose a second solution $f'(d)$ existed. Then

$$g(n) = \sum_{d|n} f'(d) \qquad (A14)$$

Let n_0 be the smallest value of n for which $f'(n_0) \neq f(n_0)$. Equations (A9) and (A14) imply that

$$0 = g(n_0) - g(n_0) = f'(n_0) - f(n_0) \tag{A15}$$

contradicting our assumption of the existence of a second solution.

Example A.1. The Euler function $\phi(n)$ is defined for positive integers as the number of integers between 1 and n that are relatively prime to n; then

$$\sum_{d|n} \phi(d) = n$$

and the use of the Möbius function shows that if $n = p_1^{q_1} p_2^{q_2} \cdots p_l^{q_l}$, then

$$\phi(n) = n \prod_{i=1}^{l} \left(1 - \frac{1}{p_i}\right) = \prod_{i=1}^{l} p_i^{q_i - 1}(p_i - 1)$$

Exercises

1. A jewelry manufacturer wishes to design a collection of patternless necklaces using black, white, and gold beads. He/she has decided that the necklace should contain 15 beads. The patternless constraint requires no necklace to use beads of all one color or to tie together three identical necklaces of length 5 or five identical necklaces of length 3, each of which would create a necklace pattern with a periodic structure. To make each necklace distinct, no two necklaces can use the same pattern. If the colored beads in one necklace are in the reverse order of the colored beads in another necklace and, if these necklaces are cyclically distinct, assume that each can be in the collection.

   (a) How large a collection can the manufacturer design? (Hint: Let $f(d)$ be the number of necklaces with period d and $g(n)$ the total number of patterns of length 15, namely, 3^{15}. Use Möbius inversion in Theorem A.2 to solve for the size of the collection.)

   (b) Repeat (a) with the additional requirement that each necklace must use all three of the available colors.

   (c) Repeat (a) with the additional requirement that each necklace contain exactly (i) 4 gold beads, (ii) 5 gold beads.

   (d) The manufacturer decides that the structural constraints in (a) are satisfactory, but he/she wishes to increase the length of the necklace so that the collection contains at least 10 million necklaces. What length necklace should he/she choose?

(e) If a 15-bead necklace is randomly strung with each color of bead equally likely, what is the probability that it will not be periodic?

2. Derive the multiplicative form of the Möbius inversion theorem: if

$$g(n) = \prod_{d|n} f(d)$$

then

$$f(m) = \prod_{n|m} g(n)^{\mu\left(\frac{m}{n}\right)}$$

where $\mu(\cdot)$ is the Möbius function.

3. The Möbius inversion in Theorem A.2 takes advantage of the fact that every integer $n = p_1^{c_1} p_2^{c_2} \cdots p_k^{c_k}$ can be described by a vector $\bar{c}$. Sums in Theorem A.3 are over certain portions of the vector space containing $\bar{c}$. Show that Theorem A.3 holds even if a dimension of the vector space containing $\bar{c}$ is dropped; that is, show that if

$$g(n) = \sum_{d \in D_n} f(d)$$

then

$$f(n) = \sum_{d \in D_n} g(d) \mu\left(\frac{n}{d}\right)$$

where

$$D_m = \{d : d|m, d \text{ not divisible by } \hat{p}\}$$

where $\hat{p}$ is a prime divisor of m.

3.13. References

E. R. Berlekamp. 1968. *Algebraic Coding Theory.* McGraw-Hill, New York: p. 28.

G. D. Birkhoff and H. S. Vandiver. 1904. "On the Integral Divisors of $a^n - b^n$." *Ann. Math.* 5: 173–80.

J. Brillhart, D. Lehmer, J. Selfridge, B. Tuckerman, and S. Wagstaff, Jr. 1983. *Factorizations of $b^n \pm 1$.* Vol. 22 in Contemporary Mathematics Series Providence, RI: American Math. Society.

L. J. Cummings. 1976. "Comma-Free Codes and Incidence Algebras." In *Combinatorial Mathematics IV*. Lecture Notes in Math. 560. Berlin: Springer-Verlag.
———. 1985. "Synchronizable Codes in the de Bruin Graph." *Ars Comb.* **19:** 73–90.
———. 1987. "Aspects of Synchronizable Coding." *Journal of Combinatorial Mathematics and Combinational Computing* **1:** 17–84.
J. S. Devitt and D. M. Jackson. 1981. "Comma-Free Codes: An Extension of Certain Enumerative Techniques to Recursively Defined Sequences." *J. Comb. Theory (A)* **30:** 1–18.
J. S. Devitt. 1983. "An Enumerative Interpretation of the Scholtz Construction for Comma-Free Codes." In *Combinatorics on Words, Progress and Perspectives*, edited by L. J. Cummings. Academic Press.
W. L. Eastman. 1965. "*On the Construction of Comma-Free Codes.*" *IEEE Trans. Inform. Theory.* **IT-11:** 263–66.
W. L. Eastman and S. Even. 1964. "On Synchronizable and PSK Synchronizable Block Codes." *IEEE Trans. Inform. Theory* **IT-10:** 351–56.
F. M. Gardner and W. C. Lindsey, eds. 1980. Special issue on Synchronization. *IEEE Trans. Commun.* **COM-28.**
E. N. Gilbert. 1960. "Synchronization of Binary Messages." *IRE Trans. Inform. Theory* **IT-6:** 470–77.
S. W. Golomb and B. Gordon. 1965. "Codes with Bounded Synchronization Delay." *Inform. Control* **8:** 355–76.
S. W. Golomb, B. Gordon, and L. R. Welch. 1958. "Comma-Free Codes." *Canad. J. of Math.* **10:** 202–9.
S. W. Golomb, L. R. Welch, and M. Delbrück. 1958. "Construction and Properties of Comma-Free Codes." *Biol. Medd. K. Dan. Vidensk. Selsk.* **23:** 1–34.
S. W. Golomb et al. 1963. "Synchronization." *IEEE Trans. Commun.* **11:** 481–91.
B. H. Jiggs. 1963. "Recent Results in Comma-Free Codes." *Canad. J. of Math.* **15:** 178–87.
W. B. Kendall and I. S. Reed. 1962. "Path-Invariant Comma-Free Codes." *IRE Trans. Inform. Theory* **IT-8:** 350–55.
V. I. Levenshtein. 1961. "Certain Properties of Code Systems." *Dokl. Akad. Nauk. SSSR* **140:** 1274–277.
W. C. Lindsey, F. Ghazvinian, W. C. Hagmann, and K. Dessouky. 1985. "Network Synchronization." *Proc. IEEE* **73,** pp. 1445–1467.
J. H. van Lint. 1985. "{0, 1, *} Distance Problems in Combinatorics." In *Surveys in Combinatorics 1985.* Invited papers for the Tenth British Combinatorial Conference, London Mathematical Society Lecture Note Series 103. Cambridge: Cambridge University Press. Pp. 113–35.
Y. Niho. 1973. "On Maximal Comma-Free Codes," *IEEE Trans. Inform. Theory* **IT-19:** 580–81.
H. Reisel. 1968. *En Bok Om Primtal.* Odense, Denmark: Studentlitteratur.
R. A. Scholtz. 1966. "Codes with Synchronization Capability." *IEEE Trans. Inform. Theory* **IT-12:** 135–42.
———. 1969. "Maximal- and Variable-Word-Length Comma-Free Codes." *IEEE Trans. Inform. Theory* **IT-15:** 300–6.
R. A. Scholtz and R. M. Storwick. 1970. "Block Codes for Statistical Synchronization." *IEEE Trans. Inform. Theory* **IT-16,** pp. 432–438.
R. A. Scholtz and L. R. Welch. 1970. "Mechanization of Codes with Bounded Synchronization Delays." *IEEE Trans. Inform. Theory* **IT-16:** 438–46.
J. J. Stiffler. 1971. *Theory of Synchronous Communications.* Prentice-Hall, New York.
B. Tang, S. W. Golomb, and R. L. Graham. 1987. "A New Result on Comma-Free Codes of Even Wordlength." *Canad. J. of Math.* **39:** 513–26.
V. K. W. Wei and R. A. Scholtz. 1980. "On the Characterization of Statistically Synchronizable Codes." *IEEE Trans. Inform. Theory* **IT-26,** pp. 733–735.

4

Infinite Discrete Sources

4.1. Agent 00111 Meets the Countably Infinite

Agent 00111 was happiest taking cases that had only a finite number of outcomes. These were well-behaved in terms of pricing and computation. Even if the finite number of outcomes were extremely large, the Shannon–McMillan Theorem (Theorem 1.2a.) often made the costing tractable. Sometimes, however, there were not a finite number of possibilities but two other possibilities. The number of outcomes could be countably or uncountably infinite. In Chapter 4, we consider infinite discrete distributions. For example, Agent 00111 may be asked to estimate how many days he would spend researching the true state of a country's secret committees and to justify the time in terms of the uncertainty eliminated. In such cases, Agent 00111's department could *theoretically* spend from here to eternity on the project, since the outcomes (in days) are countably infinite. In practice, this did not bother Agent 00111 too much—an obvious point of diminishing returns and increasing boredom caused him to quit after a period of time. However, as his scientists pointed out to him, just suppose that each day he were on a project he obtained more information than the last; how tempting it would be to stay.

This type of problem led Agent 00111 to conclude that while his practical methods were basically sound, he had to be alert for the infinite case. As an example that has an impact on Agent 00111, recall the Lempel–Ziv algorithm from Section 2.11, which transmits integers representing compressed data. These integers are positive, but they can be large; in fact, the larger the better. The amount of compression for a segment of data increases with the size of one of the transmitted parameters, i.e., L. The Lempel–Ziv algorithm has to use a reasonably compact format for transmitting large integers.

4.2. The Leningrad Paradox

Intermediate between the case of finite distributions and continuous distributions is the case of the infinite discrete distribution, an infinite sequence of probabilities

$$p_1, p_2, p_3, \ldots, p_k \geq 0 \; \forall k, \qquad \sum_{k=1}^{\infty} p_k = 1$$

These distributions were first considered seriously in the eighteenth century, after the St. Petersburg paradox achieved worldwide notoriety. The paradox involves a player in a gambling house who tosses a perfect coin successively until the first tail appears. If n heads precede the first tail, he/she receives 2^n rubles, for $n = 0, 1, 2, 3, \ldots$. What is a reasonable price for the gambling house to charge for the privilege of playing this game?

Ideally, the gambling house fee should equal the expected value of the game. With probability $1/2$, the payoff is $2^0 = 1$; with probability $1/4$, it is $2^1 = 2$; with probability $1/8$, it is $2^2 = 4$, etc. Thus, the expected value of the game is

$$1/2 \cdot 1 + 1/4 \cdot 2 + 1/8 \cdot 4 + 1/16 \cdot 8 + \cdots$$
$$= 1/2 + 1/2 + 1/2 + 1/2 + \cdots = \infty$$

and the game should have infinite value to the player. Part of the *paradox* is the simple observation that an infinite distribution can have an infinite mean. The rest of the paradox revolves around the fact that no gambling house ever has a truly infinite bankroll, and even if the gambling house limit is set quite high (say, at 1 trillion rubles), the expected value of the game is suddenly a *surprisingly small* number:

$$\sum_{k=0}^{40} \frac{2^k}{2^{k+1}} = 20.5 \text{ rubles}$$

More modest payoff functions involve less dependence on very improbable cases in evaluating the mean. For example, if the payoff for n consecutive heads before the first tail is set at n rubles, the expected value of the game is only

Infinite Discrete Sources

$$\sum_{n=1}^{\infty} n \cdot 2^{-n} = 2 \dagger$$

Let us look at another formulation of the same game. In a hypothetical primitive society, each married couple continues to produce children until the advent of the first son. We further assume that boys and girls are produced randomly, each with probability $1/2$. What is the expected family size? The brute force solution is to observe that there are n children with a probability of 2^{-n}, for an expected number of

$$\sum_{n=1}^{\infty} n 2^{-n} = 2$$

A more clever approach is to observe that each family ends up with exactly one boy, but the expected number of girls in the population equals the expected number of boys. Hence, there is an *average* of one girl per family and an expected number of two children.

Next, we compute the amount of *information*, or *entropy*, contained in the distribution $1/2, 1/4, 1/8, 1/16, 1/32, \ldots$. The brute force approach is

$$H(p_1, p_2, \cdots) = -\sum_{k=1}^{\infty} p_k \log_2 p_k = \sum_{k=1}^{\infty} k 2^{-k} = 2 \text{ bits}$$

again assuming we remember the sum of the series. A more elegant approach is to reason in terms of the coin-tossing game as follows. The first toss produces 1 bit of information. If the outcome of that toss is a tail, there is no further

† For $z < 1$, there is the result

$$\sum_{n=1}^{\infty} n z^n = \frac{z}{(1-z)^2}$$

based on

$$\sum_{n=1}^{\infty} n z^n = z \sum_{n=1}^{\infty} n z^{n-1} = z \frac{d}{dt}\left(\sum_{n=1}^{\infty} z^n\right) = z \frac{d}{dt}\left(\frac{z}{1-z}\right) = \frac{z}{(1-z)^2}$$

For $z = 1/2$, this gives

$$\frac{1/2}{(1/2)^2} = 2$$

More generally, if z is a probability p, with $1 - z = q$, the sum of this series can be written p/q^2.

uncertainty, but with probability 1/2, the first outcome is a head, after which there is as much uncertainty as there had been *ab initio*. Thus, $H = 1 + H/2$, from which $H = 2$.

This is close to Agent 00111's case discussed in Section 4.1. Suppose that Agent 00111 is on a mission. Each day, he has a half chance of having his mission terminated, either by detection or by recall. He produces one bit of information every day that he remains on the mission, i.e., after the kth day he has accumulated

$$\log_2 \frac{1}{(2^{-k})}$$

bits of information. Thus his expected total delivery of information is

$$\sum_{k=1}^{\infty} 2^{-k} \cdot k = 2 \text{ bits}$$

It is also instructive to consider ways of encoding the outcome of consecutive coin-tossing games for transmission over a binary channel. Since each outcome contains, on the average, 2 bits of information, the optimum encoding cannot allow fewer than two binary symbols per outcome. This is quite easy to achieve if we simply send 0 for *tail* and 1 for *head* for every toss in every game. Since the expected duration of each game is two tosses, this achieves the theoretical minimum in transmission. Moreover, since 0 always marks the end of a game, there is no ambiguity in decipherability concerning where the description of one game ends and the next begins. This is an ideal encoding given the fact that each toss of an ideal coin contains a full bit of information and no further information compression is possible. The entropy of the distribution in the St. Petersburg paradox is also 2 bits, since the probabilities are still $p_k = 2^{-k}$. Thus, even though the paradox involves a distribution with an infinite mean, the information contained in each outcome is small, averaging only 2 bits per game. It is natural to ask whether there are finite mean discrete distributions—and hence sequential gambling games—with infinite entropy. In Theorem 4.1, we show that the answer to this question is no, i.e., an infinite entropy implies an infinite mean. However, we substitute a weaker requirement than finite mean by requiring that with arbitrarily high probability, the outcome of a discrete distribution is finite but the entropy is infinite. Such distributions exist, e.g., Example 4.2. Thus, it is possible to describe and play sequential gambling games, where, with probability 1, any given game has only finite duration and yet with so much uncertainty about the outcome that even with optimum encoding, it is necessary to send infinitely

Infinite Discrete Sources 211

many bits on average to describe what happened! Since this situation of infinite entropy is basically a twentieth-century version of the St. Petersburg paradox, it can appropriately be called the Leningrad paradox.†

It must be emphasized that the St. Petersburg distribution just described, with a payoff of k having the probability 2^{-k}, has a very special and unusual property, since the *mean,* or *expected value,* of this distribution happens to equal the *entropy.* As we see in Theorem 4.2, this is a rare and extreme situation. The coding problem for the binary channel also becomes far less trivial when $p \neq 1/2$, as described in Sections 4.4–4.8.

4.3. Mean vs. Entropy in Infinite Discrete Distributions

We start with some formal definitions.

Definition. An infinite sequence

$$\{p_n\}_{n=1}^{\infty} \text{ with } p_n \geq 0 \text{ for } n = 1, 2, 3, \cdots, \quad \sum_{n=1}^{\infty} p_n = 1$$

is called an *infinite discrete distribution.*

Definition. If $\{p_n\}$ is an infinite discrete distribution, the following standard terminology is applicable:

$$M = \sum_{1}^{\infty} n p_n \quad (1)$$

is the mean of the distribution.

$$H = -\sum_{1}^{\infty} p_n \log_2 p_n \quad (2)$$

is the *entropy* of the distribution.

We now consider some examples concerning the convergence or nonconvergence of M and H.

---

† This could become the only context in which the name "Leningrad" will be remembered.

Example 4.1 (Finite Mean, Finite Entropy). Let $p_n = 2^{-n}$. Then

$$M = \sum_{n=1}^{\infty} n 2^{-n} = 2 \qquad H = \sum_{n=1}^{\infty} n 2^{-n} = 2$$

and both the mean and entropy are finite (and incidentally equal). This is, of course, the distribution discussed in Section 4.2.

Example 4.2 (Infinite Mean, Infinite Entropy).

$$\text{Let } \alpha = \left[1 + \sum_{n=2}^{\infty} \frac{1}{n(\log n)^2}\right]^{-1}$$

and define

$$p_1 = \alpha \qquad p_n = \frac{\alpha}{n(\log n)^2} \quad \text{for } n > 1.$$

We first note that this is an infinite discrete probability distribution

$$\sum_{n=1}^{\infty} p_n = \alpha \left[1 + \sum_{n=2}^{\infty} \frac{1}{n(\log n)^2}\right] = 1$$

since the series for α^{-1} is convergent. Next, we show the distribution has an infinite mean

$$M = \alpha + \alpha \sum_{n=2}^{\infty} \frac{1}{(\log n)^2} = \infty$$

since this series diverges.

Finally, we show that the distribution has infinite entropy

$$H = -\alpha \log \alpha - \alpha \sum_{n=2}^{\infty} \frac{(\log \alpha - \log n - 2 \log \log n)}{n(\log n)^2}$$

$$= \alpha \sum_{n=2}^{\infty} \frac{1}{n \log n} + 2\alpha \sum_{n=2}^{\infty} \frac{\log \log n}{n(\log n)^2} - \log \alpha = \infty$$

since the first sum diverges, while the remaining two terms are finite.

Infinite Discrete Sources

Example 4.3 (Infinite Mean, Finite Entropy).

$$\text{Let } p_n = \frac{6}{\pi^2 n^2}$$

Since

$$\sum_{n=1}^{\infty} p_n = \frac{6}{\pi^2} \sum_{n=1}^{\infty} \frac{1}{n^2} = 1$$

this defines an infinite discrete distribution. Compute the mean:

$$M = \frac{6}{\pi^2} \sum_{n=1}^{\infty} \frac{1}{n} = \infty$$

so that the mean is infinite. Compute the entropy:

$$H = \frac{6}{\pi^2} \sum_{n=1}^{\infty} \frac{(\log \pi^2 + 2 \log n - \log 6)}{n^2} = \log \frac{\pi^2}{6} + \frac{12}{\pi^2} \sum_{n=1}^{\infty} \frac{\log n}{n^2} < \infty$$

which is finite.

The fourth situation (*infinite entropy, finite mean*) is incapable of occurring, as Theorem 4.1 demonstrates.

Theorem 4.1.

1. An infinite distribution $\{p_n\}$ with finite mean M has finite entropy H.
2. Specifically, H (in bits) $\leq M + 3/2 + 1/e \log_2 e = M + 2.0307 \cdots$
3. This bound is easily modified for logarithmic bases other than 2. (It can also be improved on, as proved in Note 2 and Theorem 4.2.)

Proof: It is given that

$$M = \sum_{n=1}^{\infty} n p_n < \infty$$

Decompose $\{p_n\}$ into two subsequences, Q and R, as follows:

$$p_n \in Q \quad \text{if } p_n \leq 2^{-n}$$
$$p_n \in R \quad \text{if } p_n > 2^{-n} \tag{3}$$

We now bound $-p_n \log p_n$ for $p_n \in Q$, $n > 1$, and for $p_n \in R$.

For $p_n \in Q$ and $n > 1$, we note that $-x \log x$ is monotonic increasing from 0 to $(\log e)/e$ on the interval $(0, e^{-1})$, and $p_n \leq 2^{-n} < e^{-1}$. It follows that

$$-p_n \log p_n \leq -2^{-n} \log(2^{-n}) = n \cdot 2^{-n}$$

If $p_n \in R$, we note that $p_n > 2^{-n}$ implies $-\log p_n < n$. It follows that

$$-p_n \log p_n \leq -p_n \log(2^{-n}) = np_n$$

Putting these inequalities together, we obtain an upper bound for H:

$$H = -\sum_{n=1}^{\infty} p_n \log p_n$$

$$= -\sum_{p_n \in Q} p_n \log p_n - \sum_{p_n \in R} p_n \log p_n$$

$$\leq -p_1 \log p_1 + \sum_{n=2}^{\infty} n \cdot 2^{-n} + \sum_{n=1}^{\infty} np_n$$

$$\leq (\log e)/e + 3/2 + M < \infty$$

as asserted.

Note 1: For a finite discrete distribution, both the mean and the entropy are obviously finite. The probabilities $p_1, p_2, \ldots, p_n$ lead to a maximum entropy of $\log n$ for the case $p_1 = p_2 = \cdots = p_n = 1/n$ and to a maximum mean of n for the case $p_1 = p_2 = \cdots = p_{n-1} = 0$, $p_n = 1$. Taking the limit as $n \to \infty$, both of these extremal situations lead to the values $p_1 = p_2 = p_3 = \cdots = 0$, which is not a distribution. Thus, the problems of extremal mean and extremal entropy in the infinite discrete case are not solvable as direct extensions of the finite case. On the other hand, the results in Theorem 4.1 can be directly extended to continuous distributions on any semi-infinite strip $a \leq x < \infty$.

Note 2: For discrete distributions, entropy is not affected by a rearrangement of the probabilities, since it is the values that occur, and not their order,

Infinite Discrete Sources

that affect $\Sigma p_n \log p_n$. However, the mean ($\Sigma n p_n$) is quite sensitive to the *order* in which probabilities occur, and it is shown later that if $\{p_n\}$ is any distribution with infinitely many nonzero terms, it has rearrangements with infinite mean. Hence, Theorem 4.1 can be strengthened to state that if $\{p_n\}$ is an infinite discrete distribution with infinite entropy, then every rearrangement of $\{p_n\}$ has infinite mean. The converse of this statement is false, as shown by Example 4.3.

Note 3: We show that if $\{p_n\}$ is any sequence of positive terms, it has a rearrangement $\{q_n\}$ such that $\Sigma n q_n$ diverges. We separate $p_2, p_4, p_6, \cdots$ from $p_1, p_3, p_5, \ldots$. Let $\lfloor x \rfloor$ denote the greatest integer not exceeding x. Define $\{n_i\}$ as follows:

$$n_1 = \lfloor 1/p_2 + 1 \rfloor \text{ and } n_i = \max\{(n_{i-1} + 1), \lfloor 1/p_{2i} + 1 \rfloor\} \quad \text{for } i > 1$$

Note that $\{n_i\}$ is an infinite increasing sequence of integers, with $n_i p_{2i} > 1$ for all i. Let $q_{n_i} = p_{2i}$. For all subscripts k not of the type n_i, let q_k correspond to the first available p_n with odd subscript. Thus,

$$\sum_{k=1}^{\infty} k q_k \geq \sum_{i=1}^{\infty} n_i q_{n_i} > \sum_{i=1}^{\infty} 1 = \infty$$

and $\{q_k\}$ is a rearrangement of $\{p_k\}$.

Using the Lagrange multipliers method, it is possible to determine the extremal infinite discrete distribution that makes $H - M$ as large as possible for a given finite mean M, as shown in Theorem 4.2.

Theorem 4.2. Let an infinite discrete distribution have mean M and entropy computed to logarithmic base $b > 1$ of H_b. Then

$$M - H_b \geq \log_b(b - 1) \tag{4}$$

with equality only for the case

$$M = \frac{b}{b - 1}$$

and

$$p_k = \frac{1}{M - 1}\left(\frac{M - 1}{M}\right)^k = (b - 1)b^{-k} \tag{5}$$

For information measured in bits, $b = 2$ and $M - H_2 \geq 0$ with equality only for the distribution given in Example 4.1.

Proof: Since we wish to maximize

$$H = \sum_{1}^{\infty} p_k \log p_k$$

subject to the constraints

$$\sum_{1}^{\infty} p_k = 1 \quad \text{and} \quad \sum_{1}^{\infty} k p_k = M$$

we define

$$U = \sum_{k=1}^{\infty} (-p_k \log p_k + \lambda p_k + \mu k p_k)$$

For the extremal case

$$\frac{\partial U}{\partial p_k} = -(1 + \log p_k) + \lambda + \mu k = 0 \tag{6}$$

$$\log p_k = \lambda - 1 + \mu k, \; p_k = e^{\lambda - 1} e^{\mu k} \tag{7}$$

To satisfy the constraints, set

$$p_k = \frac{1}{M-1} \left(\frac{M-1}{M} \right)^k.$$

We note that this satisfies both Equation 7 and the constraints, since

$$\sum_{k=1}^{\infty} p_k = \frac{1}{M-1} \sum_{k=1}^{\infty} \left(\frac{M-1}{M} \right)^k$$

$$= \frac{1}{M-1} \cdot \frac{(M-1)/M}{[(1 - (M-1)/M)]} = \frac{1}{M[1 - (M-1)/M]} = 1 \tag{8}$$

Infinite Discrete Sources

and

$$\sum_{k=1}^{\infty} k p_k = \frac{1}{M-1} \sum_{k=1}^{\infty} k \left(\frac{M-1}{M}\right)^k$$

$$= \frac{1}{M-1} \frac{(M-1)/M}{[1-(M-1)/M]^2} = \frac{M}{M^2[1-(M-1)/M]^2} = M \quad (9)$$

Computing entropy (H_b) to the logarithmic base $b > 1$,

$$H_b = -\sum_{k=1}^{\infty} p_k \log_b p_k$$

$$= -\frac{1}{M-1} \sum_{k=1}^{\infty} \left(\frac{M-1}{M}\right)^k \left[k \log_b \frac{M-1}{M} - \log_b(M-1)\right]$$

$$= \left(\log_b \frac{M}{M-1}\right) \left[\frac{1}{M-1} \sum_{k=1}^{\infty} k\left(\frac{M-1}{M}\right)^k\right]$$

$$+ (\log_b(M-1)) \left[\frac{1}{M-1} \sum_{k=1}^{\infty} \left(\frac{M-1}{M}\right)^k\right]$$

$$= M \log_b \frac{M}{M-1} + \log_b(M-1)$$

$$= M \log_b M - (M-1) \log_b(M-1) \quad (10)$$

Now define

$$\Delta_b = H_b - M = \{M \ln M - (M-1) \ln(M-1)\} \log_b e - M.$$

To maximize Δ_b with respect to M, we note that

$$0 = \frac{d\Delta_b}{dM} = (\log_b e)[1 + \ln(M) - 1 - \ln(M-1)] - 1 = \log_b \frac{M}{M-1} - 1$$

Hence, maximization occurs when

$$\log_b\left(\frac{M}{M-1}\right) = 1 \quad \text{or} \quad \frac{M}{M-1} = b \quad M = \frac{b}{b-1} \quad (11a)$$

Entropy H_b^* for this extremal case is

$$H_b^* = \left(\frac{b}{b-1}\log_b\frac{b}{b-1} - \frac{1}{b-1}\log_b\frac{1}{b-1}\right) \tag{11b}$$

$$= \frac{b}{b-1} - \left(\frac{b}{b-1} - \frac{1}{b-1}\right)\log_b(b-1)$$

$$= M - \log_b(b-1) \tag{11c}$$

If we wish to find the distribution of greatest entropy on some measure space S, subject to a constancy constraint on the rth moment, Lagrange multipliers will yield a distribution of the form $p(t) = c_1 e^{c_2 t^r}$, where c_1 and c_2 are suitable normalization constants. Sometimes this answer will not make sense. For example, if only the 0th moment is constrained (as it always is for a probability distribution), the Lagrange multiplier method leads to a *flat* distribution, which makes sense only on compact measure spaces, since there is no flat probability distribution on $(-\infty, \infty)$ nor on the set of all positive integers. The well-known result that on $(-\infty, \infty)$ the distribution of *given variance* with greatest entropy is the Gaussian distribution is not due to the measure space being the real numbers but to the *second moment* being the highest order constraint. As a very simple example, on the space consisting of the four points $-3, -1, 1, 3$, the distribution of greatest entropy is of course $p(-3) = p(-1) = p(1) = p(3) = 1/4$. However, if the object is to maximize the ratio of entropy to variance (say, for symmetric distributions), the flat distribution achieves only the ratio

$$H/\sigma^2 = \frac{\log 4}{(1/2)(1+9)} = 0.4 \text{ bits/unit variance}$$

whereas the more Gaussianlike distribution $p(-3) = p(3) = 0, p(-1) = p(1) = 1/2$, achieves the ratio

$$H/\sigma^2 = \frac{\log 2}{1/2 + 1/2} = 1.0 \text{ bits/unit variance}$$

Exercises

1. If a biased coin with probability p for *heads* and q for *tails* ($p + q = 1$) is tossed repeatedly and we examine the number of consecutive heads preceding the first occurrence of tails, called a *run*, we obtain the *geometric distribution*

Infinite Discrete Sources 219

$$G(n, p) = p^n q \qquad n = 0, 1, 2, \ldots.$$

(a) Find the rth moment of the distribution.

(b) Find the entropy of the distribution. Can we conclude that the entropy of the distribution is the entropy per toss times the average number of tosses?

2. The Poisson distribution $P(n, \lambda)$ occurs whenever events are uniformly randomly distributed in space or time with average density λ per unit region. $P(n, \lambda)$ is the probability that exactly n events will occur in a given region.

(a) Find the rth moment of the distribution.

(b) Compute the entropy of the distribution and reduce it to simplest form.

4.4. Run-Length Encodings

Agent 00111 is back at the casino, playing a game of chance while the fate of mankind hangs in the balance. Each game consists of a sequence of favorable events (probability p), terminated by the first occurrence of an unfavorable event (probability $q = 1 - p$). More specifically, the game is roulette, and the unfavorable event is the occurrence of 00, which has a probability of $q = 1/32$. No one seriously doubts that Agent 00111 will come through again, but the Secret Service is concerned about communicating a blow-by-blow description back to Whitehall. The bartender, who is a freelance agent, has a binary channel available, but he charges a stiff fee for each binary digit sent.

The problem facing the Secret Service is how to encode the vicissitudes of the wheel while placing the least strain on the Royal Exchequer. As we observed in Section 4.2, for the case $p = q = 1/2$, the best thing to do is to use 0 and 1 to represent the two possible outcomes. However, the case at hand involves $p \gg q$, for which the direct-coding method is shockingly inefficient. Finally, a junior code clerk who has been reading up on Information Theory suggests encoding the *run lengths* between successive unfavorable events. In general, the probability of a run length of n is $p^n q$, for $n = 0, 1, 2, 3, \ldots$, which is the familiar *geometric distribution*.

If the list of possible outcomes were finite, we could list them with their probabilities, then apply Huffman coding, as in Section 2.7. However, with an infinite list, we cannot start at the bottom and work our way up. Fortunately, the fact that the probabilities follow a distribution law provides a short cut.

Let $p^m = 1/2$ or $m = -\log 2/\log p$. The results will be neater for those p for which m is an integer (viz., $p = 0.5$, $p = 0.707\cdots$, $p = 0.794\cdots$, $p = 0.849\cdots$, $p = 0.873\cdots$, etc.) and especially simple when m is a power of 2. We consider m to be an integer initially. Given that $p^m = 1/2$, a run of length $n + m$ is only half as likely as a run of length n. The respective probabilities are $p^{m+n}q = (1/2)p^n q$ and $p^n q$. Thus, on the heuristic level, we would expect the code word for run length $n + m$ to be one bit longer than the code word for run length n. This argument leads to the heuristic conclusion that there should be m code words of each possible word length $\geq m$; there are no words for the shortest word lengths, which are not used at all if $m > 1$; and possibly one transitional word length used fewer than m times. These observations are sufficient to define dictionaries for small values of m. The dictionaries for $m = 1, 2, 3$, and 4 are shown in Table 4.1, where $G(n) = p^n \cdot q$.

For larger integer values of m, we can extend the structure of the Table 4.1 examples. Recall $p^m = 1/2$ for an integer m. Let k be the smallest positive integer such that $2^k \geq 2m$. Then the code dictionary corresponding to our heuristic requirements contains exactly m words of every word length $\geq k$, as well as $2^{k-1} - m$ words of length $k - 1$. (For m a power of 2, the collection of words of length $k - 1$ is empty. Note also that m must be an integer if $2^{k-1} - m$ is to be an integer.) This result is obtained by calculating how much message space is used up by m words of every length $\geq k$.

Table 4.1. Run-Length Dictionaries for Small m

	$m = 1$			$m = 2$			$m = 3$			$m = 4$	
n	$G(n)$	Code word	n	$G(n)$	Code word	n	$G(n)$	Code word	n	$G(n)$	Code word
0	1/2	0	0	0.293	00	0	0.206	00	0	0.159	000
1	1/4	10	1	0.207	01	1	0.164	010	1	0.134	001
2	1/8	110	2	0.146	100	2	0.130	011	2	0.113	010
3	1/16	1110	3	0.104	101	3	0.103	100	3	0.095	011
4	1/32	11110	4	0.073	1100	4	0.081	1010	4	0.080	1000
5	1/64	111110	5	0.051	1101	5	0.064	1011	5	0.067	1001
6	1/128	1111110	6	0.036	11100	6	0.051	1100	6	0.056	1010
7	1/256	11111110	7	0.025	11101	7	0.041	11010	7	0.047	1011
8	1/512	111111110	8	0.018	111100	8	0.032	11011	8	0.040	11000
9	1/1024	1111111110	9	0.013	111101	9	0.026	11100	9	0.033	11001
10	1/2048	11111111110	10	0.009	1111100	10	0.021	111010	10	0.028	11010

Infinite Discrete Sources

$$\frac{m}{2^k} + \frac{m}{2^{k+1}} + \frac{m}{2^{k+2}} + \cdots = \frac{m}{2^{k-1}}$$

leaving

$$1 - \frac{m}{2^{k-1}} = \frac{2^{k-1} - m}{2^{k-1}}$$

unused, which means that $2^{k-1} - m$ words of length $k - 1$ may be adjoined.

To illustrate what happens when (1) m is not a power of 2 and (2) when m is a power of 2, we consider the cases when $m = 14$ and $m = 16$. The corresponding dictionaries are shown in Table 4.2. For the case $m = 14$, we find $k = 5$, and $2^{k-1} - m = 2$, so that there are two code words of length 4,

Table 4.2. Run-Length Dictionaries for $m = 14$ and $m = 16$

	$m = 14$				$m = 16$		
n	Code word	n	Code word	n	Code word	n	Code word
0	0000	24	101100	0	00000	24	101000
1	0001	25	101101	1	00001	25	101001
2	00100	26	101110	2	00010	26	101010
3	00101	27	101111	3	00011	27	101011
4	00110	28	110000	4	00100	28	101100
5	00111	29	110001	5	00101	29	101101
6	01000	30	1100100	6	00110	30	101110
7	01001	31	1100101	7	00111	31	101111
8	01010	32	1100110	8	01000	32	1100000
9	01011	33	1100111	9	01001	33	1100001
10	01100	34	1101000	10	01010	34	1100010
11	01101	35	1101001	11	01011	35	1100011
12	01110	36	1101010	12	01100	36	1100100
13	01111	37	1101011	13	01101	37	1100101
14	10000	38	1101100	14	01110	38	1100110
15	10001	39	1101101	15	01111	39	1100111
16	100100	40	1101110	16	100000	40	1101000
17	100101	41	1101111	17	100001	41	1101001
18	100110	42	1110000	18	100010	42	1101010
19	100111	43	1110001	19	100011	43	1101011
20	101000	44	11100100	20	100100	44	1101100
21	101001	45	11100101	21	100101	45	1101101
22	101010	46	11100110	22	100110	46	1101110
23	101011	47	11100111	23	100111	47	1101111

followed by 14 code words of lengths 5, 6, 7, etc. On the other hand, since $m = 16$ is a power of 2, the corresponding dictionary contains exactly 16 words of every word length, starting with length 5.

In a practical situation, if $m = -\log 2/\log p$ is not an integer, then the best dictionary (in the sense of meeting our requirements) oscillates between $\lfloor m \rfloor$ words of a given length and $\lfloor m \rfloor + 1$ words of another length. (Here, $\lfloor m \rfloor$ denotes the greatest integer $\leq m$.) For large m, however, there is very little penalty for selecting the *nearest* integer when designing the code. Very often, the underlying probabilities are not known accurately enough to justify picking a nonintegral value of m. For example, selecting $p = 0.95$ may involve as large a round-off error as selecting $m = 14$.

4.5. Decoding Run-Length Codes

The dictionaries in Tables 4.2 exhibit striking patterns that suggest that a simple decoding procedure may be possible. We continue to illustrate by example.

For the case $m = 16$, the following procedure for decoding is adequate:

1. Start at the beginning (left end) of the word and count the number of 1s preceding the first 0. Let this number be $A \geq 0$. Then the word consists of $A + 5$ bits.

2. Let the last 5 bits represent the ordinary binary representation of the integer R, $0 \leq R \leq 15$.

Then the correct decoding of the word is $16A + R$.

This simple decoding reveals an equally simple method of *encoding*. To encode the number N, (1) divide N by 16 to obtain $N = 16A + R$, then (2), write A 1s followed by the 5-bit binary representation of R.

The case when $m = 14$ is only slightly more complicated; the decoding procedure is

1. If a word starts in A 1s and the next 3 bits are *not all* 0s, then we consider the word to consist of $A + 5$ bits.

2. Let the last 5 bits be the binary representation of the integer R. Then the correct decoding of the code word is $14A + R - 2$.

1. If the initial A 1s are followed by three or more 0s, we consider the code word to consist of a total of $A + 4$ bits.

2. Let the last 4 bits be the binary representation of an integer R'. Then the correct decoding of the code word is $14A + R'$.

Infinite Discrete Sources

This procedure also can be inverted to describe direct encoding from ordinary numbers to code words. Generalizations are left to the reader.

Exercises

1. Is run-length coding 100% efficient for any value of m? Is it 100% efficient for all values of m?
2. Describe the best run-length code for the case $m = 2.5$.
3. Is it possible to construct a dictionary with exactly $k - 1$ code words of every length k? Why? What kind of underlying probability distribution would such a code be suited to?
4. Assume $m = 2^r$. Compare the average number of bits per message required to communicate by run-length coding. Compare this with the entropy of the corresponding geometric distribution. Evaluate the efficiency numerically for $r = 0, 1, 2, 3, 4, 5$.
5. Suppose a run-length dictionary had been designed assuming a geometric distribution with parameters p and q, but the system encountered actually has parameters p' and q'. If $q' < q$, show that the rate is less than anticipated when using the run-length-coded channel, even though the theoretical efficiency is also less.
6. Using one of the dictionaries in Table 4.2, is it possible to read messages from right to left and still perform instantaneous decoding? Does this possibility exist for all values of m?

4.6. Capacity-Attaining Codes

When Agent's 00111 casino mission was being planned, the senior cryptographer was quick to observe that although run-length coding was an improvement over no coding at all, it was less than 100% efficient. Agent 00111 recalled hearing of a technique invented by Elias and Shannon at MIT that was 100% efficient and demanded a quick briefing from his staff. At the hasty and rather acrimonious briefing, operations decided that the Elias and Shannon method required infinite computing capability and was therefore unimplementable. Thus, the run-length system was retained. However, the evaluation of system complexity was not accepted by the entire scientific staff. In fact, the decision to retain the run-length system was fortuitous; Agent 00111 managed to bribe the croupier so that the unfavorable case occurred only half as often as expected, i.e., $p \to p/2$. This, in view of Exercise 5 (Section 4.5), greatly decreased the cost of communicating.

We examine Elias and Shannon's method, which is indeed 100% efficient, for communicating an infinite sequence of events selected from a discrete distribution. Moreover, we agree with the scientific staff that the infinite-computing capability requirement is a gross overstatement. Nevertheless, operations may well have made the correct practical decision; we leave it to the reader to judge. (This method is sometimes called *Elias coding*.)

Suppose we have a memoryless source S with an n-symbol alphabet A_1, $A_2, \ldots, A_n$ and corresponding probabilities $p_1, p_2, \ldots, p_n$. Messages from this source are to be encoded for transmission over an m-symbol channel, where each of the channel symbols $g_1, g_2, \ldots, g_m$ has unit duration. The Elias and Shannon encoding scheme begins by associating all the possible infinite messages produced by the source with the points on the interval $[0, 1)$ of the real line. Moreover, this is done in such a way that the *total probability* of all infinite messages mapped into any subinterval I of $[0, 1)$ equals the length of I. More generally, the total probability of all messages mapped into any measurable set is the measure of that set.

This measure-preserving map is created as follows:

1. Divide the interval $[0, 1)$ into n subintervals $[0, p_1)$, $[p_1, p_1 + p_2)$, $[p_1 + p_2, p_1 + p_2 + p_3), \ldots, [p_1 + p_2 + \cdots + p_{n-1}, 1)$.

2. Map all messages beginning with the symbol A_i into the subinterval whose right-hand end point is $p_1 + p_2 + \cdots + p_i$. Since this is an interval of length p_i and the probability of a message beginning with A_i is p_i, the measure-preserving property holds thus far.

3. Divide each subinterval into n parts, where the ratio of the lengths of the sub-subintervals is $p_1:p_2:p_3: \cdots :p_n$, with all sub-subintervals of the form $[a, b)$.

4. Map all messages beginning $A_i A_j$ into the jth sub-subinterval of the ith subinterval. This sub-subinterval has length $p_i p_j$, which is also the probability of an arbitrary message beginning with $A_i A_j$.

5. In the same manner, divide each sub-subinterval into n parts.

6. Associate all messages beginning with the triple $A_i A_j A_k$ with the kth portion of the jth component of the ith subinterval, etc.

7. Continuing in this manner, by the Nested Interval theorem of real analysis, every *infinite* message is ultimately mapped into a single point.

8. To find the optimum encoding for an infinite message M from the source S for the m-ary channel, find the point P in $[0, 1)$ into which M is mapped.

Infinite Discrete Sources

9. Compute the infinite fractional expansion of P to the base m, which is an encoding of M (via P) into an m-symbol alphabet. This encoding is an optimum encoding for the m-ary channel, because every m-ary symbol in every position has probability $1/m$, given the correspondence between lengths of intervals on $[0, 1)$ and total probabilities of messages mapped into those intervals.

In practice, it is not necessary to have the entire infinite message before starting to encode. As soon as enough source symbols have been generated, it is evident that the message will map into a very small subinterval of $[0, 1)$, and this information will usually specify the first few digits to the base m for transmission over the m-ary channel. We illustrate this in Example 4.4.

Example 4.4 (Binary Source, Binary Channel). Suppose a source produces the two symbols A and B, with respective probabilities $p = 0.6$, $q = 0.4$. If s is any *finite* sequence of As and Bs, we write $pr(s)$ for the probability attached to all sequences that start with the sequence s. Thus, $pr(ABAA)$ is the probability of an arbitrary sequence starting with $ABAA$ and equals 0.0864.

To encode the message $AABABABB \cdots$ for transmission over the binary channel, we first observe that $pr(A) = 0.6$, $pr(AA) = 0.36$, $pr(AAB) = 0.144$, $pr(AABA) = 0.0864$, $pr(AABAB) = 0.03456$, $pr(AABABA) = 0.020736$, $pr(AABABAB) = 0.0082944$, and $pr(AABABABB) = 0.0033177$. This illustrates how the sequence is mapped into subintervals of increasingly narrow width. However, we can also state lower and upper bounds on where the intervals must start and finish. For example, any sequence beginning with AAB must be mapped into the interval $[0.216, 0.36)$, which we obtain by first narrowing down $[0, 1)$ to $[0, 0.6)$, then narrowing down $[0, 0.6)$ to $[0, 0.36)$, and then narrowing down $[0, 0.36)$ to $[0.216 = 0.6 \times 0.36, 0.36)$. Note that we select the lower or upper division of an interval according to whether the latest symbol is A or B. Noting that every number in $[0.216, 0.36)$ has a binary expansion starting with 0 implies that we have the first coded binary digit. The following data gives the results for the entire $AABABABB \cdots$ sequence. We write $a \le s < b$ to indicate that every sequence starting with s is mapped into the interval $[a, b)$.

				Raw data				Coded form
		0	$\le$	A	$<$	0.6	$<$	1	—
		0	$\le$	AA	$<$	0.36	$<$	1/2	0
0	$<$	0.216	$\le$	AAB	$<$	0.36	$<$	1/2	0
0	$<$	0.216	$\le$	$AABA$	$<$	0.3024	$<$	1/2	0
1/4	$<$	0.26789	$\le$	$AABAB$	$<$	0.3024	$<$	5/16	0100
1/4	$<$	0.26789	$\le$	$AABABA$	$<$	0.288576	$<$	5/16	0100
1/4	$<$	0.2802816	$\le$	$AABABAB$	$<$	0.288576	$<$	5/16	0100
9/32	$<$	0.28825854	$\le$	$AABABABB$	$<$	0.288576	$<$	37/128	0100100

Thus, from the first eight source symbols, it is already possible, in this typical example, for the encoder to compute the first 7 bits. From the standpoint of decoding, the 7 bits 0100100 indicate a message located in the interval from 9/32 to 37/128. Since

$$0.2802816 < 9/32 = 0.28125 < 37/128 = 0.2890625 < 0.3024$$

this is a subinterval of message space between *AABABAB* and *AABB*, i.e., the decoder can already write down *AABA*. In principle, the decoding procedure is almost identical to the coding procedure.

4.7. The Distribution of Waiting Times and Performance of Elias–Shannon Coding

Elias–Shannon coding is 100% efficient in the specific sense that every binary symbol transmitted conveys a full bit of information. To prove this, we first observe that the encoding scheme is a measure-preserving mapping from the set of all infinite messages in the alphabet A, B [where $pr(A) = p$ and $pr(B) = q = 1 - p$] to the unit interval $[0, 1)$, since the *length* of any subinterval $[a, b)$ of $[0, 1)$ is the total probability of all messages mapped into the subinterval. Thus, the probability that a message begins with A is p, and the set of messages beginning with p corresponds to the subinterval $[0, p)$, with length (i.e., measure) equal p. By construction, the same result holds for the set of messages beginning with any specified finite string of As and Bs. More generally, the set of messages mapped into any measurable subset of $[0, 1)$ has probability equal to the measure of that subset, although we do not invoke the formal measure theory necessary to prove the assertion. Next, we note that binary encoding assigns the first code symbol to be 0 or 1 based on whether the message will map into $[0, 1/2)$ or $[1/2, 1)$, which are two equally likely situations, each having probability measure $1/2$. Similarly, each succeeding binary code symbol carries a full bit of information, representing a choice between two equiprobable alternatives. There is no guarantee that the next channel symbol will be computable from the source message in any specified number of steps, although we could construct an example where the subinterval of $[0, 1)$ becomes narrower and narrower but never ceases to contain the demarcation point between sending 0 and 1 as the first transmitted symbol. However, such extreme cases are quite rare.

We now show that the *expected waiting time* required to compute the next channel symbol is finite and that *all the moments* for the distribution of waiting times are finite. For clarity of exposition, the proof will be demon-

Infinite Discrete Sources

strated for the case of the binary source (probabilities p and q) being encoded for the binary channel (symbols 0 and 1), where moreover p (and thus q also) is not of the form $k/2^s$, with k and s both integers. For this restricted case, fairly strong upper bounds on the moments of the waiting-time distribution are obtained.

Suppose that the first k symbols ($k \geq 0$) from the source are already known, say, $A_1 A_2 \cdots A_k$ and the corresponding subinterval $[P, Q)$ of $[0, 1)$ has been chosen. All the *bits* that can be assigned to this message are then transmitted, and we wish to know how many *more symbols* are needed to determine the next bit. To obtain the next bit, there is a binary demarcation point B with the interval $[P, Q)$, i.e., with $P < B < Q$, and when the subinterval is narrowed to fit within either $[P, B)$ or $[B, Q)$, the next bit can be transmitted. As soon as the next source symbol is considered, it will choose between two subintervals of $[P, Q)$, say, $[P, D)$ and $[D, Q)$, where $(D - P)/(Q - D) = p/q$. Since $D \neq B$ by the assumption of the nonbinary nature of p and q, *either* $[P, D)$ is a subinterval of $[P, B)$, *or* $[D, Q)$ is a subinterval of $[B, Q)$, *but not both*. Thus, there is a probability p_1 (which is either p or q but cannot be specified between them *a priori*) that the very next symbol determines the next bit. On the other hand, with probability $\overline{p_1} = 1 - p_1$, the next symbol narrows the subinterval without excluding the binary demarcation point B, in which case the problem looks as it originally did. Thus, with probability $\overline{p_1} p_2$ (where p_2 is also chosen between p and q), the *second* symbol determines the next bit. Generalizing, the kth symbol determines the next bit with probability $\overline{p_1 p_2 p_3} \cdots \overline{p_{k-1}} p_k$, where each of $p_1, p_2, p_3 \cdots$ is chosen from the set consisting of p and q.

Let E be the expected waiting time for determining the next bit and assume $p > q$. Define

$$P_k = \overline{p_1 p_2 p_3} \cdots \overline{p_{k-1}} p_k \leq p^k$$

from which

$$E = \sum_{k=1}^{\infty} k P_k \leq \sum_{k=1}^{\infty} k p^k = \frac{p}{q^2}.$$

Moreover, if the rth moment of the waiting time is M_r,

$$M_r = \sum_{k=1}^{\infty} k^r P_k \leq \sum_{k=1}^{\infty} k^r p^k < \infty$$

which proves that all the moments are finite, as asserted.

A more stringent upper bound for the moments M_r can be found as follows:

$$\text{Let } S_1 = p_1 + \overline{p_1}p_2 + \overline{p_1p_2}p_3 + \overline{p_1p_2p_3}p_4 + \cdots$$

and in general, let

$$S_i = p_i + \overline{p_i}p_{i+1} + \overline{p_ip_{i+1}}p_{i+2} + \overline{p_ip_{i+1}p_{i+2}}p_{i+3} + \cdots$$

Note that for each i, S_i is the 0th moment of a hypothetical distribution, i.e., $S_i = 1$ for $i = 1, 2, 3 \cdots$

Next we rearrange terms in M_r to yield an expression in terms of the S_i

$$\begin{aligned}
M_r &= \sum_{k=1}^{\infty} k^r P_k \\
&= p_1 + 2^r \overline{p_1}p_2 + 3^r \overline{p_1p_2}p_3 + 4^r \overline{p_1p_2p_3}p_4 + \cdots \\
&= S_1 + (2^r - 1)\overline{p_1}(p_2 + \overline{p_2}p_3 + \overline{p_2p_3}p_4 + \overline{p_2p_3p_4}p_5 + \cdots) \\
&\quad + (3^r - 2^r)\overline{p_1p_2}S_3 + \cdots \\
&= \sum_{k=1}^{\infty} [k^r - (k-1)^r]\overline{p_1p_2} \cdots \overline{p_{k-1}}S_k \\
&= \sum_{k=1}^{\infty} [k^r - (k-1)^r]\overline{p_1p_2} \cdots \overline{p_{k-1}} \\
&\leq \sum_{k=1}^{\infty} [k^r - (k-1)^r]p^{k-1} \\
&= \sum_{k=1}^{\infty} k^r p^{k-1} - \sum_{k=1}^{\infty} (k-1)^r p^{k-1} \\
&= q \sum_{k=1}^{\infty} k^r p^{k-1} \quad (12)
\end{aligned}$$

To repeat, we obtain

$$M_r \leq q \sum_{k=1}^{\infty} k^r p^{k-1} \quad (13)$$

Infinite Discrete Sources

which improves the previous upper bounds by a factor of q/p. In particular,

$$E = M_1 \leq \frac{q}{p} \sum_{k=1}^{\infty} kp^k = \frac{q}{p} \times \frac{p}{q^2} = \frac{1}{q}$$

a rather good upper bound.

The fact that all the moments of the waiting-time distribution are finite suggests that it may even be practical to consider using a coding scheme of this sort. The risk of an inordinate waiting time exists in many of the practical mass communication systems in use today, but this should not automatically exclude the method. In support of this view, there is reason to believe that the expected backlog, defined as the average number of symbols lost in the double process of encoding n symbols from the A, B alphabet to n' symbols in the 0, 1 alphabet, and then decoding back to n'' symbols in the A, B alphabet, remains bounded as n increases. That is, there is evidence to suggest that for large n, $E(n - n'') \leq B$, where E denotes expected value and B is a finite constant independent of n. This is a far stronger assertion than the expected waiting time for obtaining each additional code symbol is finite. If this assertion is true, it represents another way in which the Elias–Shannon coding is 100% efficient.

In the crucial issue of computational complexity, however, it is unfortunately the case that the amount of computation required per bit of encoded message, or per symbol of decoded message, grows in an unbounded manner with the distance from the starting point of the message. For this reason, these codes seem to be of theoretical rather than practical importance. [See, however, Jelinek (1968), who uses the Elias–Shannon procedure to approximate the Huffman code words for the nth extension of a source.]

Exercises

1. Suppose A and B are source symbols with respective probabilities of 0.7 and 0.3. Encode the message

    $$AABAAABBABAABAA \cdots$$

    as far as possible with the Elias–Shannon method. Then decode this encoding as far as possible. Suppose that the second symbol in the encoded sequence is erased before reception. What does the receiver know about the transmitted message?

2. Suppose the memoryless source alphabet is A, B, C, with probabilities 5/9, 1/3, and 1/9, respectively. Use the Elias–Shannon method to encode $BAABABACBAABBACABAC \cdots$ for the decimal channel.

3. With the coding in Problem 2, write a 10-symbol source message for which no digits can be transmitted.

4. Obtain a still better bound for the distribution of waiting times (see Section 4.4.4) by assuming statistical independence of the various p_is.

5. If $p = q = 1/2$, the expected waiting time to send the next bit is only one symbol (since the encoding is direct: $A = 0, B = 1$). Show, however, that if $p = 1/2 + \epsilon$, then the expected waiting time approaches two symbols as $\epsilon \to 0$. (That is, the expected waiting time is a discontinuous function of p, with a large discontinuity at $p = 1/2$.)

6. Consider the first-order Markov source having the state diagram shown in Figure 4.1.

Figure 4.1. A first-order Markov source state diagram.

Assuming that the source is started with its stationary distribution, encode

$$aabbbaba \cdots$$

using the Elias–Shannon coding procedure. Be sure that the encoding procedure you use is a measure-preserving map.

4.8. Optimal, Asymptotically Optimal, and Universal Codes

In many cases, Agent 00111 could not tolerate preparing a new code book for each new countably infinite discrete distribution; however, his ideal solution was not so ambitious as the Lempel–Ziv approach. Agent 00111 wanted a code book that given an *a priori* probability distribution, could be adapted to work nearly as well as any other. It is worth mentioning that we have already given examples that approach this ideal: The class of codes in

Infinite Discrete Sources

Section 4.4 specified as ideal for geometric distributions arising from run lengths. However, the exact value of p in the geometric distribution need not be specified, and the class of codes is equally good across *all* possible geometric distributions. (This is not to say that individual codes are equally efficient.) Agent 00111 wanted a class of codes, or even a single code, that is good across an even wider class of distributions. Solutions to this problem were provided by Elias; however, we must first establish exactly what we are looking for.

Let M be any countably infinite set of outcomes. We assume that M has been ordered so that the most probable events are labeled first

$$Pr(m_1) \geq Pr(m_2) \geq \cdots \geq Pr(m_{j-1}) \geq Pr(m_j) \geq \cdots \quad (14)$$

For any set X, let X^* denote the set consisting of all sequences of elements selected from X.

We assume that we have a finite alphabet B, and we wish to consider possibilities for a uniquely decodable countably infinite codebook $C \subset B^*$, i.e., given a one-to-one mapping of M into C we ask that this map induce a one-to-one mapping from C^* to M^*.

Given a candidate code C, we order the elements of C in terms of increasing length

$$L(c_1) \leq L(c_2) \leq \cdots \leq L(c_j) \leq \cdots \quad (15)$$

where $L(\ )$ is the length of the enclosed sequence of elements in B.

It is obvious that the most efficient use of C, in terms of average codeword length, is to map the most probable messages to the code words of least length. That is, with the preceding labeling, we map m_i to c_i. We call this use of C the *minimal encoding,* and for the remainder of Section 4.8, we assume that all codes are used with their minimal encodings.

There are some immediate bounds for minimal encodings. The average code-word length of such a code is a function of the length function and the probability distribution alone. We write $E_P(L)$ for this average and refer to the probability distribution as P. Several of the results in Chapter 2 extend without trouble to the countably infinite case, e.g., the Kraft inequality (Theorem 2.1) and its converse. The results we need are stated without proof in Theorem 4.3.

Theorem 4.3. Let B be a finite set and let (M, P) be a countable source with entropy $H(P) = H$ (entropy computed to base $|B|$). Then

1. If $C \subset B^*$ is uniquely decipherable with length function L and size $|C| \geq |M|$,

$$E_P(L) \geq \begin{cases} 0 & H = 0 \\ \max\{1, H\} & 0 < H \leq \infty \end{cases} \quad (16)$$

2. There is a prefix code $C(P) \subset B^*$ with a length function L_P given by

$$L_P(c_j) = \log\left(\frac{1}{\Pr(m_j)}\right) \quad 1 \leq j \leq |C| \quad (17)$$

and minimal average code-word length

$$E_P(L_P) \leq \begin{cases} 0 & H = 0 \\ 1 + H & 0 < H < \infty \end{cases} \quad (18)$$

3. Define

$$R_P(H) = \begin{cases} 1 + H & 0 < H \leq 1 \\ 1 + \dfrac{1}{H} & 1 \leq H < \infty \end{cases}$$

For the preceding prefix code, the ratio of the average code-word length to the minimum possible obeys

$$\frac{E_P(L_P)}{\max\{1, H\}} \leq R_P(H) \quad (19)$$

4. With the preceding notation, $R_P(H) \leq 2$ and

$$\lim_{H \to \infty} R_P(H) = \lim_{H \to 0} R_P(H) = 1 \quad (20)$$

Proof: The preceding statements represent a standard extension of the results in Chapter 2.

Theorem 4.3 shows that for a particular P, there exists a good code C_P. However, we want to take a particular code, use it with minimal encoding on (M, P), with P variable, and examine how well it performs. Let $L_C(\ )$ denote the length function of C. We write $L = L_C$ if there is no fear of ambiguity. Noting the result in Equation (19), we define the following:

Infinite Discrete Sources

A code C is *universal* if and only if there is a constant K for which

$$\frac{E_P(L)}{\max\{1, H(P)\}} < K \tag{21}$$

for all P for which $0 < H(P) < \infty$.

A universal code C is called *asymptotically optimal* if and only if the ratio is bounded above by a function of $H(P)$ that tends to 1 as $H(P)$ tends to infinity: Formally,

$$\frac{E_P(L)}{\max\{1, H(P)\}} \leq R(H(P)) \leq K \tag{22}$$

with

$$\lim_{H \to \infty} R(H) = 1 \tag{23}$$

These definitions define a code that possesses some of the properties of the very best codes matched to (M, P), but the code does not depend on P.

Example 4.5 (A Universal Code for the Positive Integers). Let a positive integer j be expanded in a standard $|B|$-ary notation; e.g., $|B| = 2$ implies binary notation, $|B| = 8$ implies octal, *etc.* Then the length of j is $1 + \lfloor \log j \rfloor$ (logs to the base $|B|$). Now insert 0 between each pair of symbols and place 1 after the last symbol in the sequence (the least significant bit). Thus, j is encoded to an element of $(B - \{0\})(0B)^*1$, where $(\ )^*$ denotes repetition of $(\ )$ an arbitrary* of times.

Decoding is easy. A device reads symbols, printing them out at odd times, continuing when there is a 0 at even times and stopping when there is a 1 at an odd time. The fact that the code is universal follows from Lemma 4.1, an inequality due to Wyner.

Lemma 4.1. Provided $Pr(m_1) \geq Pr(m_2) \geq \cdots \geq Pr(m_j) \geq \cdots$, then $E_P[\log(j)] \leq H(P)$.

Proof

$$1 \geq \sum_{i=1}^{j} Pr(m_i) \geq j Pr(m_j) \tag{24}$$

which implies

$$\log(j) \le \log\left[\frac{1}{Pr(m_j)}\right] \tag{25}$$

Averaging with respect to P gives the desired result.

Now, for the code of Example 4.5

$$E_P(L) = \sum_{j=1}^{(\infty)} Pr(j)|L(j)| = 2\sum_{j=1}^{\infty} Pr(j)[1 + \lfloor\log(j)\rfloor]$$

$$\le 2 + 2E_P[\lfloor\log(j)\rfloor]$$

$$\le 2 + 2E_P[\log(j)]$$

$$\le 2 + 2H(P) \tag{26}$$

Thus, the code is universal with

$$E_P(L) \le 2 + 2H(P) \tag{27}$$

$$\frac{E_P(L)}{\max\{1, H(P)\}} \le 4 \tag{28}$$

Note: This code is not asymptotically optimal.

Example 4.6. It should be apparent that code C in Example 4.5 is not very efficient; however, it is easily extendable. Let N^+ denote the set of positive integers. We may expand an integer $j \in N^+$ in standard $d = |B|^k$ notation and apply construction 1 to the alphabet B^k. Less formally, we insert a 0 between blocks of k symbols from B and terminate with a 1. We call this code C_k. We define

$$U_k(h) = 1 + k + \left(1 + \frac{1}{k}\right)h$$

$$K_k = 2 + k + \frac{1}{k}$$

and

$$R_k(h) = \frac{U_k(h)}{\max\{1, h\}}$$

Infinite Discrete Sources 235

By an argument used for C in Example 4.5, it can be shown that C_k is a universal code with

$$E_P(L_k) \leq U_k[H(P)] \tag{29}$$

$$\frac{E_p(L_k)}{\max\{1, H(P)\}} \leq R_k[H(P)] \tag{30}$$

$$R_k[H(P)] \leq K_k \tag{31}$$

$$\lim_{H \to \infty} R_k(H) = 1 + \frac{1}{k} \tag{32}$$

Note, this implies that no single C_k is asymptotically optimal, but as k increases, the codes come closer to being asymptotically optimal.

Example 4.7 (An Asymptotically Optimal Code for the Positive Integers). For a positive integer j, we expand j in standard $|B|$ notation to obtain a sequence of length $l(j) = 1 + \lfloor \log_B(j) \rfloor$. We then compute the integer $n = n(j)$ for which

$$\binom{n+1}{2} \geq l(j) > \binom{n}{2}$$

and set

$$m = \binom{n+1}{2} - l(j)$$

We encode j by taking its $|B|$ expansion and:

1. Padding the left with m zeroes to give a length sequence

$$\binom{n+1}{2}$$

2. Splitting the length $\binom{n+1}{2}$ sequence into a number of segments so that there is precisely one segment of length k, $1 \leq k \leq n$. Note:

$$\sum_{i=1}^{n} k = \binom{n+1}{2}$$

3. Inserting 0 between each segment and placing 1 at the end.

Thus, if $|B| = 2$, $j = 75$, the binary representation of j is

$$1\ 0\ 0\ 1\ 0\ 1\ 1 \qquad l(j) = 7$$

$n = 4$, since

$$10 = \binom{5}{2} \geq 7 > \binom{4}{2}$$

Note, $m = 3$ and by Step 1, we encode 75 to 0001001011; by Step 2, we encode to 0|00|100|1011, and finally by Step 3, we add padding to obtain

$$0|0|0\ 0|0|1\ 0\ 0|0|1\ 0\ 1\ 1|1$$
$$\ \ x\ \ \ \ \ x\ \ \ \ \ \ \ \ x\ \ \ \ \ \ \ \ \ \ \ \ x$$

where x indicates the symbols added for padding. Observe that the process is easily reversible, i.e., decoding is possible. Given the beginning of the sequence, synchronization is achieved by looking for a 1 in a position reserved for padding. The padding locations are identified by referring to an incremental counter that is reset for each symbol. Reading between the padding gives the original integer.

Using Wyner's inequality (Lemma 4.1) and some generalizations, Elias (1975) proved that

$$E_P(L) \leq 1 + H(P) + \sqrt{1 + 8H(P)}$$

$$\frac{E_p(L)}{\max\{1, H(P)\}} \leq R_p[H(P)]$$

where

Infinite Discrete Sources

$$R_p[H(P)] \leq 5$$

and

$$\lim_{H \to \infty} R_p[H(P)] = 1$$

Note that this code provides a solution to the Lempel–Ziv algorithm (see section 2.10.2) requirement for transmitting large positive integers in a compact format (see Section 2.10). In fact, the exact probability distribution induced from the positive integers by operating the Lempel–Ziv algorithm on a stationary ergodic information source does not seem to have been studied beyond the level of average values (see Section 2.10.3).

4.9. The Information-Generating Function of a Probability Distribution

The moment-generating function of classical probability theory is both a convenient means of calculating the moments of a distribution and an effective embodiment of the properties of the distribution for various analytical purposes. In this section, a generating function is defined whose derivatives, evaluated at a certain place, yield the moments of the self-information (entropy) of the distribution. The first of these moments is the entropy of the distribution. (As a formalism, this technique works equally well for discrete and continuous distributions, although we stress the discrete.) Moreover, the information-generating function summarizes those aspects of the distribution that are invariant beyond measure-preserving rearrangements of the probability space.

Let $P = \{p(n)\}$ be a probability distribution with $n \in N$, where N is a discrete sample space. We define the *information-generating function* $T(u)$ for this distribution by:

$$T(u) = \sum_{n \in N} p^u(n) \tag{33}$$

where u is a real (or complex) variable. Clearly, $T(1) = 1$, and since $0 \leq p(n) \leq 1$ for all $n \in N$, the expression for $T(u)$ in Equation 33 is convergent for all $|u| \geq 1$.

Observe that the derivative $T'(u)$ is given by

$$T'(u) = \sum_{n \in N} \log p(n) \cdot p^u(n) \tag{34}$$

since $p^u(n) = \exp\{u \cdot \log p(n)\}$ and uniform absolute convergence of Equation 33 justifies term-by-term differentiation for all $|u| > 1$. If the sum in Equation 34 converges at $u = 1$, it must equal $T'(1)$, and we have

$$-T'(1) = -\sum_{n \in N} p(n) \log p(n) = H(P) \tag{35}$$

the entropy of the probability distribution P.

More generally, we obtain the rth derivative of $T(u)$ evaluated at $u = 1$ to be

$$T^{(r)}(1) = \sum_{n \in N} p(n) \log^r p(n) \tag{36}$$

provided that this sum converges. Except for a factor of $(-1)^r$, this is the rth moment of the self-information of the distribution.

In Table 4.3, we give $T(u)$ for several examples of discrete distributions—the uniform distribution, the geometric distribution, and the inverse α-power distribution. In each of these cases, $T(u)$ is readily computed and generally has a simpler form than the expression for entropy itself. The third case in-

Table 4.3. The Information-Generating Functions for Three Kinds of Discrete Distributions

Sample space N	Probability function $p(n)$	Information-generating function $T(u)$	Entropy $-T'(1)$
$n = 1, 2, \ldots, A$	$p(n) = \dfrac{1}{A}$ for $n \in N$	$T(u) = A^{1-u}$	$\log A$
$n = 0, 1, 2, 3, \ldots$	$p(n) = qp^n,\ p + q = 1$	$T(u) = \dfrac{q^u}{1 - p^u}$	$-\dfrac{1}{q}(p \log p + q \log q)$
$n = 1, 2, 3, \ldots$	$p_\alpha(n) = \dfrac{n^{-\alpha}}{\zeta(\alpha)}$	$T_\alpha(u) = \dfrac{\zeta(\alpha u)}{\zeta^u(\alpha)}$	$\log \zeta(\alpha) - \alpha \dfrac{\zeta'(\alpha)}{\zeta(\alpha)}$
	$\zeta(\alpha) = \sum\limits_{n=1}^{\infty} n^{-\alpha},\ \alpha > 1$		

Infinite Discrete Sources

volves the Riemann Zeta Function as the normalization constant for the distribution, which leads to an expression for entropy reminiscent of analytic number theory.

The binomial distribution and the Poisson distribution are not shown in Table 4.3 because the authors were unable to reduce $T(u)$ to closed form in these cases. The expressions for entropy in these two cases are also unknown in closed form.

4.9.1. Uniqueness of the Inverse

Given the information-generating function $T(u)$, how accurately can we determine the probability distribution corresponding to it? We can write Equation 33 in the continuous form

$$T(u) = \int_S p^u(s) \, d\mu \tag{37}$$

where S is the sample space with measure μ and $s \in S$. The value of $T(u)$ is left unchanged by any measure-preserving rearrangement of $p(s)$ on S. In the discrete case, this simply means a permutation of the order of probabilities, while in the continuous case, it allows subintervals, *etc.*, to be reshuffled. Hence, we can at most reconstruct $p(s)$, given $T(u)$ modulo the group of measure-preserving rearrangements on S. If S is only semi-infinite, we can pick the canonical $p(s)$ modulo the group of rearranged transformations to be the monotonically nonincreasing distribution.

For the discrete case, we can invert by a number of techniques. Using measure theory, since

$$T(u)^{1/u} = \left(\sum p^u \right)^{1/u} = \|p(n)\|_u$$

is the *norm* of the distribution in the L_u metric, we know that

$$\lim_{u \to \infty} T(u)^{1/u} = p_{\max} \tag{38}$$

where $p_{\max}$ is the maximum probability in the distribution [see, for example, Loomis (1953), p. 39].

Thus, the *largest* probability of the distribution (which is the first value in canonical form) is uniquely recovered by the inversion formula in Equation 38. If we then define $T_1(u) = T(u) - p_1^u$, where $p_1 = p_{\max}$ and more generally, $T_n(u) = T_{n-1}(u) - p_n^u$ for all $n > 1$, we have the recursively defined inversion

$$\lim_{u \to \infty} T_n(u)^{1/u} = p_{n+1} \qquad (39)$$

In this manner, the distribution can be reconstructed unambiguously (except for rearranging the sample space) from the information-generating function. Analogous techniques enable inversion in the continuous case.

As an alternative to measure theory and infinite limits, we can evaluate $T(1) = 1$, $T(2)$, $T(3)$, ..., $T(N)$, where N is at least twice the size of the discrete distribution, i.e., N is at least twice the number of outcomes with nonzero probability. These evaluations are analogous to the *syndromes* in coding theory and in particular obey the Newton identities [e.g., MacWilliams and Sloane (1977), p. 244]. It follows that the T sequence obeys a recursion rule.† After computing the shortest possible recursion, the inverse roots of the recursion polynomial give the values of the probabilities. Since the roots of a polynomial are unordered, the probabilities are necessarily unordered.

4.9.2. Composition of Generating Functions

Given two independent random variables X and Y, the product of their moment-generating functions is well known to be the moment-generating function for the sum distribution. The analogous result for information-generating functions follows.

If $T_X(u)$ and $T_Y(u)$ are the information-generating functions of the independent random variables X and Y, then the product $T_X(u)T_Y(u)$ is the information-generating function of the Cartesian product random variable $Z = X \times Y$, whose events are the ordered pairs (x, y), with $x \in X$, $y \in Y$, and probability $p_Z(x, y) = p_X(x)p_Y(y)$. That is,

$$T_Z(u) = T_X(u)T_Y(u) \qquad (40)$$

The proof, in both the discrete and the continuous cases, is an immediate consequence of the trivial identity

$$p_Z^u(x, y) = (p_X(x)p_Y(y))^u = p_X^u(x)p_Y^u(y) \qquad (41)$$

Another form of distribution composition occurs if we must choose between sampling from distribution X or from distribution Y, where the probability of X is α, and the probability of Y is $\beta = 1 - \alpha$. For this case, the

---

† This type of solution is familiar to anyone who has studied Reed–Solomon decoding. Either Euclid's or Berlekamp's algorithm can be used to find the shortest possible recursion.

Infinite Discrete Sources 241

compound distribution W (choosing between X and Y and then choosing the appropriate event) has the information-generating function

$$T_W(u) = \alpha^u T_X(u) + \beta^u T_Y(u) \tag{42}$$

Again, proof of this identity is immediate.

We can describe Equations 40 and 42 as representing the effect of composing two independent distributions in series and in parallel, respectively.

If we are concerned about the *structure* of the space A consisting of all possible information-generating functions $T(u)$ over an underlying measure space, we may note that from Equation 40, A is closed with respect to multiplication and from Equation 42, A is closed under pseudolinear averaging. If as in Section 4.9.2, we are concerned with probability distributions on the integers, the space A contains a maximum member $T(u) \equiv 1$, which is the information-generating function for the distribution where one of the outcomes has probability 1 (a delta function). The members of space A are lower bounded by

$$L(u) = \begin{cases} 1 & \text{for } u = 1 \\ 0 & \text{for } u > 1 \end{cases} \tag{43}$$

which is not the transformation of any admissible distribution but may be regarded heuristically as corresponding to the (fictitious) uniform distribution on the integers.

4.10. Notes

For a modern discussion of the St. Petersburg paradox, see Feller (1950). The discussion of run-length coding follows Golomb (1966a). The optimal-coding scheme evidently originated with Elias and Shannon, though we can quote no original reference to it. Billingsley (1961) uses the Elias–Shannon mapping procedure to map symbol sequences into points on the unit interval, and then uses Hausdorff dimension theory to prove that the most efficient coding of a sequence for an r-input channel is the r-adic expansion of the point representing the sequence. Elias (1975) introduced the idea of universal and asymptotically optimal codes, as well as the examples described. Information-generating functions were introduced in Golomb (1966b). Finally, we have chosen not to define or expose the measure theory underlying much of this chapter. We recommend Cramer (1946) and Loomis (1953) as background references.

4.11. References

P. Billingsley. 1961. "On the Coding Theorem for the Noiseless Channel." *Ann. Math. Stat.* **32**: 576–601.

H. Cramer. 1946. *Mathematical Methods of Statistics.* Princeton, NJ: Princeton University Press.

L. D. Davisson. 1973. "Universal Noiseless Coding." *IEEE Trans. Inform. Theory* **IT-19**: 783–95.

P. Elias. 1975. "Universal Code-Word Sets and Representations of the Integers." *IEEE Trans. Inform. Theory* **IT-21**: pp. 194–203.

W. Feller. 1957. *An Introduction to Probability Theory and Its Applications.* Wiley, New York.

R. G. Gallager and D. C. van Vorhis. 1975. "Optimal Source Codes for Geometrically Distributed Integer Alphabets." *IEEE Trans. Inform. Theory* **IT-21**: 228–30.

S. W. Golomb. 1966a. "Run-Length Encodings." *IEEE Trans. Inform. Theory* **IT-12**: pp. 309–401.

———. 1966b. "The Information-Generating Function of a Probability Distribution." *IEEE Trans. Inform. Theory* **IT-12**: pp. 75–7.

———. 1970. "A Class of Probability Distributions on the Integers." *J. Number Th.* **2**: 189–92.

———. 1980. "Sources Which Maximize the Choice of a Huffman Coding Tree." *Inform. Contr.* **45**: 263–72.

D. A. Huffman. 1951. "A Method for the Construction of Minimum Redundancy Codes." *Proc. IRE* **40**: pp. 1098–1101.

F. Jelinek, 1968. "Buffer Overflow in Variable Length Coding of Fixed Rate Sources." *IEEE Trans. Inform. Theory* **IT-14**:490–501.

L. H. Loomis. 1953. *Abstract Harmonic Analysis.* Van Nostrand.

F. J. MacWilliams and N. J. A. Sloane. 1977. *The Theory of Error-Correcting Codes.* North-Holland.

5

Error Correction I: Distance Concepts and Bounds

5.1. The Heavy-Handed Cancellation Problem

Agent 00111 was worried! He was firmly established on foreign soil and had been communicating with Whitehall regularly in code on the back of postage stamps. The system worked beautifully until the local post office changed clerks. Using lots of ink and a heavy hand, the new clerk had managed to render illegible parts of 13 consecutive posted messages by canceling the stamps! While the messages were not completely obliterated, some symbols were impossible to read.

The code clerk at Whitehall performed a statistical analysis on the damaged stamps and determined that the symbol $x(k)$ appearing at a specific position k in the message would be obliterated with probability 0.05 independent of what happened to any other letters in the message. The instruction went out to Agent 00111 immediately: "Use auxiliary binary code (16, 11). We will use erasure decoding." Whitehall had done it again!

What had Whitehall done? They ordered Agent 00111 to switch to a code with 16-symbol words, which could be read without error if three or fewer symbols in each word were obliterated. There were 2^{11} words in the code. The probability that a code word would be readable for this code and channel is

$$\binom{16}{0}(0.95)^{16} + \binom{16}{1}(0.95)^{15}(0.05) + \binom{16}{2}(0.95)^{14}(0.05)^2$$
$$+ \binom{16}{3}(0.95)^{13}(0.05)^3 \doteq 0.9930$$

Since Agent 00111's messages are about 100 bits long (say, nine words or 99 bits) the messages should now get through the channel with a probability of $(0.9930)^9 = 0.9387$, which is a significant improvement over the original 99-bit binary message correct transmission probability of $(0.95)^{99} \doteq 0.00623$.

The code clerk at Whitehall had correctly performed certain analytic steps. The first step in analyzing a situation where errors occur is to study the error statistics. With this information, it is possible to derive optimal decoding algorithms and measure the resulting performance for a given choice of code. The performance measure can then (in theory) be maximized by the choice of code. Chapter 5 follows this outline for a particularly simple type of error-prone channel, one where geometric intuition can aid in understanding the concept of error correction, and then concludes with a discussion of more general channels and more general bounds on what can be achieved.

5.2. Discrete Noisy Channels

When discussing noiseless channels, the simple statement that channel input can always be determined from channel output was a sufficient description of the channel. Channels not satisfying this condition are described by the obvious antonym *noisy*. Consider what is involved in the complete statistical description of a channel with input alphabet **X** and output alphabet **Y**. Repeated use of the channel requires that for each choice of

Sequence length $= n$

Input sequence $= x = (x_1, x_2, \ldots, x_n)$, where $x \in X^n$

Output sequence $= y = (y_1, y_2, \ldots, y_n)$, where $y \in Y^n$

the time t at which the input sequence is transmitted, the probability

$$Pr(Y_t = y | X_t = x) \qquad (1)$$

is specified, where Y_t represents the received random n-tuple beginning at time t and X_t represents the transmitted n-tuple beginning at time t.

This unwieldly description makes many real-life problems quite difficult. However, there are many real channels for which the following simplifying assumptions can be made. Let X_{t+k-1} and Y_{t+k-1} be the random variables equal to the value of the kth component of X_t and Y_t, respectively.

Error Correction I

1. The channel is *stationary;* i.e.,

$$Pr(Y_{t_1} = y | X_{t_1} = x) = Pr(Y_{t_2} = y | X_{t_2} = x) \qquad (2)$$

for all t_1, t_2, y, x, and n of interest. More succinctly, the statistical structure of the channel does not change with time.

2. The channel is *memoryless;* i.e.,

$$Pr(Y_t = y | X_t = x) = \prod_{k=1}^{n} Pr(Y_{t+k-1} = y_k | X_{t+k-1} = x_k) \qquad (3)$$

for all t, y, x, and n of interest. Equivalently, the random output variables in the vector Y_t are conditionally independent, given the input sequence. A more physical interpretation of memoryless channels is that the channel interference, which generates an output vector from an input vector, acts independently on each element of the input vector.

3. The channel is *synchronized,* i.e., the transmitter and receiver have synchronized clocks available so that the time location of a transmitted n-tuple and the corresponding received n-tuple may be specified *a-priori*.

Unless otherwise specified, all noisy channels will be assumed to be stationary, memoryless, and synchronized. We define

$$P(y|x) \triangleq Pr(Y = y | X = x) \qquad (4)$$

for all $y \in Y$, $x \in X$, where Y represents the random channel output when X is the input. This information is most easily displayed in matrix form

$$P = [P(y_i | x_j)] \qquad (5)$$

where y_i for $i = 1, \ldots, |Y|$ are the elements in the output alphabet and x_j for $j = 1, \ldots, |X|$ are elements in the input alphabet. The $|Y| \times |X|$ matrix P is called a *channel matrix*.

When a channel is stationary, memoryless, and synchronized, a complete statistical description of the channel requires knowledge of only the channel matrix. Whenever a channel is specified by simply stating its channel matrix, it may be assumed that the channel is stationary, memoryless, and synchronized.

The channel matrix of a noiseless channel is characterized by only one nonzero entry in each row. Another simple channel, the *deterministic channel,* has only one nonzero entry in each column; i.e., for any given x_i, $P(y_j | x_i)$

= 1 for one value of j and zero for all other elements in the ith row of P. It is interesting to note that deterministic channels can be noisy.

Example 5.1. Consider the two channel mappings shown in Figure 5.1.

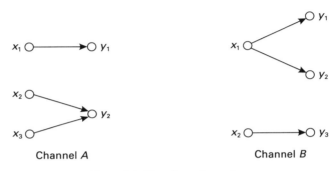

Figure 5.1. Two channel mappings.

$$P_A = \begin{bmatrix} 1 & 0 & 0 \\ 0 & 1 & 1 \end{bmatrix} \quad P_B = \begin{bmatrix} p & 0 \\ 1-p & 0 \\ 0 & 1 \end{bmatrix}$$

In channel A, only one arrow leaves every input node, implying that one output is associated with each input, and thus the channel is deterministic. However, if y_2 is the observed output of channel A, the input cannot be uniquely determined, and therefore channel A is *not* noiseless. On the other hand, channel B is noiseless, since only one arrow arrives at each output node, indicating that the input can always be uniquely determined from the output. Channel B is not deterministic, since the path departing from input note x_1 must be chosen according to a probability distribution.

Many channels representing physical communication systems have special symmetry properties in their channel matrix. A channel is *input uniform* if the elements of each column in the channel matrix are a permutation of the first column's elements. Likewise, a channel is *output uniform* if the elements of each row of the channel matrix are a permutation of the first row's elements. An example of a widely used input- and output-uniform channel is the *binary symmetric channel* (BSC) with channel matrix given by

$$\Pi_{BSC} = \begin{bmatrix} 1-p & p \\ p & 1-p \end{bmatrix} \quad (6)$$

Assuming that the input and output alphabets are identical, p denotes the probability of an error in the transmission of an input symbol through the channel. An example of a channel that is only input uniform is the *binary erasure channel* (BEC), where p represents the probability that an input symbol will be erased while being transmitted through the channel.

$$\Pi_{BEC} = \begin{bmatrix} 1-p & 0 \\ p & p \\ 0 & 1-p \end{bmatrix} \quad (7)$$

It is worth noting that the Agent 00111 example in Section 5.1 was a BEC with $p = 0.05$. It was assumed in the analysis that the Whitehall clerks could observe which bits were obliterated. Examples of channels having output uniformity but not input uniformity are rare, in practice.

Exercises

1. For channel B in Example 5.1, calculate $H(\mathbf{X})$, $H(\mathbf{Y})$, and $I(\mathbf{X}; \mathbf{Y})$, assuming $P(x_1) = P(x_2)$. When observing the output y, on the average, we obtain $H(\mathbf{Y})$ bits of information per symbol, of which $I(\mathbf{X}; \mathbf{Y})$ bits are about the channel input symbol x. Explain what the remaining $H(\mathbf{Y}) - I(\mathbf{X}; \mathbf{Y})$ bits/symbol are about.

2. Let Π_1 and Π_2 be the channel matrices of two independent channels connected serially; i.e., output symbols of Π_1 are input symbols of Π_2. Derive the channel matrix P of the composite channel in terms of Π_1 and Π_2. Let $Q_1, Q_2, \ldots, Q_{|\mathbf{X}|}$ denote the *a priori* probability of the input symbols of the first channel. If $Q = (Q_1, Q_2, \ldots, Q_{|\mathbf{X}|})^t$ find the probabilities of the output symbols of the second channel in terms of Π_1, Π_2, and Q.

3. Using the results of Problem 2, find the composite channel matrix of two serially connected binary symmetric channels. Can you generalize this result to the case of N identical binary symmetric channels in series? What are the entries in the resulting matrix?

4. Consider a stationary, memoryless, synchronized channel whose input alphabet $\mathbf{X}$ is identical to its output alphabet.

(a) For what types of channel matrices is it possible to view **X** as a group under an operation $\&$ and think of the output symbol y as

$$y = x \ \& \ e$$

where x is the input symbol and e is an error symbol selected according to the probability distribution $Pr(e)$, with e *statistically independent of x?*

(b) How is $Pr(e)$ related to the channel matrix in this case?

(c) Channels satisfying the constraints in (a) are called *additive error channels*. Is the binary erasure channel an additive error channel?

5.3. Decoding Algorithms

Let **C** denote the subset of the collection $\mathbf{X}^n$ of all possible input n-tuples that the encoder and decoder have agreed to use as code words. The problem confronting the decoder in any communication system is deciding which code words were transmitted when the only information available is the code **C** used by the transmitter, a statistical description of the channel, and the (word-synchronized) channel output symbol stream. Let $\mathbf{X}^n$ denote the set of all possible channel input n-tuples and let $\mathbf{Y}^n$ denote the set of all possible channel output n-tuples. If $x \in \mathbf{X}^n$ and $y \in \mathbf{Y}^n$ are typical input and output sequences, respectively, assume that the channel is described by the conditional probability distribution $Pr(y|x)$. The object of the decoder is to deduce the transmitted sequence as reliably as possible, i.e., to minimize the probability of error P_E in determining the channel input.

Suppose that the decoder has observed the channel output n-tuple $y_0 \in \mathbf{Y}^n$. It is now possible to calculate the *a posteriori* probability that an input n-tuple $x \in \mathbf{X}^n$ was transmitted, namely, $Pr(x|y_0, \mathbf{C})$. Notice that this probability depends on the code used, not just on the particular input n-tuple x for which it is being evaluated. By applying Bayes' law and shortening notation, we have

$$Pr(x|y_0, \mathbf{C}) = \frac{Pr(y_0, x|\mathbf{C})}{Pr(y_0|\mathbf{C})}$$

$$= \frac{Pr(y_0|x, \mathbf{C}) \times Pr(x|\mathbf{C})}{\sum_{x' \in \mathbf{X}^n} Pr(y_0|x', \mathbf{C})Pr(x'|\mathbf{C})} \tag{8}$$

Error Correction I

$Pr(x|y_0, \mathbf{C})$ is called the *a posteriori* probability of x given y_0 and given the code $\mathbf{C}$. When the channel input x is given, then the output probability does not depend on $\mathbf{C}$, since $\mathbf{C}$ merely specifies what input vectors may be used. Hence,

$$Pr(x|\mathbf{C}) \triangleq \begin{cases} Pr(x) & x \in \mathbf{C} \\ 0 & x \notin \mathbf{C} \end{cases} \quad (9)$$

and for $x \in \mathbf{C}$,

$$Pr(y_0|x, \mathbf{C}) = Pr(y_0|x) \quad (10)$$

reducing Equation 8 to

$$Pr(x|y_0, \mathbf{C}) = \begin{cases} \left(\dfrac{Pr(y_0|x)Pr(x)}{\sum_{x' \in \mathbf{C}} Pr(y_0|x')Pr(x')} \right) & \text{if } x \in \mathbf{C} \\ 0 & \text{if } x \notin \mathbf{C} \end{cases} \quad (11)$$

Of course, the *a posteriori* probability of x given y_0 still depends on how code words are assigned to message sequences and the probabilistic description of the message sequences.

After observing y_0, suppose that the decoder decides x_0 was transmitted. The probability of error P_E is simply the probability that some other sequence was transmitted.

$$P_E = \sum_{x \in \mathbf{C}, x \neq x_0} P(x|y_0, \mathbf{C}) \quad (12)$$

$$= 1 - P(x_0|y_0, \mathbf{C}) \quad (13)$$

To minimize P_E, it is apparent that the decoder should state the following:

"x_0 was sent" if $P(x_0|y_0, \mathbf{C}) > P(x|y_0, \mathbf{C})$ for all $x \in \mathbf{C}$ $x \neq x_0$ (14)

or equivalently from Equation 11,

"x_0 was sent" if $P(y_0|x_0)P(x_0)$

$$> P(y_0|x)P(x) \quad \text{for all } x \in \mathbf{C} \quad x \neq x_0 \quad (15)$$

In other words, the decoder should decide on x_0 if it is more probable than any other alternative, given that **C** was used and y_0 was received. If there exist two values of x that maximize $Pr(x|y_0, \mathbf{C})$, then either may be chosen for the decoder output. This method of making a decoding decision is sometimes termed a *conditional maximum-likelihood decision rule*. Its use is somewhat tempered by the fact that the decoder requires subjective information about the *a priori* probability distribution of messages to be transmitted.

If the probability distribution of the transmitted symbol sequence is unknown, a *maximum-likelihood decision rule* is often used, which states

$$x_0 \text{ was sent if } P(y_0|x_0) > P(y_0|x) \quad \text{for all } x \in \mathbf{C} \quad x \neq x_0 \quad (16)$$

where y_0 is the observed output sequence. In other words, the decoder decides on x_0 if this maximizes the probability of receiving y_0. As seen from Equation 15, the maximum-likelihood decision rule is identical to the conditional maximum-likelihood decision rule when all input code n-tuples are *a priori* equally likely. When the *a priori* distribution $Pr(x)$ is unknown, the maximum-likelihood decision rule minimizes probability of error as follows. We can choose the probability distribution $Pr(x)$ that expresses maximum uncertainty and use it in the conditional maximum-likelihood decision rule. Since the distribution of maximum uncertainty from an information-theoretic viewpoint is simply the distribution having maximum entropy, in the absence of other constraints, we should choose the *a priori* distribution $Pr(x)$ to be a constant for all x. Hence, the conditional maximum-likelihood decision rule reduces to the maximum-likelihood decision rule. If $Pr(x)$ is known but is *not* an equally likely distribution, the maximum-likelihood decision rule is a legitimate though suboptimal way of making decisions under a P_E criterion.

Let us see what happens to the maximum-likelihood decision rule when a binary symmetric channel with symbol error probability p is employed. Then **X** is the same as **Y** and

$$Pr(y|x) = \prod_{i=1}^{n} Pr(y_i|x_i) \quad (17)$$

where

$$y = (y_1, y_2, \ldots, y_n)$$
$$x = (x_1, x_2, \ldots, x_n) \quad (18)$$

and for a BSC,

Error Correction I

$$Pr(y_i | x_i) = \begin{cases} 1 - p & y_i = x_i \\ p & y_i \neq x_i \end{cases} \quad (19)$$

Let

$$d(x, y) = |\{i : x_i \neq y_i, 1 \leq i \leq n\}| \quad (20)$$

That is, $d(x, y)$ is the number of positions in which x and y differ. Equation 17 simplifies to

$$Pr(y|x) = p^{d(x,y)}(1 - p)^{n-d(x,y)} \quad (21)$$

$$= (1 - p)^n \left(\frac{p}{1 - p}\right)^{d(x,y)} \quad (22)$$

Notice that $Pr(y|x)$ depends on x and y only through the value of $d(x,y)$. When $p < 1/2$, $Pr(y|x)$ is a *monotone decreasing* function of $d(x, y)$, and therefore the maximum-likelihood decision rule when y_0 was received reduces to

$$x_0 \text{ was sent} \Leftrightarrow d(x_0, y_0) < d(x, y_0) \quad \text{for all } x \in C \quad x \neq x_0 \quad (23)$$

It is apparent that a unique minimum for $d(x, y_0)$ over $x \in C$ may not exist. When this happens, the decoder may decide that either one of the minimizing code words was transmitted, or it may announce that an error has been detected, and not make a decision.

The quantity $d(x, y)$ defined by Equation 20 is called the *Hamming distance between x and y*. As its name implies, the Hamming distance function $d(\cdot, \cdot)$ satisfies all the properties of a metric on the space of n-tuples X^n. The maximum-likelihood decision rule for the BSC states that the code word x_0 closest in the Hamming metric to the received n-tuple y_0 was transmitted.

Example 5.2. Consider the code in Table 5.1 that is to be used in a BSC with error rate $p < 1/2$.

Table 5.1. A Code Example

Label (x_i)	Binary code
$x_1 =$	1 1 1 1 1
$x_2 =$	1 1 0 0 0
$x_3 =$	0 0 1 1 0
$x_4 =$	0 1 1 0 1

The 32 binary 5-tuples can be tabulated in columns according to the word to which they are closest. The decoding procedure is to state that x_i was sent if the received 5-tuple is in the column headed by x_i. The result is shown in Table 5.2.

Table 5.2. A Decoding Table

d	x_1	x_2	x_3	x_4	x_1, x_4	x_2, x_3
0	11111	11000	00110	01101		
1	10111	01000	10110	00101	01111	
	11011	10000	01110	01001	11101	
	11110	11100	00010	01100		
		11010	00100			
		11001	00111			
2	10011	10001	00011	00001	01011	10100
					10101	10010
						01010
						00000

When a pair of binary code words are separated by an even Hamming distance, it is possible to find symbol sequences that are the same distance from each code word. If no other code word is closer to the sequences having this equidistant property, a choice of either code word is acceptable. This is the case in Table 5.2 for words in columns x_1, x_4, and x_2, x_3.

A code is said to have the capability of correcting k_c errors if *regardless of the transmitted code word,* a minimum Hamming distance criterion results in correct decoding whenever k_c or fewer errors are made. To ensure this property, it is necessary for all code words to be at least $2k_c + 1$ Hamming distance units apart, since this guarantees that received words of distance k_c units or less from the transmitted word are at least $k_c + 1$ units from any other code word and will thus be properly decoded. This characteristic is a direct result of the triangle inequality property of the Hamming metric, i.e.,

$$d(x, y) + d(y, z) \geq d(x, z)$$

Figure 5.2 abstractly represents two code words with Hamming distance 4 between them and some of the possible received words. For $k_c = 1$, $2k_c + 1 < 4$, a code having minimum distance 4 between code words can always decode words with single errors correctly.

Error Correction I

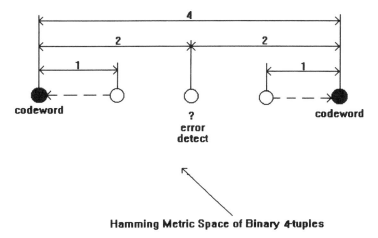

Figure 5.2. Decoding algorithm (-->) for two code words, 4 units apart, and the intermediate received n-tuples.

However, if two errors are made during transmission, it is possible that the received word may be equidistant between two code words. In this minimum distance 4 case, proper decoding of two errors cannot be guaranteed, but double errors can be detected. If three or more errors are made, the received word may be distance 1 or 0 from some nontransmitted code word, and the decoding procedure may yield an incorrect decision. Thus, a code with minimum distance between code words equal to 4 can be termed a single-error-correcting, double-error-detecting code.

An equally valid description of a minimum distance 4 code would be that it is a triple-error-detecting code. Since all code words are at least distance 4 apart, it is a simple matter to observe the received word and note when three or less errors have been made. However, this code is *not* a single-error-correcting, triple-error-detecting code, since it is impossible to guarantee differentiating between a single error (which is corrected) and a triple error (which is detected) when two code words are distance 4 apart. The (16, 11) code of Section 5.1 had a minimum distance of 4 and was used for triple-error detection.

Definition. Let **C** be a code with minimum distance between words equal to $d_{\min}$. We say that **C** is a k_c error-correcting, k_d error-detecting code, $k_c \le k_d$, if

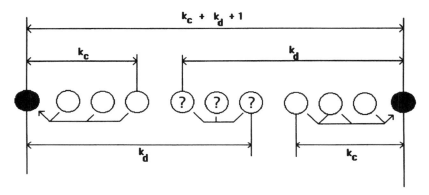

Figure 5.3. General decoding algorithm (-->) for a code with minimum distance $k_c + k_d + 1$.

$$k_c + k_d + 1 = d_{\min} \qquad (24)$$

Notice that there may be more than one solution to Equation 24 for a given $d_{\min}$ and therefore **C** may be described in several different ways. A diagram of the general decoding algorithm is shown in Figure 5.3, with ? indicating an error detection.

It is possible to define and search for codes in relation to the Lee metric rather than the Hamming metric. See Problem 5 of the Exercises.

Exercises

1. Let x, y, and z represent elements in a space. A metric is a function into the nonnegative real numbers defined on pairs of elements in the space, having the following properties:

    (a) $d(x, y) \geq 0$ with equality if and only if $x = y$
    (b) $d(x, y) = d(y, x)$
    (c) $d(x, z) \leq d(x, y) + d(y, z)$

Error Correction I

Show that the Hamming distance is a metric defined on the space of all n-tuples over an arbitrary alphabet **X**.

2. Let the words of a code using binary symbols (0 and 1) have length n. Consider these words to be n-dimensional vectors and relate the Hamming distance between words to the usual Euclidean distance between points in n-space.

3. Tabulate the minimum-distance decoding algorithm for the following ternary code: 000, 012, 201, 210, 111. Describe the code in terms of k_c and k_d.

4. In Section 5.3, we demonstrated that minimum Hamming distance decoding provided a minimum probability of error when communicating through a binary symmetric channel.

   (a) Are there any other binary-input binary-output channels for which this decoding scheme minimizes error probability for arbitrary n? If so, describe them.

   (b) Repeat (a) assuming that n is fixed.

   (c) Give examples of $|\mathbf{X}|$-input $|\mathbf{X}|$-output channels for which minimum Hamming distance decoding minimizes error probability. Assume $|\mathbf{X}| > 2$, n arbitrary.

5. Consider an $r \times r$ channel matrix with the following properties: Each row is a cyclic shift one digit to the right of the previous row; the matrix is symmetric; and the diagonal entry of the matrix is the largest element in the matrix.

   (a) If x denotes a channel input vector of length n and y denotes a channel output vector of length n, what properties of the (x, y) vector pair must be known to compute $Pr(y|x)$?

   (b) Describe the maximum-likelihood decoding procedure in terms of the properties in (a).

   (c) Consider an alphabet of symbols $x_1, \ldots, x_{|\mathbf{X}|}$, where the Lee distance $L(x_i, x_j)$ between two symbols is defined as

   $$L(x_i, x_j) = \min(i \ominus j, j \ominus i)$$

   where $\ominus$ denotes modulo $|\mathbf{X}|$ subtraction. We define the Lee distance between y and x as the sum of the distances between elements:

$$L(y, x) = \sum_{i=1}^{n} L(x_i, y_i)$$

(i) Show that the Lee distance $L(y, x)$ is a metric.

(ii) Under what conditions on the channel matrix is the Lee distance a maximum-likelihood decoding metric?

6. Suppose that a code with minimum Hamming distance d_{min} is used over a binary erasure channel. How many erased symbols per word can be reincarnated?

7. What minimum Hamming distance is required to guarantee the simultaneous correction of k_c errors and reincarnation of k_e erasures?

8. The table of the largest binary code with distance d_{min} and word length n gives results only for odd values of d_{min}. The reason for this is evident from the following result: If d_{min} is odd and $|C(d_{min}, n)|$ denotes the size of the largest dictionary having parameters d_{min} and n, then

$$|C(d_{min}, n)| = |C(d_{min} + 1, n + 1)|$$

Prove this result. (*Hint:* Consider extending each code word in the dictionary with odd minimum distance in the following way: If the code word contains an odd number of 1s, append an additional 1 to the code word; otherwise, append a 0. Determine the minimum distance of this new code.)

5.4. A Hamming Distance Design Theorem

Suppose we try to communicate through a BSC with a code **C** of n-tuples chosen to have Hamming distance at least d_{min} between words. Since the BSC is memoryless, the probability $Pr(e)$ of exactly e errors occurring during code-word transmission is given by the binomial distribution

$$Pr(e) = \binom{n}{e} p^e (1 - p)^{n-e} \tag{25}$$

for $0 \le e \le n$, where p is the symbol error probability of the BSC. The expected value and variance of the number of errors during code-word transmission through the BSC are then

Error Correction I

$$E\{e\} = np \tag{26}$$

$$\mathrm{Var}\{e\} = np(1-p) \tag{27}$$

While the number of errors e is not always close to np, the fractional number of errors e/n occurring in long code words is very close to p. This is verified by applying Chebychev's inequality [Feller (1950)] to the binomial distribution as follows:

$$Pr\{|e - np| > n\epsilon\} = Pr\left\{\left|\frac{e}{n} - p\right| > \epsilon\right\} < \frac{np(1-p)}{(n\epsilon)^2} \tag{28}$$

Hence, we can state that for any $\epsilon > 0$,

$$Pr\left\{\frac{e}{n} > p + \epsilon\right\} < \frac{p(1-p)}{n\epsilon^2} \tag{29}$$

That is, a small probability of even a small deviation ($>\epsilon$) from the expected fraction of errors can be insured for very large n.

Now suppose that the code in question has the property that the fractional number of errors that it can guarantee correcting is greater than p, i.e.,

$$\frac{1}{n}\left\lfloor\frac{d_{\min} - 1}{2}\right\rfloor = p + \epsilon \tag{30}$$

where this equation serves to determine the positive number ϵ. For this ϵ, the left side of Equation 29 can be interpreted as an upper bound on the probability of a decoding error. Hence, using Equations 29 and 30 yields

$$Pr\{\text{decoding error}\} < \frac{p(1-p)}{n\left(\frac{1}{n}\left\lfloor\frac{d_{\min}-1}{2}\right\rfloor - p\right)^2} \tag{31}$$

where the quantity we square is required to be positive. Obviously, the probability of decoding error can be made arbitrarily small by choosing codes with arbitrarily large values of n, each having the additional property that

$$\frac{d_{\min}}{2n} - p \geq \epsilon$$

for some fixed positive number ϵ.

Example 5.3. Suppose the BSC error rate is $1/4$. Consider the simple code consisting of two words

$$C = \{(000\cdots 0), (111\cdots 1)\}$$

for which

$$n = d_{min}$$

Equation 31 indicates that

$$Pr\{\text{decoding error}\} < \frac{3}{16n\left(\frac{1}{n}\left\lfloor\frac{n-1}{2}\right\rfloor - 1/4\right)^2}$$

Values for this bound are given in Table 5.3.

Table 5.3. Upper Bounds on the Probability of a Decoding Error

n	Upper bound on $Pr\{\text{decoding error}\}$
5	1.666667 (!)
11	0.407407
21	0.174515
51	0.063723
101	0.030915
1001	0.000309
10001	0.000003

It should be clear that this example code is very inefficient in terms of redundancy.

For the purpose of efficiency in communication, when choosing codes with larger values of n, it is desirable to choose codes containing many words. Since the space of n-tuples over a finite alphabet is finite, the characteristics of having both a large number of code words and large distances between code words are at odds with each other. The quantities of interest can be defined precisely as follows. Let $C(n, d_{min})$ be a code with words of length n and minimum distance between words d_{min}. Let $|C(n, d_{min})|$ be the number

Error Correction I

of code words in $\mathbf{C}(n, d_{\min})$. Then the size $C_n(D)$ of the largest dictionary with n-tuples having fractional minimum distance D is given by

$$C_n(D) = \max_{\{\mathbf{C}(n,d_{\min}):(\frac{d_{\min}}{n}) \geq D\}} |\mathbf{C}(n, d_{\min})| \qquad (32)$$

It is easily verified that for a fixed n, $C_n(D)$ is a monotone nonincreasing function of D. We demonstrate in the following sections that for a fixed D, $C_n(D)$ apparently increases exponentially with n. Therefore, the quantity of interest should be the asymptotic exponential growth rate of $C_n(D)$, which we define to be the *asymptotic transmission rate*

$$R(D) = \lim_{n \to \infty} \frac{1}{n} \log C_n(D) \qquad (33)$$

Unfortunately, the existence of this limit has not been demonstrated, and it will be demonstrated only that, for large n, the values of

$$\frac{1}{n} \log C_n(D)$$

can be bounded to a certain interval. In more detail, we define

$$\bar{R}(D) = \limsup_{n \to \infty} \frac{1}{n} \log C_n(D) \qquad (34a)$$

$$\underline{R}(D) = \liminf_{n \to \infty} \frac{1}{n} \log C_n(D) \qquad (34b)$$

and observe that for large values of n, achievable transmission rates must fall somewhere between $\underline{R}(D)$ and $\bar{R}(D)$. As an example, given a BSC with error rate p, $p < 1/2$, and arbitrary $\epsilon_1 > 0$, $\epsilon_2 > 0$, and $\epsilon_3 > 0$, we may combine Equations 31, 34a, and 34b to find a code $\mathbf{C}$ with the following properties

$$Pr\{\text{decoding error}\} < \epsilon_1 \qquad (35)$$

and the transmission rate R of the code in the range

$$\underline{R}(2p + \epsilon_2) - \epsilon_3 < R = \frac{1}{n} \log |\mathbf{C}| < \bar{R}(2p + \epsilon_2) + \epsilon_3 \qquad (36)$$

Obviously for the code $\mathbf{C}(n, d_{\min})$ in question,

$$\frac{d_{\min}}{n} \geq D = 2p + \epsilon_2 \tag{37}$$

and there is sufficient distance between code words to permit a high probability of correcting all errors when n is large. The remaining question is to determine $\underline{R}(D)$ and $\bar{R}(D)$. This has been achieved or approximated with a number of bounds, each improving on the earlier results in some way. These bounds are discussed in Sections 5.5–5.8, 5.11, and 5.12. Interesting classes of codes, e.g., perfect and equidistant codes, can be defined by the property of meeting some of the bounds with equality. These code classes are discussed in Sections 5.9 and 5.10.

Exercise

1. Repeat the bound derivation in Section 5.4, this time for the binary erasure channel. Evaluate your bound as a function of n for $\mathbf{C} = \{(00\cdots0), (11\cdots1)\}$ when the erasure probability is $1/4$.

5.5. Hamming Bound

The best known bound on code dictionary size $|\mathbf{C}|$ is the Hamming (or sphere-packing) bound, which upper bounds $|\mathbf{C}|$ for a given n, $d_{\min}$, and alphabet size $|\mathbf{X}|$. Its derivation results from a simple counting argument. All n-tuples within Hamming distance r of an n-tuple x are said to lie in a *Hamming sphere* of radius r around x. If c is a code word, then the number of words differing from c in r places or less, i.e., the number N of n-tuples in a Hamming sphere of radius r around c is given by

$$N = \sum_{i=0}^{r} (|\mathbf{X}| - 1)^i \binom{n}{i} \tag{38}$$

where the number of words of Hamming distance i from c is counted by the $(|\mathbf{X}| - 1)^i$ errors that could occur in a given i places times the number

Error Correction I

$$\binom{n}{i}$$

of ways that the i error positions could be chosen. Now let **C** be any constructable code having minimum distance $d_{\min}$, where

$$d_{\min} \geq 2r + 1 \tag{39}$$

Hamming spheres of radius r around the $|\mathbf{C}|$ code words of **C** are mutually nonintersecting, and therefore the space of n-tuples over the alphabet **X** must contain $N|\mathbf{C}|$ n-tuples. Since there are a total of $|\mathbf{X}|^n$ n-tuples that can compose these spheres,

$$|\mathbf{C}| \sum_{i=0}^{r} (|\mathbf{X}| - 1)^i \binom{n}{i} \leq |\mathbf{X}|^n \tag{40}$$

or equivalently from Equations 39 and 40, using the largest possible value of r,

$$|\mathbf{C}| \leq \frac{|\mathbf{X}|^n}{\sum_{i=0}^{d'} (|\mathbf{X}| - 1)^i \binom{n}{i}}, \quad d' = \left\lfloor \frac{d_{\min} - 1}{2} \right\rfloor \tag{41}$$

where $\lfloor z \rfloor$ denotes the largest integer not exceeding z. Equation 41 is known as the *Hamming bound* on code dictionary size $|\mathbf{C}|$.

When **X** is a binary alphabet, then the Hamming bound can be simplified using the following result:

Lemma 5.1. If r and n are integers for which $0 < r < n/2$, then

$$\left[8n\left(\frac{r}{n}\right)\left(1 - \frac{r}{n}\right)\right]^{-1/2} 2^{nH\left(\frac{r}{n}, 1-\frac{r}{n}\right)} \leq \sum_{i=0}^{r} \binom{n}{i} \leq 2^{nH\left(\frac{r}{n}, 1-\frac{r}{n}\right)} \tag{42}$$

where $H(\,,\,)$ is the entropy function whose arguments are probabilities and whose units are bits/symbol.

Proof: The derivation of Equation 42 stems from Stirling's approximation to $n!$. (See MacWilliams and Sloane (1977), pp. 309–10.)

Using Equation 42 to simplify the Hamming bound for the binary case gives

$$|\mathbf{C}| \le \left[8n\left(\frac{r}{n}\right)\left(1-\frac{r}{n}\right)\right]^{1/2} 2^{n\left[1-H\left(\frac{r}{n},1-\frac{r}{n}\right)\right]} \qquad (43)$$

where from Equation 39

$$r = \left\lfloor \frac{d_{\min}-1}{2} \right\rfloor \qquad (44)$$

Since Equations 43 and 44 must hold for all binary dictionaries with specified $d_{\min}$, n, the right side of Equation 43 must be an upper bound on the maximum dictionary size $C_n(D)$ over all dictionaries having word length n and fractional distance D

$$D = \frac{d_{\min}}{n} = \frac{2r + \left\{\begin{matrix}1\\2\end{matrix}\right\}}{n} \qquad (45)$$

where the choice of 1 or 2 in Equation 45 depends on whether $d_{\min}$ is odd or even. Therefore, for large n,

$$C_n(D) \le \left[9n\left(\frac{D}{2}\right)\left(1-\frac{D}{2}\right)\right]^{1/2} 2^{n\left[1-H\left(\frac{D}{2},1-\frac{D}{2}\right)\right]} \qquad (46)$$

and the attainable information rate discussed in Section 5.4 is upper bounded by

$$\bar{R}(D) = \limsup_{n\to\infty} \frac{1}{n}\log_2 C_n(D) \qquad (47)$$

$$\le \lim_{n\to\infty}\left\{\frac{1}{2}\frac{\log_2 n}{n} + \frac{1}{2n}\log_2\left[\frac{9D}{2}\left(1-\frac{D}{2}\right)\right]\right\}$$

$$+ 1 - H\left(\frac{D}{2}, 1-\frac{D}{2}\right) \qquad (48)$$

As n approaches infinity, the first two terms disappear, and we are left with the asymptotic version of the Hamming bound for binary codes

Error Correction I

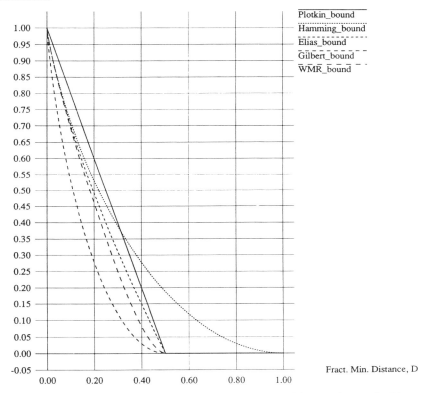

Figure 5.4. Bounds on asymptotic code rates versus fractional minimum distance for binary codes.

$$\bar{R}(D) \le 1 - H\left(\frac{D}{2}, 1 - \frac{D}{2}\right) \tag{49}$$

This bound is sketched in Figure 5.4.

Codes that achieve equality in Equation 40, i.e., attain the Hamming bound, are called *perfect* codes. These codes are discussed in Section 5.9.

Exercises

1. Consider a code where every sequence of n symbols is distance d or less from exactly one code word. Such a code is called a perfect (close-packed or lossless) code.

(a) How would you modify Equation 40 for perfect codes?

(b) Can a perfect code with even minimum distance $d_{\min}$ exist?

(c) Can you find examples (i.e., values of $|\mathbf{C}|$, $|\mathbf{X}|$, n, and $d_{\min}$) for which the modified equation in (a) is satisfied?

(d) Can you construct an example of a perfect code?

2. Consider a code with alphabet $(0, 1)$, word length n, and odd minimum Hamming distance $d_{\min}$ between words. A code of word length $n + 1$ can be constructed from the code of word length n by adding 0 or 1 to the original code words to make the number of 1s in each word even. Show that the new code has minimum distance $d_{\min} + 1$.

3. Let $d(x, y)$ be any distance measure on n-tuples that has the additive property

$$d(x, y) = \sum_{i=1}^{n} d(x_i, y_i)$$

Both the Hamming and the Lee distance satisfy this relation. Let

$$A_c^{(n)}(z) = \sum_d A_c^{(n)}(d) z^d$$

where $A_c^{(n)}(d)$ is the number of n-tuples that are exactly distance d from c.

(a) Show that for distance measures having the additive property,

$$A_c^{(n)}(z) = \prod_{i=1}^{n} A_{c_i}^{(1)}(z)$$

where $c = (c_1, c_2, \ldots, c_n)$.

(b) For the Hamming and Lee metrics, $A_{c_i}^{(1)}(z)$ is independent of c_i. State $A^{(1)}(z)$ for the

   (i) Hamming metric with alphabet $\mathbf{X}$

   (ii) Lee metric with $|\mathbf{X}|$ even

(iii) Lee metric with $|X|$ odd

(c) Using the preceding results, tabulate the volume of a Lee sphere with $n = 8$ and $|X| = 5$ as a function of its radius.

(d) Using sphere-packing arguments, determine an upper bound on the dictionary size of a code with $|X| = 5$ and $n = 8$ that will correct all errors causing the channel output n-tuple to differ from the channel input n-tuple by at most a Lee distance equal to unity.

5.6. Plotkin's Bound

Another upper bound on $d_{\min}$ can be derived using an average-distance concept. Let $d(c_i, c_j)$ denote the Hamming distance between the code words c_i and c_j. The average distance d_{av} between words, which is an upper bound on the minimum distance $d_{\min}$, is given by

$$d_{\min} \leq d_{av} = \left[\frac{|C|(|C|-1)}{2}\right]^{-1} \sum_{i=2}^{|c|} \sum_{j=1}^{i-1} d(c_i, c_j) \qquad (50)$$

where $[|C|(|C|-1)/2]$ is the number of terms in the sum in Equation 50. Since an upper bound on $d_{\min}$ is called for, we must upper bound the sum in Equation 50. This sum can be rewritten as

$$\sum_{i>j}\sum d(c_i, c_j) = \sum_{i>j}\sum \sum_{k=1}^{n} d(c_{ik}, c_{jk})$$

$$= \sum_{k=1}^{n} \sum_{i>j}\sum d(c_{ik}, c_{jk}) \qquad (51)$$

where c_{ik} denotes the kth symbol in the word c_i and $d(c_{ik}, c_{jk})$ is 1 if c_{ik} and c_{jk} differ and 0 otherwise. According to Plotkin, Equation 51 implies

$$\sum_{i>j}\sum d(c_i, c_j) \leq \sum_{k=1}^{n} \max_{\{c_{ik}, i=1,\ldots,|C|\}} \left\{\sum_{i>j}\sum d(c_{ik}, c_{jk})\right\} \qquad (52)$$

The problem as stated by Equation 52 is simply to choose c_{ik}, $i = 1, \ldots, |C|$ from an alphabet X to maximize the contribution

$$\sum\sum_{i>j} d(c_{ik}, c_{jk})$$

of the kth symbol position to the average code-word distance. It can be shown that

$$\max_{\{c_{ik}, i=1,\ldots,|\mathbf{C}|\}} \sum\sum_{i>j} d(c_{ik}, c_{jk}) \leq \left(\frac{|\mathbf{C}|}{|\mathbf{X}|}\right)^2 \frac{|\mathbf{X}|(|\mathbf{X}|-1)}{2} \quad (53)$$

where each symbol occurs approximately an equal number of times in the maximal choice of symbols. Substituting Equation 53 into Equation 52 gives

$$d_{\min} \leq n\left(\frac{|\mathbf{C}|}{|\mathbf{C}|-1}\right)\left(\frac{|\mathbf{X}|-1}{|\mathbf{X}|}\right) \quad (54)$$

Under the condition that

$$\frac{d_{\min}}{n} > \frac{|\mathbf{X}|-1}{|\mathbf{X}|} \quad (55)$$

Equation 54 can be rewritten as a bound on dictionary size

$$|\mathbf{C}(n, d_{\min})| \leq \frac{d_{\min}/n}{(d_{\min}/n) - [(|\mathbf{X}|-1)/|\mathbf{X}|]} \quad (56)$$

We refer to Equations 55 and 56 as *Plotkin's bound*.

The asymptotic attainable transmission rate for a given fractional distance D can be bounded by noting that Equations 55 and 56 are interpretable as follows:

$$\left\{D > \frac{|\mathbf{X}|-1}{|\mathbf{X}|}\right\} \Rightarrow \left\{C_n(D) \leq \frac{D}{D - [(|\mathbf{X}|-1)/|\mathbf{X}|]}\right\}$$

$$\Rightarrow \bar{R}(D) = \limsup_{n\to\infty} \frac{1}{n} \log C_n(D) = 0 \quad (57)$$

Thus, the exponential rate of growth of dictionary size $C_n(D)$ with n for a given fractional distance D is zero when D is greater than $(|\mathbf{X}|-1)/|\mathbf{X}|$. Hence, Plotkin's bound, as stated in Equation 56, is a low-rate or large-distance bound.

Error Correction I

Plotkin's bound can be used with a recurrence relation on dictionary size to obtain a meaningful bound when $d_{\min}/n$ is smaller than the value allowed by Equation 55. Notice that

$$|\mathbf{C}(n, d_{\min})| = \sum_{x \in \mathbf{X}} |\mathbf{C}_x(n, d_{\min})| \tag{58}$$

where $\mathbf{C}(n, d_{\min})$ is any code consisting of n-tuples with minimum distance at least $d_{\min}$ and $\mathbf{C}_x(n, d_{\min})$ consists of all n-tuples in $\mathbf{C}(n, d_{\min})$ beginning with the symbol x. Since there are only $|\mathbf{X}|$ terms in Equation 58,

$$|\mathbf{C}(n, d_{\min})| \leq |\mathbf{X}| |\mathbf{C}_x(n, d_{\min})| \tag{59}$$

for some choice of x. A code $\mathbf{C}(n-1, d_{\min})$ of word length $n-1$ and minimum distance at least $d_{\min}$ can be constructed from $\mathbf{C}_x(n, d_{\min})$ by simply removing the first symbol x in each code word; hence,

$$|\mathbf{C}(n, d_{\min})| \leq |\mathbf{X}| |\mathbf{C}(n-1, d_{\min})| \tag{60}$$

Repeating this construction several times indicates that

$$|\mathbf{C}(n, d_{\min})| \leq |\mathbf{X}|^{n-k} |\mathbf{C}(k, d_{\min})| \tag{61}$$

When k is small enough, the Plotkin bound in Equations 55 and 56 can be applied to $\mathbf{C}(k, d_{\min})$; therefore,

$$|\mathbf{C}(n, d_{\min})| \leq \frac{|\mathbf{X}|^{n-k}(d_{\min}/k)}{(d_{\min}/k) - [(|\mathbf{X}| - 1)/|\mathbf{X}|]} \tag{62}$$

whenever

$$\frac{d_{\min}}{k} > \frac{|\mathbf{X}| - 1}{|\mathbf{X}|} \tag{63}$$

The iterated bound in Equation 62 may hold for many values of k, but by choosing k to be the largest integer satisfying

$$\frac{d_{\min}}{k} - \frac{1}{|\mathbf{X}|k} \geq \frac{|\mathbf{X}| - 1}{|\mathbf{X}|} \tag{64}$$

it is possible to arrive at a relatively simple asymptotic result. Solving for k in Equation 64 implies

$$k + r = \frac{|\mathbf{X}| d_{\min} - 1}{|\mathbf{X}| - 1} \tag{65}$$

where $0 \leq r < 1$. Substituting k given by Equation 65 into Equation 62 yields

$$|\mathbf{C}(n, d_{\min})| \leq \frac{|\mathbf{X}|^{n-[(|\mathbf{X}|d_{\min}-1)/(|\mathbf{X}|-1)]r+1} d_{\min}}{(|\mathbf{X}| - 1)r + 1} \tag{66}$$

The factor $|\mathbf{X}|^r [(|\mathbf{X}| - 1)r + 1]^{-1}$ is upper bounded by unity for $0 \leq r < 1$, and therefore,

$$|\mathbf{C}(n, d_{\min})| \leq |\mathbf{X}|^{n-[(|\mathbf{X}|d_{\min}-1/|\mathbf{X}|-1)]} d_{\min} \tag{67}$$

Again by fixing the fractional Hamming distance D and letting n become large, the asymptotic transmission rate can be bounded by

$$\bar{R}(D) \leq \log|\mathbf{X}| \left(1 - \frac{|\mathbf{X}|}{|\mathbf{X}| - 1} D \right) \tag{68}$$

where the base of the logarithm indicates the units in which information is measured. This extended version of Plotkin's bound is shown in Figure 5.4 for binary alphabets.

If Plotkin's bound is met, i.e., $d_{\min} = d_{av}$ in Equation 50, every two distinct code words differ in the same number of places. Codes with this property are discussed in Section 5.10.

Exercises

1. In Section 5.6, we computed the bound

$$|\mathbf{C}(n, d_{\min})| \leq |\mathbf{X}|^{n-k} |\mathbf{C}(k, d_{\min})|$$

Suppose we apply a bound to $\mathbf{C}(k, d_{\min})$ such that

$$|\mathbf{C}(k, d_{\min})| \leq B(k, d_{\min})$$

[$B(k, d_{\min})$ is not necessarily the Plotkin low-rate bound], where

Error Correction I

$$\lim_{k \to \infty} \frac{1}{k} B(k, D'k) = R_B(D')$$

Assuming that $d_{\min}$, n, and k go to infinity in such a way that

$$\lim_{n \to \infty} \frac{k}{n} = \alpha \qquad \lim_{n \to \infty} \frac{d_{\min}}{n} = D$$

show that

$$\bar{R}(\alpha D') \leq (1 - \alpha) \log |\mathbf{X}| + \alpha R_B(D')$$

where α is any real number in the range $0 \leq \alpha \leq 1$.

Note: This result tells us that any given bound $R_B(D')$ at fractional distance D' can be extended to lower fractional distances by drawing a straight line between $[D', R_B(D')]$ and $(0, \log |\mathbf{X}|)$ in the $\bar{R}, D$ plane.

2. (a) Verify

$$\max_{\{c_{ik}: i=1,2,\ldots,|\mathbf{C}|\}} \sum\sum_{i>j} d(c_{ik}, c_{jk}) \leq \frac{|\mathbf{C}|^2}{2} \cdot \frac{|\mathbf{X}| - 1}{|\mathbf{X}|}$$

where $d(x, y)$ is the Hamming distance between x and y.

(b) Can you derive an upper bound on

$$\sum\sum_{i>j} L(c_{ik}, c_{jk})$$

when $L(x, y)$ is the Lee distance between x and y? Use this result to determine the equivalent of Plotkin's bound in the Lee metric.

5.7. Elias Bound

Elias combined the sphere-packing concept of Hamming and the average-distance concept of Plotkin to generate a bound that is better than either of those two bounds. Sphere-packing concepts are used to find a portion of a code where words tend to cluster, and the average distance bound is applied to this subset of words. For simplicity, we derive the Elias bound for binary alphabets, though the result can be generalized.

Let **C** be a code containing binary n-tuples. Then there are

$$\sum_{i=0}^{d} \binom{n}{i}$$

n-tuples within distance d of each code word, giving a total of

$$|\mathbf{C}| \sum_{i=0}^{d} \binom{n}{i}$$

(not necessarily distinct) n-tuples in the Hamming spheres about the $|\mathbf{C}|$ code words. Let $N_d(x)$ be the number of code words within distance d of an n-tuple x. Then counting $|\{(x, c): x \in \mathbf{X}^n, c \in \mathbf{C}, d(x, c) \leq d\}|$ in two ways by first picking x and then picking c, we obtain

$$\sum_{x \in \mathbf{X}^n} N_d(x) = |\mathbf{C}| \sum_{i=0}^{d} \binom{n}{i} \tag{69}$$

Since $\mathbf{X}^n$ contains 2^n n-tuples, it follows that for some value of x, $N_d(x)$ must contain at least the average value of $N_d(x)$, i.e.,

$$N_d(x) \geq \left\lceil 2^{-n} |\mathbf{C}| \sum_{i=0}^{d} \binom{n}{i} \right\rceil \tag{70}$$

In the remainder of the derivation, we assume that $d < n/2$ and d and $|\mathbf{C}|$ are large enough to make the lower bound on $N_d(x)$ at least 2.

Let $c_1, c_2, \ldots, c_{N_d}$ be code words in **C** that are within Hamming distance d of the n-tuple x satisfying Equation 70. Then consider the difference vectors $a_1, \ldots, a_{N_d}$, given by

$$a_i = c_i \ominus x, \quad i = 1, 2, \ldots, N_d \tag{71}$$

where $\ominus$ denotes modulo 2 subtraction (addition) of the vectors, element by element. Let **A** equal a new code composed of the n-tuples $a_1, a_2, \ldots, a_{N_d}$. Notice that the Hamming distance between c_i and c_j is the same as the distance between a_i and a_j. In more detail,

$$a_i \ominus a_j = (c_i \ominus x) \ominus (c_j \ominus x) = c_i \ominus c_j \tag{72}$$

Error Correction I

and a_i and a_j differ in the same positions that c_i and c_j differ. Therefore from the construction of $\mathbf{A}$, we have obtained the following relations:

1. $|\mathbf{A}| \geq \lceil 2^{-n}|\mathbf{C}| \sum_{i=0}^{d} \binom{n}{i} \rceil$

2. $d_{\text{minA}} \geq d_{\text{minC}}$, where the subscripts denote the code to which the minimum distances apply

3. $w(a_i) \leq d$ for all i, where $w(a_i)$ denotes the *Hamming weight* of the n-tuple a_i, i.e., the number of nonzero elements in the n-tuple a_i

Notice that either $|\mathbf{A}|$ or d may be viewed as a free parameter in these relations. Neither is a parameter that can be determined from the original code $\mathbf{C}$.

We now apply the average-distance Plotkin bound to the localized code $\mathbf{A}$; hence,

$$d_{\text{minC}} \leq d_{\text{minA}} \leq d_{\text{avgA}} = \left[\frac{|\mathbf{A}|(|\mathbf{A}|-1)}{2}\right]^{-1} \sum\sum_{i>j} d(a_i, a_j) \qquad (73)$$

To make this bound independent of the choice of the code $\mathbf{A}$, the quantity on the right side of Equation 73 must be maximized over all choices of $\mathbf{A}$ satisfying the Relations 1–3. We first perform the maximization for a fixed value of $|\mathbf{A}|$ satisfying Relation 1 but only under the restriction that

4. $$\sum_{a_i \in \mathbf{A}} w(a_i) \leq |\mathbf{A}| d$$

Since any code satisfying Relations 2 and 3 will satisfy Relation 4, we have enlarged the set of possible $\mathbf{A}$ over which maximization is performed, yielding a valid upper bound to the right side of Equation 73.

Let z_k be the number of code words in $\mathbf{A}$ having a 0 in the kth position. Then we must maximize

$$\sum\sum_{i>j} d(a_i, a_j) = \sum_{k=1}^{n} z_k(|\mathbf{A}| - z_k) \qquad (74)$$

subject to the constraint of Relation 4 that

$$\sum_{k=1}^{n} (|\mathbf{A}| - z_k) \leq |\mathbf{A}| d \qquad (75)$$

It is easily shown that the right side of Equation 74 is maximized under the constraint in Equation 75 by setting

$$z_k = \frac{|\mathbf{A}|\,d}{n} \qquad (76)$$

Hence, combining Equations 73, 74, and 76 yields

$$\frac{d_{\min C}}{n} \leq \frac{2d}{n}\left(1 - \frac{d}{n}\right)\frac{|\mathbf{A}|}{|\mathbf{A}| - 1} \qquad (77)$$

where $|\mathbf{A}|$, d, and $|\mathbf{C}|$ are related only by Relation 1. Equation 77 and Relation 1 are the finite form of the *Elias bound*. The bound must hold for all choices of the free parameter, say, d, with Relation 1 used to determine the range of $|\mathbf{A}|$, and both d and Relation 4 used in Equation 77 to determine the bound.

The asymptotic version of the Elias bound assumes that $|\mathbf{A}|$ and d vary linearly with n and hence that

$$|\mathbf{A}| \approx \alpha n$$
$$d \approx \lambda n \qquad (78)$$

where α and λ are constants. Then asymptotically Equations 77 and 78 have the form

$$\frac{d_{\min C}}{n} \leq 2\lambda(1 - \lambda) \qquad (79)$$

$$|\mathbf{C}| \leq [8n\lambda(1 - \lambda)]^{1/2}\alpha n\, 2^{n[1 - H(\lambda, 1-\lambda)]} \qquad (80)$$

where Equation 80 has been determined from Relation 1 and Equation 42. Dropping the subscript $\mathbf{C}$ on minimum distance and computing an upper bound on asymptotic transmission rate yield the asymptotic version of the Elias bound

$$D \leq 2\lambda(1 - \lambda) \qquad (81)$$

$$\bar{R}(D) \leq 1 - H(\lambda, 1 - \lambda) \qquad (82)$$

where D is the fractional distance and λ can take on any value in the range $0 < \lambda < 1$. This parametrically described bound is plotted in Figure 5.4.

5.8. Gilbert Bound

In previous sections, we derived upper bounds on dictionary size for fixed word length n and minimum Hamming distance d_{min}. The following constructive algorithm supplies a simpler lower bound to dictionary size $|\mathbf{C}|$ for fixed n, d_{min}, and alphabet size $|\mathbf{X}|$.

1. Let the set $\mathbf{S}^n$ of allowable n-tuples initially consist of all possible $|\mathbf{X}|$-ary n-tuples $\mathbf{S}^n = \mathbf{X}^n$.
2. Choose a code word from $\mathbf{S}^n$.
3. Remove all n-tuples from $\mathbf{S}^n$ that are within Hamming distance $d_{min} - 1$ of the code word chosen in Step 2.
4. If $\mathbf{S}^n$ is not empty, return to Step 2; otherwise stop.

Notice that $\mathbf{S}^n$ initially contains $|\mathbf{X}|^n$ elements. Each time a code word is selected in Step 2, Step 3 removes at most

$$\sum_{i=0}^{d_{min}-1} (|\mathbf{X}| - 1)^i \binom{n}{i} \quad n\text{-tuples}$$

from $\mathbf{S}^n$. Hence, we can *guarantee* that the algorithm will not stop after $m - 1$ code-word selections if

$$(m - 1) \sum_{i=0}^{d_{min}-1} (|\mathbf{X}| - 1)^i \binom{n}{i} < |\mathbf{X}|^n \qquad (83)$$

The largest m satisfying Equation 83 is a lower bound on dictionary size; therefore, the attainable dictionary size $|\mathbf{C}|$ must satisfy

$$|\mathbf{C}| > |\mathbf{X}|^n \left[\sum_{i=0}^{d_{min}-1} (|\mathbf{X}| - 1)^i \binom{n}{i} \right]^{-1} \qquad (84)$$

The asymptotic form of Equation 84 in terms of asymptotic transmission rate and fractional distance can be determined from Equation 42.

$$R(D) = \liminf_{n \to \infty} \frac{1}{n} C_n(D) \geq 1 - H(D, 1 - D) \qquad (85)$$

where $C_n(D)$ represents the maximum attainable dictionary size for word length n and minimum distance at least D_n. The Gilbert bound in Equation 84 has been improved slightly, but the asymptotic result in Equation 85 (see Figure 5.4) has not been improved in an abstract sense. However, in recent years, Russian researchers have discovered classes of nonbinary codes with asymptotic properties superior to those of the Gilbert lower bound in some regions. These codes are defined using algebraic geometric methods.

Exercises

1. In the algorithm on which the Gilbert bound is based, selection of the first code word is followed by removal of

$$\sum_{i=0}^{d_{\min}-1} (|\mathbf{X}| - 1)^i \binom{n}{i}$$

   n-tuples

   (a) Show that during any subsequent code-word selection, there exists an n-tuple in $\mathbf{S}^n$ that is distance $d_{\min}$ from a previously selected code word.

   (b) Show that if the n-tuple in (a) is selected as the code word, then the number of elements in $\mathbf{S}^n$ is reduced in Step 3 by at most

$$\sum_{i=0}^{d_{\min}-1} (|\mathbf{X}| - 1)^i \binom{n}{i} - N(n, d_{\min}, |\mathbf{X}|)$$

   Hence, $N(n, d_{\min}, |\mathbf{X}|)$ is the number of n-tuples in the intersection of two Hamming spheres of radius $d_{\min} - 1$ with centers Hamming distance $d_{\min}$ apart.

   (c) Evaluate $N(n, d_{\min}, |\mathbf{X}|)$.

   (d) Derive an improved bound based on the preceding results.

2. Suppose that the symbol alphabet contains four elements and the *Lee* distance is in fact the design metric. Can you derive the Gilbert bound for this metric?

5.9. Perfect Codes

The Hamming bound in Section 5.5 states that a code $\mathbf{C}$, $\mathbf{C} \subseteq \mathbf{X}^n$, with minimum distance $d_{\min}$ has the property that spheres of radius r,

$$r = \left\lfloor \frac{d_{\min} - 1}{2} \right\rfloor \tag{86}$$

centered at code words, do not intersect. If in addition, these spheres exhaust the space $\mathbf{X}^n$, i.e., every n-tuple in $\mathbf{X}^n$ is packed perfectly with spheres, then the code $\mathbf{C}$ is called *close packed* or *perfect*.

If the metric employed is Hamming distance, a perfect code must satisfy the Hamming bound with equality

$$|\mathbf{C}| = \frac{|\mathbf{X}|^n}{\sum_{i=0}^{r} \binom{n}{i}(|\mathbf{X}| - 1)^i} \tag{87}$$

If we randomly choose values for $|\mathbf{X}|$, n, and r, it is doubtful that the right side of Equation 87 will be an integer and hence that equality will hold. Furthermore, if the right side of Equation 87 is an integer, then the numerology is correct, but there is *no guarantee* that the space $\mathbf{X}^n$ can be perfectly packed by Hamming spheres of radius r. The following are perfect codes in the Hamming metric:

1. $|\mathbf{X}|$ arbitrary, $r = n$, $|\mathbf{C}| = 1$

2. $|\mathbf{X}|$ arbitrary, $r = 0$, $|\mathbf{C}| = 2^n$

3. $|\mathbf{X}| = 2$, $n = 2r + 1$, $|\mathbf{C}| = 2$ (binary repetition)

4. Hamming single-error-correcting codes, i.e., $|\mathbf{X}| = q$, q a prime power, $r = 1$,

$$n = (q^m - 1)/(q - 1)$$
$$|\mathbf{C}| = q^{n-m}, m \geq 2$$

5. Nonlinear codes with the same parameter as Case 3.

6. Two Golay codes, i.e., $|\mathbf{X}| = 2$, $r = 3$, $n = 23$, $|\mathbf{C}| = 2^{12}$; and $|\mathbf{X}| = 3$, $r = 2$, $n = 11$, $|\mathbf{C}| = 3^6$

Cases 1–3 are trivial cases; the perfect codes of major interest are the Hamming and Golay codes. Examples of these codes are given in the Appendix to Chapter 7.

As already indicated, perfect single-error-correcting codes in the Hamming metric are usually referred to as *Hamming codes.* The relationship between word length and alphabet size for the known Hamming codes is even more precise than indicated by Equation 87, which for $r = 1$ is

$$[1 + n(|\mathbf{X}| - 1)] \Big| |\mathbf{X}|^n \tag{88}$$

If the alphabet size is a power of a prime, i.e.,

$$|\mathbf{X}| = p^m \tag{89}$$

then for a code to be a Hamming code, Relation 88 implies that

$$1 + n(|\mathbf{X}| - 1) = p^a |\mathbf{X}|^b \tag{90}$$

for some integers a and b with $0 \le a < m$. Solving for n gives

$$n = \frac{p^a |\mathbf{X}|^b - 1}{|\mathbf{X}| - 1} = p^a \cdot \frac{|\mathbf{X}|^b - 1}{|\mathbf{X}| - 1} + \frac{p^a - 1}{p^m - 1} \tag{91}$$

The first term in Equation 91 is an integer, but the second term is an integer only if a is zero. Therefore, when $|\mathbf{X}|$ is a power of a prime, the word length of a perfect Hamming code must satisfy

$$n = \frac{|\mathbf{X}|^b - 1}{|\mathbf{X}| - 1} \tag{92}$$

for some integer b. In fact, a Hamming code exists for each n of the form given in Equation 92 when $|\mathbf{X}|$ is a power of a prime.

The question of the existence of other perfect codes is answered to a great extent by Theorem 5.2.

Theorem 5.2 (Tietäväinen). When $|\mathbf{X}|$ is a prime or a prime power, all perfect codes are examples of Case 1–6 perfect codes.

The proof of Theorem 5.1, which will not be given here, is based on several specialized results and the Elias bound, which indicate that the Ham-

Error Correction I

ming bound cannot be achieved for n large and rates significantly less than unity. Very little is known about the existence of perfect codes when $|X|$ is not the power of a prime.

The concept of sphere packing is not limited to the Hamming metric; it can be extended to other metrics, including the Lee metric (see Question 5 in Section 5.3). The known perfect codes in the Lee metric have the following parameters:

1. r arbitrary, $n = 2$, $|X| = m(2r^2 + 2r + 1)$ where m is any positive integer
2. $r = 1$, n arbitrary, $|X| = 2n + 1$

In both cases, the alphabet size $|X|$ is larger than the radius r of the Lee sphere and larger than the word length n.

When $|X|$ is greater than $2r + 1$, it is possible to count the number N of vectors in a Lee sphere of radius r:

$$N = \sum_{k \geq 0} 2^k \binom{n}{k} \binom{r}{k} \tag{93}$$

In Equation 93, we regard the n components of a code word as boxes, with r "error balls" to distribute among these boxes. For any $k \leq r$, we consider the problem of distributing r error balls to *exactly* k boxes. There are

$$\binom{n}{k}$$

ways of choosing k of the n boxes to contain all the balls. Each of these k boxes must be designated as either containing a *positive* or *negative* deviation, for a factor of 2^k. There are

$$\binom{r}{k}$$

ways of distributing up to r balls to k boxes so that no box is empty. Multiplying these three factors together and then summing over k leads directly to the formula for N.

In the two-dimensional case, the volume of a Lee sphere of radius r is given by $2r^2 + 2r + 1$. The perfect code described in (1) therefore packs the space of $|X|^2$ 2-tuples with $m|X|$ code words. Figure 5.5 geometrically indicates a perfect double-Lee error-correcting code over a 13-symbol alphabet.

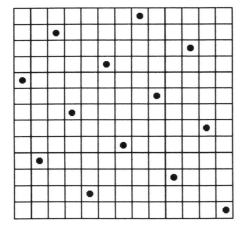

Figure 5.5. A close-packed double-Lee error-correcting code.

In the single error-correcting Lee code case (2) with n arbitrary, it is more difficult to visualize the packing geometrically, but it is easily described mathematically. The unit radius spheres contain $2n + 1$ points, and the corresponding perfect codes contain $|\mathbf{X}|^{n-1}$ code words. Packing may be accomplished with the following algorithm.

As centers of the spheres, use the set $\mathbf{C}$ of all points $(c_1, c_2, \ldots, c_n)$ of $\mathbf{X}^n$ that satisfy

$$\sum_{i=1}^{n} i c_i \equiv 0 \,(\bmod\, 2n + 1) \tag{94}$$

The number of solutions to this congruence is clearly $|\mathbf{X}|^{n-1}$, since any choice of $c_2, c_3, \ldots, c_n$ may be made, and there is a unique value of c_1 modulo $|\mathbf{X}|$ to satisfy the congruence. Also, every point in $\mathbf{X}^n$ is within a Lee distance of 1 from some point in this set. If $b = (b_1, b_2, \ldots, b_n)$ is any point of $\mathbf{X}^n$, we compute

$$\sum_{i=1}^{n} i b_i \equiv k (\bmod\, 2n + 1) \tag{95}$$

where $-n \le k \le +n$. If $k = 0$, then b is a member of the set $\mathbf{C}$. If $k > 0$, we change b_k to $b_k - 1$ to move from b to a member r of $\mathbf{C}$ at Lee distance 1. If $k < 0$, we change $b_{|k|}$ to $b_{|k|} + 1$ to move from b to a member of $\mathbf{C}$ at Lee distance 1.

Error Correction I

Each point **C** has only $2n$ neighbors at a Lee distance of 1. Thus, spheres around these points can account for at most $|\mathbf{X}|^{n-1}(2n+1) = |\mathbf{X}|^n$ points if the spheres are all disjoint. However, since every point in $\mathbf{X}^n$ is within distance 1 from some point **C**, the spheres must be disjoint and fill the space. Thus, the code is close packed.

Exercises

1. Sketch
   (a) A typical radius 3 Lee sphere in two dimensions for alphabet size 7
   (b) A typical radius 3 Lee sphere in two dimensions for alphabet size 5
   (c) A Hamming sphere of radius 1 in three dimensions for a ternary alphabet
   (d) A typical radius 1 Lee sphere in three dimensions

2. Give an example of a perfect, single-error-correcting code in both the Hamming and Lee metrics.

3. Consider the memoryless channel whose input and output alphabets **X** and **Y** are $\{0, 1, 2, 3, 4\}$ and whose channel matrix has the form

$$\begin{bmatrix} 1/2 & 0 & 0 & 0 & 1/2 \\ 1/2 & 1/2 & 0 & 0 & 0 \\ 0 & 1/2 & 1/2 & 0 & 0 \\ 0 & 0 & 1/2 & 1/2 & 0 \\ 0 & 0 & 0 & 1/2 & 1/2 \end{bmatrix}$$

   (a) If a code **C** is selected from $\mathbf{X}^2$ and the code word (c_1, c_2) is transmitted, what are the possible received code words?
   (b) How many code words can the code **C** in (a) contain if correct decoding is guaranteed with probability 1? Note that this question can be interpreted in terms of a packing problem, but the shapes being packed are not spheres whose centers are code words but the shapes described in (a).

4. Design a perfect single error-correcting Lee code in two dimensions over a five-symbol alphabet.

5. Show that the number of ways of putting up to r balls into k boxes so that no box is empty is

$$\binom{r}{k}$$

5.10. Equidistant Codes

In Section 5.10, we are interested in codes that meet the Plotkin bound in Section 5.6. Such a code has the property that two distinct code words differ in a constant number of places. The Plotkin bound for the Hamming metric states that

$$|C(n, d_{\min})| \le \frac{d_{\min}/n}{(d_{\min}/n) - [(|X| - 1)/|X|]} \qquad (96)$$

where the denominator in Equation 96 must be positive. A review of the derivation in Section 5.6 shows that equality holds in Equation 96 when the following conditions hold:

1. $d_{\min} = d_{av}$
2. $|X| \,|\, |C|$

Hence, a code that achieves the Plotkin bound with equality must have

$$d(c_i, c_j) = d_{\min} \qquad (97)$$

for all choices of c_i and c_j, $i \ne j$. Codes that achieve the Plotkin bound are called *maximal equidistant codes*. Condition 2, the alphabet size $|X|$ must divide the code size $|C|$, provides the correct numerology so that the simple upper bound on d_{av} that leads to Equation 96 is in fact tight. Many codes achieve the Plotkin bound, but there is no general definitive statement about the values of $|X|$, n, and $d_{\min}$ for which equidistant codes exist.

Two of the best known equidistant code families are

1. Codes derived from Hadamard matrices

$|X| = 2$, $|C| = m$, $n = m - 1$, $d_{\min} = m/2$, $m = 2$ or a multiple of 4

Error Correction I

2. Codes derived from linear feedback shift registers

$|\mathbf{X}| = p^m$, p prime, $n = |\mathbf{X}|^k - 1$, $|\mathbf{C}| = |\mathbf{X}|^k$, $d_{\min} = |\mathbf{X}|^{k-1}(|\mathbf{X}| - 1)$

By changing the existence question slightly, some definitive results can be obtained. Notice that the bound on $|\mathbf{C}|$ in Equation 96 depends only on $d_{\min}/n$ and $|\mathbf{X}|$, and for maximal equidistant codes,

$$\frac{d_{\min}}{n} = \frac{|\mathbf{C}|}{(|\mathbf{C}| - 1)} \frac{(|\mathbf{X}| - 1)}{|\mathbf{X}|} \tag{98}$$

Furthermore, the equality constraint (condition 2) implies that code size is a multiple of alphabet size, i.e.,

$$|\mathbf{C}| = m|\mathbf{X}| \tag{99}$$

for some integer m. Substituting Equation 99 into Equation 98 we see that an equidistant code must satisfy the relation

$$\frac{d_{\min}}{n} = \frac{m(|\mathbf{X}| - 1)}{(m|\mathbf{X}| - 1)} \tag{100}$$

We will now demonstrate by construction that for each value of m and $|\mathbf{X}|$, there exists a maximal equidistant code satisfying the relations in Equations 99 and 100.

Construction Algorithm

1. List all possible columns containing each of the $|\mathbf{X}|$ symbols exactly m times.
2. As code words, choose the rows of the column array.

Example 5.4. Let $m = 3$, $|\mathbf{X}| = 2$. Column list:

```
0 0 0 0 0 0 0 0 0 0 1 1 1 1 1 1 1 1 1 1
0 0 0 0 1 1 1 1 1 1 1 1 1 1 0 0 0 0 0 0
0 1 1 1 0 0 0 1 1 1 1 0 0 0 1 1 1 0 0 0
1 0 1 1 0 1 1 0 0 1 0 1 0 0 1 0 0 1 1 0
1 1 0 1 1 0 1 0 1 0 0 0 1 0 0 1 0 1 0 1
1 1 1 0 1 1 0 1 0 0 0 0 0 1 0 0 1 0 1 1
```

$$d_{\min} = d_{av} = 12, \qquad n = 20$$

$$\frac{d_{\min}}{n} = \frac{3}{5}, \qquad |\mathbf{C}| = 6$$

These parameters satisfy Equations 99 and 100.

In general, the construction provides n columns, each of length $|\mathbf{C}| = m|\mathbf{X}|$. The number of distinct columns n is given by the multinomial coefficients

$$n = \binom{|\mathbf{C}|}{m, m, \ldots, m} = \frac{|\mathbf{C}|!}{(m!)^{|\mathbf{X}|}} = \frac{(m|\mathbf{X}|)!}{(m!)^{|\mathbf{X}|}} \qquad (101)$$

Exercise

Prove Equation (101).

Two rows (code words) have the same symbol in the same column exactly

$$\frac{(|\mathbf{C}| - 2)!}{(m!)^{|\mathbf{X}|-1}(m-2)!}$$

times, corresponding to the number of distinct ways the remaining $|\mathbf{C}| - 2$ column positions can be filled with the remaining symbols. Hence, the distance $d_{\min}$ for the construction is given by

$$\begin{aligned}d_{\min} &= n - |\mathbf{X}| \frac{(|\mathbf{C}| - 2)!}{(m!)^{|\mathbf{X}|-1}(m-2)!} \\ &= n\left(1 - \frac{m-1}{|\mathbf{C}| - 1}\right) = \frac{nm(|\mathbf{X}| - 1)}{m|\mathbf{X}| - 1}\end{aligned} \qquad (102)$$

which obviously satisfies Equation 100 with equality.

The construction may be modified to yield smaller values of n by noting that columns equivalent under renaming the alphabet, e.g.,

$$\begin{pmatrix}0\\0\\1\\0\\2\end{pmatrix} = \begin{pmatrix}1\\1\\0\\1\\2\end{pmatrix} = \begin{pmatrix}2\\2\\1\\2\\0\end{pmatrix} = \begin{pmatrix}0\\0\\2\\0\\1\end{pmatrix} = \begin{pmatrix}1\\1\\2\\1\\0\end{pmatrix} = \begin{pmatrix}2\\2\\0\\2\\1\end{pmatrix}$$

all have the same Hamming distances between elements. Hence, the columns listed in Step 1 of the construction algorithm can be separated into equivalence classes under the alphabet-renaming operation, with each class containing

Error Correction I

$|\mathbf{X}|!$ columns. If exactly one column from each equivalence class of columns is used to construct the code, then both n and d_{min}, as given by Equations 101 and 102, will be reduced by a factor of $|\mathbf{X}|!$, and the resultant parameters will satisfy Equations 99 and 100.

Example 5.5. In Example 5.4, the last ten columns were the complement of the first ten columns. Thus, the first ten columns are inequivalent under the $0 \leftrightarrow 1$ renaming operation. Hence, the code can be reduced to

$$\begin{matrix} 0 & 0 & 0 & 0 & 0 & 0 & 0 & 0 & 0 & 0 \\ 0 & 0 & 0 & 0 & 1 & 1 & 1 & 1 & 1 & 1 \\ 0 & 1 & 1 & 1 & 0 & 0 & 0 & 1 & 1 & 1 \\ 1 & 0 & 1 & 1 & 0 & 1 & 1 & 0 & 0 & 1 \\ 1 & 1 & 0 & 1 & 1 & 0 & 1 & 0 & 1 & 0 \end{matrix}$$

where

$$|\mathbf{C}| = 6 \qquad n = 10 \qquad d_{min} = 6$$

still satisfy the bound, and the code given by the preceding rows is still maximal equidistant.

The preceding construction indicates that maximal equidistant codes with specified $|\mathbf{C}|$, $|\mathbf{X}|$, and d_{min}/n exist if and only if these parameters satisfy Equations 99 and 100. Hence, the existence question reduces to determining values of n to be used in the specified fractional distance d_{min}/n. By reducing the denominator on the right side of Equation 100 to its smallest integer value, noting that m cannot divide $m|\mathbf{X}| - 1$, it follows that n must take the form

$$n = j \frac{m|\mathbf{X}| - 1}{\text{g.c.d.}(|\mathbf{X}| - 1, m|\mathbf{X}| - 1)} \qquad (103)$$

where j is any positive integer. Thus, the existence question for maximal equidistant codes is reduced to determining which of the possible values of n described in Equation 103 actually yield a solution.

Exercises

1. Design a maximal equidistant code with parameters $|\mathbf{C}| = 8$, $|\mathbf{X}| = 2$, $n = 7$.

2. Suppose that $|\mathbf{C}|$, $|\mathbf{X}|$, and d_{min}/n satisfy the constraints in Equations 99 and 100, and furthermore codes with these parameters can be constructed for $n = n_1$ and $n = n_2$. For what other word lengths n can codes with the parameters $|\mathbf{C}|$, $|\mathbf{X}|$, and d_{min}/n be constructed?

3. Prove that the code constructed in the following manner is maximal equidistant and determine its parameters.

   (a) $\mathbf{X} = \{0, 1, 2, \ldots, p - 1\}$, p prime
   (b) $\mathbf{C} = \{c^{(0)}, c^{(1)}, \ldots, c^{(p-1)}\}$, where

   $$c^{(j)} = (j, 2j, 3j, \ldots, (p-1)j)$$

   All arithmetic is modulo p. (Note that addition and multiplication modulo p on $\mathbf{X}$ describes a *field,* and therefore, all elements of $\mathbf{X}$ have additive and multiplicative inverses, except 0, which has no multiplicative inverse.)

4. Prove that the following code $\mathbf{C}'$ is a maximal equidistant code. $\mathbf{C}'$ contains $(p + 1)$-tuples, p prime, over alphabet $\mathbf{X} = \{0, 1, \ldots, p - 1\}$, of the following forms:

   (a) $(x, x, \ldots, x, 0)$, $x \in \mathbf{X}$
   (b) $(\Pi^i c^{(j)}, j)$, $0 \leq i < p$, $1 \leq j < p$

   where

   $$c^{(j)} = [0, j, 2j, \ldots, (p-1)j]$$

   and Π^i denotes an i-unit cyclic-shift operator.

5.11. Hamming Distance Enumeration

When codes are small, it is relatively simple to enumerate all distances between code words; for example, the code $\mathbf{C}$ with

Error Correction I

$$c_1 = 11111$$
$$c_2 = 11000$$
$$c_3 = 00110$$
$$c_4 = 01101$$

has Hamming distance properties that are easily enumerated in Table 5.4.

Table 5.4. Hamming Distance Properties

$d(c_i, c_j)$	c_1	c_2	c_3	c_4
c_1	0	3	3	2
c_2	3	0	4	3
c_3	3	4	0	3
c_4	2	3	3	0

Now, consider the amount of effort required to construct a similar table for a code containing thousands or even millions of very long code words. We are going to develop an alternative method for evaluating the number a_d of ordered pairs (c_i, c_j) of code words separated by a Hamming distance exactly equal to d. In the preceding example

$$a_0 = 4$$
$$a_1 = 0$$
$$a_2 = 2$$
$$a_3 = 8$$
$$a_4 = 2$$

Notice that every entry in the table is counted once, with the pair (c_1, c_2) considered distinct from the pair (c_2, c_1).

In the following analysis, we consider the code alphabet to be the group of integers under addition modulo $|\mathbf{X}|$:

$$\mathbf{X} = \{0, 1, 2, \ldots, |\mathbf{X}| - 1\} \tag{104}$$

(This is no restriction on the results, since any alphabet can be mapped one-to-one onto an alphabet $\mathbf{X}$ of the form given in Equation 104, and the Hamming distance properties of the alphabet and corresponding code will be preserved.) We assume that the code $\mathbf{C}$ is a collection of n-tuples from $\mathbf{X}^n$ and use the following transform relations for *real-valued* functions defined on $\mathbf{X}^n$. Let $f(x)$ be a real-valued function on $\mathbf{X}^n$ and define

$$F(y) = \sum_{x \in \mathbf{X}^n} f(x) \exp\{-ix^t y / |\mathbf{X}|\} \qquad (105)$$

where x and y are viewed as column vectors from $\mathbf{X}^n$ and $(\ )^t$ denotes transpose. Notice that either real or modulo $|\mathbf{X}|$ arithmetic can be used to compute $x^t y$, since either answer gives the same result when used as an exponent on an $|\mathbf{X}|^{th}$ root of unity.

The reader may verify that the following relations are valid:

1. Inversion theorem:

$$f(x) = \frac{1}{|\mathbf{X}|^n} \sum_{y \in \mathbf{X}^n} F(y) \exp\{+ix^t y / |\mathbf{X}|\} \qquad (106)$$

for all $x \in \mathbf{X}^n$.

2. Convolution theorem:
   If

$$h(z) = \sum_{x \in \mathbf{X}^n} f(x) g(x - z) \qquad (107)$$

then

$$H(y) = F(y) G^*(y) \qquad (108)$$

for all $y \in \mathbf{X}^n$, where $*$ denotes conjugation of the complex valued transform.

3. Inner-Product theorem:

$$\sum_{x \in \mathbf{X}^n} f(x) g(x) = \frac{1}{|\mathbf{X}|^n} \sum_{y \in \mathbf{X}^n} F(y) G^*(y) \qquad (109)$$

All of these results are similar to the generally more familiar inversion, convolution, and Parseval or Plancherel theorems of Fourier transforms, the main

Error Correction I

difference being that the space on which the present functions are defined is finite. Now let us return to the distance enumeration problem.

We define the indicator function $f(x)$ of a code **C** as

$$f(x) = \begin{cases} 1 & \text{if } x \in \mathbf{C} \\ 0 & \text{if } x \notin \mathbf{C} \end{cases} \quad (110)$$

The correlation function $g(x)$ of $f(x)$ then has a simple interpretation:

$$g(y) \triangleq \sum_{x \in \mathbf{X}} f(x)f(x - y) = |\{(x, y): x \in \mathbf{C}, x - y \in \mathbf{C}\}|$$

$$= |\{(x, x'): x \in \mathbf{C}, x' \in \mathbf{C}, x - x' = y\}|$$

$$= \text{Number of ordered code-word pairs } (x, x')$$

$$\text{for which } x - x' = y \quad (111)$$

The number of ordered pairs of code words separated by a given Hamming distance can be extracted from Equation 111 as follows. Let

$$h_z(y) = z^{w(y)} \quad (112)$$

where z is an indeterminate and the Hamming weight $w(y)$ of y is the number of nonzero elements of y. The identity

$$d(x, x') = w(x - x') \quad (113)$$

follows immediately from the definition of Hamming distance. Then the polynomial in z defined by

$$\sum_{y \in \mathbf{X}^n} h_z(y)g(y) = \sum_{d=0}^{n} \sum_{y: w(y)=d} z^{w(y)} g(y)$$

$$= \sum_{d=0}^{n} z^d \left(\sum_{y: w(y)=d} g(y) \right) = \sum_{d=0}^{n} a_d z^d \quad (114)$$

has coefficients a_d of z^d, which correspond to the total number of ordered code-word pairs separated by Hamming distance d.

Computationally, there is no difference between performing the correlation in Equation 114 and enumerating a distance table, as we did at the beginning of Section 5.11; however, transform theory gives us another expression for Equation 114 that may be easier to evaluate. Applying the Inner-Product theorem in Equation 109 and then the Convolution theorem to Equation 114 gives

$$\sum_{d=0}^{n} a_d z^d = \frac{1}{|\mathbf{X}|^n} \sum_{y \in \mathbf{X}^n} H_z(y) G^*(y) = \frac{1}{|\mathbf{X}|^n} \sum_{y \in \mathbf{X}^n} H_z(y) |F(y)|^2$$

since

$$G(y) = F(y) \cdot F^*(y) = |F(y)|^2 \quad (115)$$

where $H_z(y)$ and $F(y)$ are transforms of the weight enumeration function in Equation 112 and the code indicator function in Equation 110. $H_z(y)$ can be evaluated by expanding each $x \in X^n$:

$$H_z(y) = \sum_{x \in \mathbf{X}^n} h_z(x) \exp\{-ix^t y/|\mathbf{X}|\}$$

$$= \sum_{x \in \mathbf{X}^n} z^{\sum_{j=1}^{n} w(x_j)} e^{-\frac{i}{|\mathbf{X}|} \sum_{j=1}^{n} x_j y_j}$$

$$= \prod_{j=1}^{n} \left[\sum_{x_j \in \mathbf{X}} z^{w(x_j)} e^{-\frac{i}{|\mathbf{X}|} x_j y_j} \right] \quad (116)$$

Using the fact that the Hamming weight of the symbol 0 is zero and for all other symbols it is 1, Equation 116 reduces to

$$H_z(y) = \prod_{j=1}^{n} \left[1 - z + z \left(\sum_{x_j \in \mathbf{X}} e^{-\frac{i}{|\mathbf{X}|} x_j y_j} \right) \right] \quad (117)$$

In more detail,

$$\left[\sum_{x_j \in \mathbf{X}} z^{w(x_j)} e^{-\frac{i}{|\mathbf{X}|} x_j y_j} \right] - \left(z \sum_{x_j \in \mathbf{X}} e^{-\frac{i}{|\mathbf{X}|} x_j y_j} \right)$$

$$= \sum_{x_j \in \mathbf{X}} [z^{w(x_j)} - z] e^{-\frac{i}{|\mathbf{X}|} x_j y_j} = 1 - z \quad (118)$$

Error Correction I

Furthermore, when y_j is *not* zero, the sum over $x_j \in \mathbf{X}$ in Equation 118 is zero. When y_j is zero, the sum is $|\mathbf{X}|$; hence,

$$H_z(y) = (1 - z + z|\mathbf{X}|)^{n-w(y)}(1 - z)^{w(y)} \qquad (119)$$

Substituting Equation 119 into Equation 116 gives

$$\sum_{d=0}^{n} a_d z^d = \left(\frac{1 - z + z|\mathbf{X}|}{|\mathbf{X}|}\right)^n \sum_{y \in \mathbf{X}^n} \left(\frac{1 - z}{1 - z + z|\mathbf{X}|}\right)^{w(y)} |F(y)|^2 \qquad (120)$$

Now if we define

$$b_w = \sum_{y:w(y)=w} |F(y)|^2 \qquad (121)$$

then finally

$$\sum_{d=0}^{n} a_d z^d = \left(\frac{1 - z + z|\mathbf{X}|}{|\mathbf{X}|}\right)^n \sum_{w=0}^{n} b_w \left(\frac{1 - z}{1 - z + z|\mathbf{X}|}\right)^w \qquad (122)$$

Equation 122 or an equivalent set of equations relating the coefficients of z^k on each side of Equation 122 is usually referred to as the *MacWilliams identities*. If the coefficients b_w, $w = 0, 1, \ldots, n$, can be determined, then the quantities of interest, namely, a_d, $d = 0, 1, \ldots, n$, can be derived using the MacWilliams identities.

Using the transform definition in Equation 105 and Equations 110 and 121, an explicit equation for the coefficients b_w can be written

$$b_w = \sum_{y:w(y)=w} \left|\sum_{x \in \mathbf{X}^n} f(x) \exp\{-ix^t y/|\mathbf{X}|\}\right|^2$$

$$= \sum_{y:w(y)=w} \left|\sum_{c \in \mathbf{C}} \exp\{-ic^t y/|\mathbf{X}|\}\right|^2 \qquad (123)$$

The weights b_w have no simple interpretation for arbitrary codes. As we see next, they have a very specific interpretation if the code $\mathbf{C}$ is a linear code.

Example 5.6. Consider a code containing a single overall check symbol; that is, the last symbol of the code word c is determined by

$$c_n = \sum_{i=1}^{n-1} c_i \bmod |\mathbf{X}|$$

where the beginning of the code word $(c_1, c_2, \ldots, c_{n-1})$ can be any of the $|\mathbf{X}|^{n-1}$ possible n-tuples. This code contains

$$|\mathbf{C}| = |\mathbf{X}|^{n-1}$$

possible words. We now determine the values of b_w; first,

$$b_0 = \left| \sum_{c \in \mathbf{C}} 1 \right|^2 = |\mathbf{C}|^2 = |\mathbf{X}|^{2(n-1)}$$

For the other coefficients, we have to consider the inner sum in Equation 123:

$$s_y = \sum_{c \in \mathbf{C}} \exp\{ic^t y/|\mathbf{X}|\}$$

The sum is over all possible code-word beginnings, since the last entry is a function of the beginning; hence,

$$s_y = \sum_{c_1 \in \mathbf{X}} \cdots \sum_{c_{n-1} \in \mathbf{X}} \exp\left\{\frac{i}{|\mathbf{X}|} \left(\sum_{j=1}^{n-1} c_j y_j + y_n \sum_{j=1}^{n-1} c_j\right)\right\}$$

$$= \prod_{k=1}^{n-1} \left(\sum_{c_k \in \mathbf{X}} \exp\left\{\frac{i}{|\mathbf{X}|} c_k(y_k + y_n)\right\}\right)$$

The sum over $c_k \in \mathbf{X}$ is zero unless $y_k + y_n$ is zero. If $y_k + y_n$ is zero, the sum is $|\mathbf{X}|$; hence,

$$\sum_{c \in \mathbf{C}} \exp\{ic^t y/|\mathbf{X}|\} = \begin{cases} |\mathbf{X}|^{n-1} & \text{if } y_k = -y_n \text{ for all } k < n \\ 0 & \text{otherwise} \end{cases}$$

Thus, it appears that the only nonzero contributions to the b_w occur when

$$y = (-y_n, -y_n, \ldots, -y_n, y_n)$$

for which

Error Correction I

$$w(y) = \begin{cases} 0 & y_n = 0 \\ n & y_n \neq 0 \end{cases}$$

Thus, we obtain $|\mathbf{X}| - 1$ ($y_n \neq 0$) contributions of $|\mathbf{X}|^{2(n-1)}$ to b_n. Summarizing:

$$b_0 = |\mathbf{X}|^{2(n-1)}$$
$$b_w = 0, \quad 0 < w < n$$
$$b_n = (|\mathbf{X}| - 1)|\mathbf{X}|^{2(n-1)}$$

Applying the MacWilliams identities in Equation 122 and using the binomial expansion gives:

$$\sum_{d=0}^{n} a_d z^d = \frac{|\mathbf{X}|^{2(n-1)}}{|\mathbf{X}|^n} [(1 - z + z|\mathbf{X}|)^n + (|\mathbf{X}| - 1)(1 - z)^n]$$

$$= |\mathbf{X}|^{n-2} \sum_{j=0}^{n} \binom{n}{j} [(|\mathbf{X}| - 1)^j + (|\mathbf{X}| - 1)(-1)^j] z^j$$

from which it follows that

$$a_d = |\mathbf{X}|^{n-2} \binom{n}{d} [(|\mathbf{X}| - 1)^d + (|\mathbf{X}| - 1)(-1)^d]$$

The reader should realize that

$$b_0 = |\mathbf{C}|^2 \tag{124}$$
$$a_0 = |\mathbf{C}| \tag{125}$$

for all codes with distinct words. Equation 125 may be used as a check on the computation of $\{b_w\}$ and the application of the MacWilliams identities.

Exercises

1. Verify the identity

$$\sum_{c \in \mathbf{X}} \exp\{ic^t y/|\mathbf{X}|\} = \begin{cases} |\mathbf{X}| & y = 0 \\ 0 & y \neq 0 \end{cases}$$

for $y \in \mathbf{X}$, where $\mathbf{X}$ is the integers under addition and multiplication modulo $|\mathbf{X}|$.

2. Derive the Inversion theorem, Correlation theorem, and Inner-Product theorem for the finite Fourier transform defined in Equation 105.

3. Let $\mathbf{X}$ be a binary alphabet and consider a code $\mathbf{C}$ whose words c take the form

$$c = (c_1, c_2, c_3, c_4, c_5, c_6)$$

where

$$(c_1, c_2, c_3, c_4) \in \mathbf{X}^4$$

and

$$c_5 = c_1 \oplus c_2$$
$$c_6 = c_3 \oplus c_4$$

with $\oplus$ denoting modulo 2 addition.

(a) Determine b_w, $0 \leq w \leq 6$.

(b) Use the MacWilliams identities to evaluate the number of ordered pairs of code words separated by distance d.

(c) Check your answer by direct evaluation.

4. A binary Hamming code $\mathbf{C}$ with words

$$c = (c_1, c_2, c_3, c_4, c_5, c_6, c_7)$$

is described by

$$(c_1, c_2, c_3, c_4) \in \mathbf{X}^4$$

where $\mathbf{X}$ is the binary alphabet and

$$c_5 = c_1 + c_2 + c_4$$

Error Correction I

$$c_6 = c_1 + c_3 + c_4$$
$$c_7 = c_2 + c_3 + c_4$$

Determine the number a_d of ordered pairs of code words from **C** that are separated by distance d, $0 \le d \le n$.

5. Let the distance enumeration polynomial of this section be defined by

$$a(z) = \sum_{d=0}^{n} a_d z^d$$

Assume that a code **C** with polynomial $a(z)$ is transmitted over a binary symmetric channel having error rate p. If a pure error detection scheme is used, determine the quantities P_C, P_D, and P_E (see Section 5.14) in terms of $|\mathbf{X}|$, n, p, and $a(z)$.

6. Determine an explicit expression for a_d in terms of b_w, $0 \le w \le n$.

5.12. Pless Power Moment Identities and the Welch, McEliece, Rodemich, and Rumsey (WMR) Bound

In Section 5.12, we consider only binary codes, in which case the MacWilliams identities reduce to

$$\sum_{d=0}^{n} a_d z^d = \frac{1}{2^n} \sum_{w=0}^{n} b_w (1+z)^{n-w}(1-z)^w \tag{126}$$

Equation 126 is linear in both the variables a_d, $0 \le d \le n$, and b_w, $0 \le w \le n$. We now derive a set of equations that can be solved for s unknown values of a_d when the remaining values of a_d and $b_0, b_1, \ldots, b_{s-1}$ are known.

Multiplying both sides of Equation 126 by e^{nt} and substituting e^{-2t} for z gives:

$$\sum_{d=0}^{n} a_d e^{(n-2d)t} = \sum_{w=0}^{n} b_w (\cosh t)^{n-w}(\sinh t)^w \tag{127}$$

In more detail, we use the identities

$$2\cosh t = e^{-t} + e^{+t} = e^t\{1 + e^{-2t}\}$$
$$2\sinh t = e^{+t} - e^{-t} = e^t\{1 - e^{-2t}\}$$

So

$$\frac{1}{2^n}(1+z)^{n-w}(1-z)^w, \text{ for } z = e^{-2t}$$

equals

$$(\cosh t)^{n-w}(\sinh t)^w e^{-nt}$$

Differentiating both sides of Equation 127 r times with respect to t and evaluating at t equals 0 yields

$$\sum_{d=0}^{n} a_d(n-2d)^r = \sum_{w=0}^{n} b_w F_r^{(w)}(n) \tag{128}$$

where

$$F_r^{(w)}(n) = \frac{d^r}{dt^r}[(\cosh t)^{n-w}(\sinh t)^w]\big|_{t=0} \tag{129}$$

Notice that when $w > r$, the rth derivative in Equation 129 contains a $\sinh t$ factor, and hence, as $\sinh 0 = 0$:

$$F_r^{(w)}(n) = 0 \quad w > r \tag{130}$$

As a result, Equation 128 reduces to

$$\sum_{d=0}^{n} a_d(n-2d)^r = \sum_{w=0}^{r} b_w F_r^{(w)}(n) \tag{131}$$

Equation 131 for $r = 0, 1, 2, \cdots$ is known as the *Pless power-moment identities*. If $b_0, b_1, \ldots, b_{s-1}$ and all but s of the values of a_d are known, then the Pless power-moment identities for $r = 0, 1, \ldots, s-1$ provide s linear equations in s unknowns that can be solved for the unknown values of a_d.

The coefficients $F_r^{(w)}(n)$ in general are messy expressions; however, they have been used twice to obtain asymptotic bounds for binary codes (we omit the proofs). McEliece *et al.* (1974) used these methods to obtain an upper bound for $\bar{R}(D)$ that *for low-rate codes* is superior to the Elias bound. This bound is

$$\bar{R}(D) \le \min_{0<\alpha<D} [1 - H(\alpha, 1 - \alpha)]$$
$$\times \left[\frac{-\log(1 - 2D)}{\log(1 - 2\alpha) - \log(1 - 2D)} \right], \quad D \le 1/2 \quad (132)$$

McEliece *et al.* (1977) improved the bound in Equation 132 to obtain a bound uniformly tighter than the Elias bound. (The bound is called either the WMR bound or the Jet Propulsion Laboratories (JPL) bound or in eastern Europe, the four Americans' bound.) The result follows.

$$\bar{R}(D) \le \min_{0<u<1-2D} [1 + g(u^2) - g(u^2 + 2Du + 2D)], \quad D \le 1/2 \quad (133)$$

where

$$g(x) = h_2((1 - \sqrt{1 - x})/2) \quad (134)$$

and $h_2(\ )$ is the binary entropy function, i.e., $h_2(y) = H(y, 1 - y)$.

This last upper bound for $\bar{R}(D)$ remains the best known in general. The WMR bound is shown in Figure 5.4. (For D in the range $0.273 < D < 0.5$, a closed-form expression for the minimum exists.) The methods used in the WMR bound have been extended by Kabatiansky and Levenshtein (1978) to give analogous bounds for sphere packings.

The proof of the WMR bound involves a class of orthogonal polynomials known as *Krawtchouk* polynomials. The mathematical principles have been generalized by Delsarte (1973) into an elegant area of combinatorial mathematics known as *association schemes*.

5.13. Finite State Channels

To prevent the reader from getting the wrong impression, we state unequivocally that the communication world is not composed solely of memoryless channels. Unfortunately, channel modeling becomes a formidable task for many real-world channels with memory. The concept of a *channel state* s, which is in effect the memory of the channel, provides the following format for describing the channel. Let channel input sequence $= (x_1, x_2, \ldots, x_n) = x$ (135)

channel output sequence $= (y_1, y_2, \ldots, y_n) = y$ (136)

channel state sequence $= (s_1, s_2, \ldots, s_n) = s$ (137)

and denote by s_{t-1} the state of the channel at the time x_t is transmitted. Assuming s_0 is the initial state, then we formally write

$$Pr(y, s | x, s_0) = \prod_{t=1}^{n} Pr(y_t, s_t | y_{t-1}, y_{t-2}, \ldots, y_1,$$

$$s_{t-1}, s_{t-2}, \ldots, s_1, s_0, x_n, x_{n-1}, \ldots, x_1) \quad (138)$$

Under the assumptions that

1. Channel output is independent of future inputs when the present input is known,
2. Channel output is independent of prior outputs and prior inputs when the most recent channel state is known (a result of the definition of a channel state),

the conditional probabilities in Equation 138 simplify to

$$Pr(y, s | x, s_0) = \prod_{t=1}^{n} Pr(y_t, s_t | x_t, s_{t-1}) \quad (139)$$

Under these assumptions, the description of the channel has been simplified from describing $Pr(y, s | x, x_0)$ for all $y \in \mathbf{Y}^n$, $s \in \mathbf{S}^n$, $x \in \mathbf{X}^n$, and $s_0 \in \mathbf{S}$ (here, $\mathbf{S}$ is the set of channel states) to describing $Pr(y_t, s_t | x_t, s_{t-1})$ for $y_t \in \mathbf{Y}$, $s_t \in \mathbf{S}$, $x_t \in \mathbf{X}$, and $s_0 \in \mathbf{S}$ and each value of t. The description is further simplified if the Markov chain in question is homogeneous so that there is no dependence on the time index t. When $|\mathbf{S}|$ is finite, then Equation 139 and the complete description of $Pr(y_t, s_t | x_t, s_{t-1})$ for all possible values of its arguments constitute a complete description of a *finite-state channel.*

Our purpose in this exposition is not to present a complete theory of coding for finite-state channels, but simply to illustrate what can happen when the channel is not memoryless. Using this approach, we assume henceforth that the finite-state channel is homogeneous and has the additional property that

$$Pr(y_i, s_i | x_i, s_{i-1}) = Pr(y_i | x_i, s_{i-1}) Pr(s_i | x_i, s_{i-1}) \quad (140)$$

i.e., the channel output and the next state are independent when conditioned on knowing the channel input and most recent state. When the present output

Error Correction I

of the channel depends in some manner on previous inputs as well as the present input, then *intersymbol interference* is said to be present. If the state s_{i-1} depends statistically on the previous inputs in any way, intersymbol interference is present. The channel is *free of intersymbol interference* if the statistical description $Pr(s_i | x_i, s_{i-1})$ of the homogeneous Markov channel-state sequence does not depend on x_i, i.e.,

$$Pr(s_i | x_i, s_{i-1}) = Pr(s_i | s_{i-1}) \quad \text{for all } s \in S, s_0 \in S, x \in X \quad (141)$$

The following two examples illustrate the preceding channel types.

Example 5.7 (Pure Intersymbol Interference). Consider a channel having

$$X = Y, S = X^m$$

whose operation can be described in block-diagram form, as shown in Figure 5.6.

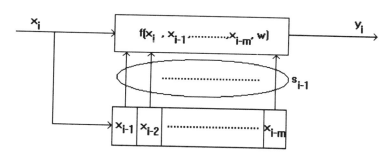

Figure 5.6. m-State shift register memory.

Here, the most recent state s_{i-1} is simply the contents of the shift register

$$s_{i-1} = (x_{i-1}, \ldots, x_{i-m})$$

and

$$Pr(s_i | x_i, s_{i-1}) = \begin{cases} 1 & \text{if } s = (x_i, x_{i-1}, \ldots, x_{i-m+1}) \\ 0 & \text{otherwise} \end{cases}$$

The function $f(x_i, x_{i-1}, \ldots, x_{i-m}, \omega)$ indicates that the ith channel output symbol is a random function of the present channel input symbol x_i and the previous m channel inputs $x_{i-1}, x_{i-2}, \ldots, x_{i-m}$. The variable ω, which denotes a point in a probabilistic sample space, has been inserted to emphasize that the function can be random.

Example 5.8 (Pure Fading Channel). A pure fading channel does not have intersymbol interference, and the channel-state Markov chain is stationary and irreducible. The second requirement prevents anomalous channel behavior, such as permanent breakdowns, *etc.* The block diagram for this type of channel is shown in Figure 5.7a.

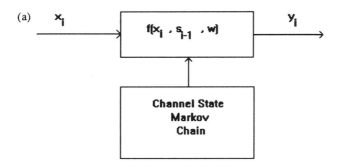

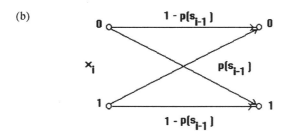

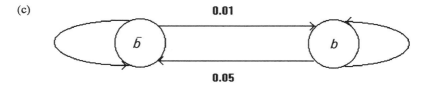

Figure 5.7. Pure fading channel examples.

Error Correction I

As a special case, consider the *binary symmetric fading* channel shown in Figure 5.7b, where $S = \{b, \bar{b}\}$, b indicates burst error state, and $\bar{b}$ indicates random error state. The state diagram for the channel-state Markov chain is shown in Figure 5.7c. The BSC error rate p is the following function of the channel state:

$$p(b) = 0.4$$

$$p(\bar{b}) = 0.01$$

Equivalently, in the burst error state, the probability of error when transmitting a symbol is very high; in the random error state, the probability of error is quite low. Because of the transition probabilities chosen in the channel-state diagram, the next state is quite likely to be the same as the present state, therefore, the states tend to persist. This means that transmission errors tend to occur in bunches (or *bursts*) corresponding to when the channel is in state b.

The class of all finite-state channels is so large and the general description so complicated that no standard coding techniques have been developed that are universally applicable. However, the following coding techniques are worth mentioning.

To equalize pure intersymbol interference channels: Assuming that the channel has finite memory size m, a simple automated method for estimating the input sequence from the output sequence can be developed. One *equalization* method is based on Viterbi's decoding algorithm for convolutional codes. Further coding may be required to reduce errors not eliminated in the equalization process.

To correct burst error for pure fading channels:

1. Use pure error detection with retransmission when errors occur, which is a viable technique in many situations (see Section 5.14).

2. Use *interleaving*, a standard technique when errors tend to occur in bursts (see Figure 5.8). This involves writing B code words from a t random error-correcting code with words of length n as *rows* of a $B \times n$ matrix. The matrix is then read out in columns for transmission. Hence, symbols from the same code word are spaced B symbols apart in the transmitted symbol stream. If the channel is in the burst state for at most tB symbols and no errors are made when it is not in the burst state, then each code word contains at most t errors. After the receiver reverses the interleaving operation, these errors are correctable by the decoder.

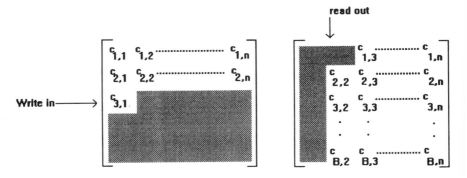

(a) Interleaving Operation

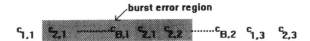

(b) Transmission through Fading Channel

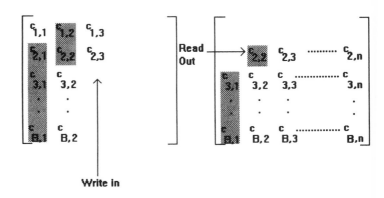

(c) Interleaving Inversion

Figure 5.8. Burst spreading by interleaving.

Error Correction I

3. Use the *supersymbol* concept by viewing the channel input alphabet **X** and the alphabet in which the code is written as separate entities for the purposes of coding. The code alphabet is taken to be the set $\mathbf{X}^B$ of all B-tuples over **X**. A t random error-correcting code over the alphabet $\mathbf{X}^B$ will, among other things, correct t consecutive errors in symbols from $\mathbf{X}^B$, which may be caused by the channel being in the burst error mode for as few as $B(t-1) + 1$ symbols. Hence, any received sequence containing a "bunch" of errors spanning at most $B(t-1) + 1$ symbols can be corrected by this coding technique. The Reed–Solomon codes (see Chapter 7) may be useful as a supersymbol code.

The reader should realize at this point that a coding problem approach is critically dependent on a relatively complete knowledge of the statistical behavior of the channel. Or, adopting Agent 00111's motto, "Be ye admonished that the first commandment of good code design is *know thy channel!*"

Exercises

1. If a received n-tuple contains errors, the first in position m and the last in position $m + b - 1$, and possibly some in between these extremes, the n-tuple is said to contain a *burst error* of length b.

   (a) In an additive error channel with input alphabet **X**, what is the number $N(n, b)$ of distinct burst errors of length b that can occur in an n-tuple?

   (b) Develop a sphere-packing upper bound on the number $|C|$ of words in a code **C**, selected from $\mathbf{X}^n$, that can correct all burst errors of length b or less.

   (c) Develop a lower bound on the number $|C|$ of words in a code **C**, **C** $\subseteq \mathbf{X}^n$, that can detect all burst errors of length b or less.

   (d) Repeat (c) under the stronger constraint that **C** must be able to correct all bursts of length b or less.

2. The interleaving and supersymbol concepts are methods for adapting codes with random-error-correcting capability to fading channels. Using the notion of a burst error in Problem 1 and the Gilbert bound for random-error-correcting codes (see Section 5.8), determine lower bounds on the

size of a code **C**, $C \subseteq X^n$, that can correct burst errors of length b or less when **C** is constructed by

(a) The interleaving method.
(b) The supersymbol method.

3. Design a binary burst-error-correcting code **C** with words of length 70 that can correct error bursts spanning up to ten symbols in the code. Try to maximize dictionary size in your design.

4. Repeat Problem 3 but require only burst-error-detection capability in the code.

5.14. Pure Error Detection

There was a time when Agent 00111 thought all of his communication problems could be solved by error-correcting codes. The realization that this was not true came to him one cold wintry day on a mountain top where he was communicating by semaphore code with a cohort several miles away. The channel alphabet consisted of 32 possible flag-pair positions. For security purposes, the transmitted sequence was garbled so that errors in reading the semaphore could easily be undetected. To alleviate this problem, Agent 00111 employed a double error-correcting code of word length 15. After receiving each 15-tuple, he would sit down and go through the long and tedious process of decoding and correcting his mistakes. It took him 1 minute to determine if the received 15-tuple was a code word and 10 minutes to correct it if it was not. Fifty percent of the time, Agent 00111 read the semaphore without error. The 20 code words he received took him, on average,

$$20 \times 1 \text{ min} + 10 \times 10 \text{ min} = 2 \text{ hr}$$

to decode. Later, as he thawed out beside a roaring fire, he thought, "I'm going to get frost bite if I have to spend 2 hours decoding outside again. Tomorrow I'll just ask for a retransmit instead of trying to correct errors."

The next day, he received another 20 15-tuples; ten were received correctly the first time, four were repeated once, two were repeated twice, and four were repeated three times. Since each retransmit request and subsequent retransmission took about 1 minute plus an extra minute for decoding, his total retransmission and decoding time was

Error Correction I

$$20 \times 1 \text{ min} + 10 \times 2 \text{ min} + 6 \times 2 \text{ min} + 4 \times 2 \text{ min} = 1 \text{ hr}$$

Certainly, there are many situations where the decoder has only to verify that received data are correct. The performance of a given code is easily determined from three quantities:

P_C = probability of correct code-word transmission

P_D = probability of a detectable incorrect code-word transmission

P_E = probability of a nondetectable incorrect code-word transmission

Since a transmission error is detectable when the received n-tuple is not a code word, the only nondetectable transmission errors occur when the channel converts the transmitted code word into a *different* received code word.

To evaluate the quantities P_C, P_D, and P_E when the communication channel is assumed to be stationary and memoryless with all errors equally likely, create the following channel matrix for input and output alphabet **X**:

$$\begin{bmatrix} 1-p & q & \cdots & q \\ q & 1-p & \cdots & q \\ \vdots & \vdots & & \vdots \\ q & q & \cdots & 1-p \end{bmatrix}$$

where p is the probability of a channel symbol transmission error and

$$q = \frac{p}{|\mathbf{X}| - 1} \tag{142}$$

This channel is usually referred to as an $|\mathbf{X}|$-ary symmetric channel with error rate p. It follows immediately that

$$P_C = (1 - p)^n \tag{143}$$

where n is the code-word length. Error probability is also computed as

$$P_E = \sum_{c \in \mathbf{C}} Pr(c \text{ is transmitted}) \sum_{\substack{c' \in \mathbf{C} \\ c' \neq c}} Pr(c' \text{ is received} \mid c \text{ is transmitted})$$

$$= \sum_{c \in \mathbf{C}} Pr(c \text{ is transmitted}) \sum_{\substack{c' \in \mathbf{C} \\ c' \neq c}} [(1-p)^{n-d(c',c)}][p^{d(c',c)}] \tag{144}$$

where d is the Hamming distance metric. Assuming that all code words are equally likely to be transmitted, we can write

$$P_E = \left[-1 + \frac{1}{|C|} \sum_{c \in C} \sum_{c' \in C} \left(\frac{q}{1-p} \right)^{d(c,c')} \right] (1-p)^n \quad (145)$$

Finally

$$P_C + P_D + P_E = 1 \quad (146)$$

implies that

$$P_D = 1 - \frac{(1-p)^n}{|C|} \sum_{c \in C} \sum_{c' \in C} \left(\frac{q}{1-p} \right)^{d(c,c')} \quad (147)$$

The remaining problem is to find a simple method for evaluating double sums in Equations 145 and 147.

Example 5.9. Consider the ternary code containing the words shown in Table 5.5

Table 5.5. A Ternary-Code Example

000, 111, 222
102, 021, 210
012, 201, 120

Code words are transmitted over a ternary symmetric channel with error rate $1/5$. As can be verified by the reader, for each choice of c, there are

- Two choices of c' for which $d(c, c') = 3$
- Six choices of c' for which $d(c, c') = 2$

$$P_C = (1-p)^3 = 0.512$$
$$P_D = 1 - (1-p)^3 - 2q^3 - 6(1-p)q^2 = 0.438$$
$$P_E = 2q^3 + 6(1-p)q^2 = 0.050$$

Error Correction I

Notice that the received n-tuple is erroneous almost 50% of the time, but as an undetected error it occurs only 5% of the time. It follows that the probabilities P_C, P_D, and P_E can be found precisely if the Hamming distance distribution can be enumerated; i.e., if the results in Section 5.11 are applicable.

Exercises

1. Let **C** be a maximal equidistant (unshortened) code over an alphabet **X**, with $|\mathbf{C}| = m|\mathbf{X}|$.

   (a) For fixed m and $|\mathbf{X}|$, determine the behavior of P_C, P_D, and P_E as a function of n.

   (b) Sketch your results in (a) as a function of n for $|\mathbf{X}| = 2$, $m = 1, 2, 3$, and $p = 1/5$.

   (c) Do any of the probabilities P_C, P_D, and P_E asymptotically decrease exponentially with n? If so, determine the rate r of decrease:

   $$r = \frac{1}{n} \log P(n)$$

2. Consider an error-detection scheme for a memoryless channel for which the quantities P_C, P_D, and P_E are known. When an error is detected, retransmission of the code word is requested.

   (a) What is the probability that a code word must be transmitted exactly j times before the decoder accepts the transmission?

   (b) What is the expected value of j in (a)?

   (c) What is the probability of eventually decoding the transmission correctly?

   (d) In Problem 2 and Section 5.14, we have assumed that there is no cost associated with repeating a request and no possibility that the request will be misinterpreted. Equivalently, we have assumed that a perfect feedback channel is available. Repeat Problem 2 assuming that a BSC with error rate ϵ is available as a feedback channel and after each code-word reception, 0 is transmitted over the feedback channel if the reception is not acceptable. Consider the correct transmission of a single code word to be the objective of this exercise and carefully state the algorithm agreed on by transmitter and receiver before repeating (a), (b), and (c).

5.15. Notes

The Hamming distance concept is the standard technique for studying error-correcting codes. The error-correcting codes were developed by R. W. Hamming, who worked with Shannon at AT&T Bell Telephone Laboratories in the 1940s. Hamming's well-known paper (1950) introduces the Hamming distance concept, sphere packing in this metric, the perfect single-error-correcting codes, and a parity check to improve distance properties. The only major paper preceding Hamming's on the subject of error correction is that of Golay (1949), which describes the perfect binary triple-error-correcting code. The study of perfect codes continues in the work of van Lint (1971) and Tietäväinen (1973). Golomb and Welch (1968) contains basic work on perfect codes in the Lee metric.

Since Hamming's bound proved to be unattainable for most d_{min} and n, a succession of upper bounds on attainable dictionary sizes followed: Plotkin's low-rate bound (1960), Elias's bound (unpublished) discovered independently by Bassalygo (1965), the McEliece–Rumsey–Welch bound (1974), and the McEliece–Rodemich–Rumsey–Welch bound (1977). (The last two bounds are often referred to as the JPL bounds.) Johnson (1971) made minor improvements in these bounds for finite values of n, d_{min} and tabulated the tightest known upper bounds for d_{min} up to 21 and values of n up to and exceeding 50.

Gilbert (1952) gave a basic lower bound to dictionary size, which in its asymptotic form has not been exceeded to this date. Varshamov (1957) refined the finite version of the Gilbert bound. Tables of the largest known binary code at the time it appeared for a given n and d_{min} are given in an updated version of a paper by Sloane (1972) appearing in Berlekamp's survey (1974) of famous papers in coding theory. An updated table is also given in Verhoeff (1987).

MacWilliams (1963) first developed the method in Section 5.12 for evaluating distances between the words of a linear code. Zierler (unpublished) first demonstrated the approach used in Section 5.12, which applies more generally to all fixed-word-length codes. Pless (1963) developed the equivalent power moment identities that eventually served as the basis for the WMR bound. Delsarte (1973) generalized the underlying combinatorial mathematics into the interesting and powerful subject of *Association Schemes*. Kabatiansky and Levenshtein (1978) extended the WMR work to sphere packings and beyond. Conway and Sloane (1988) present an interesting survey of this work.

There are many situations where forward error correction (FEC), i.e., error correction on a one-way communication channel, is not competitive with pure error detection in an automatic-repeat-request system (ARQ). For a discussion of the merits of these types of systems for telephone data communication, see Burton and Sullivan (1972). A discussion of the Viterbi algorithm (see Chap. 7) and its application to intersymbol interference problems is given in Forney (1972) and (1973). An excellent collection of papers discussing problems encountered in matching codes to real-world channels is presented in a special issue of *IEEE Transactions on Communication Technology*, edited by Stiffler (1971).

References

L. A. Bassalygo. 1965. "New Upper Bounds for Error-Correcting Codes." *Probl. Peredachi. Inf.* **1**: 41–45.

E. R. Berlekamp, ed. 1974. *Key Papers in the Development of Coding Theory.* IEEE Press, New York.

H. O. Burton and D. D. Sullivan, 1972. "Error and Error Control." *Proc. IEEE* **60**: 1293–1301.

J. H. Conway and N. J. A. Sloane. 1988. *Sphere Packings, Lattices and Groups.* Springer-Verlag, New York.

P. Delsarte. 1973. *An Algebraic Approach to the Association Schemes of Coding Theory.* Philips Research Reports Supplements No. 10, Eindhoven.

W. Feller. 1950. An Introduction to Probability Theory and its Applications Vol. 1: Wiley, New York.

G. D. Forney, Jr. 1972. "Maximum-Likelihood Sequence Estimation of Digital Sequences in the Presence of Intersymbol Interference." *IEEE Trans. Inform. Theory* **18**: 363–78.

———. 1973. "The Viterbi Algorithm." *Proc. IEEE* **61**: 268–78.

E. N. Gilbert. 1952. "A Comparison of Signalling Alphabets." *Bell Sys. Tech. J.* **31**: 504–22.

M. J. E. Golay. 1949. "Notes on Digital Coding." *Proc. IRE* **37**: 657.

S. W. Golomb and L. R. Welch. 1968. "Algebraic Coding in the Lee Metric." In *Error-Correcting Codes*, edited by H. B. Mann. Wiley, New York.

R. W. Hamming. 1950. "Error-Detecting and Error-Correcting Codes." *Bell Sys. Tech. J.* Vol. 29: 147–60.

S. M. Johnson. 1971. "On Upper Bounds for Unrestricted Binary-Error-Correcting Codes." *IEEE Trans. Inform. Theory* **17**: 466–78.

G. A. Kabatiansky and V. I. Levenshtein. 1978. "Bounds for Packings on a Sphere and in Space." *PPI* **14**: pp. 3–25.

J. H. van Lint. 1971. *Coding Theory.* Lecture Notes in Mathematics 201, Springer-Verlag, New York.

F. J. MacWilliams. 1963. "A Theorem on the Distribution of Weights in a Systematic Code." *Bell Sys. Tech. J.* **42**: 79–84.

F. J. MacWilliams and N. J. A. Sloane. 1977. *The Theory of Error-Correcting Codes.* North-Holland, Amsterdam.

R. J. McEliece, H. Rumsey, Jr., and L. R. Welch. 1974. "A Low-Rate Improvement on the Elias Bound." *IEEE Trans. Inform. Theory* **20**: 676–78.

R. J. McEliece, E. R. Rodemich, H. Rumsey, Jr., and L. R. Welch. 1977. "New Upper Bounds on the Rate of a Code via the Delsarte-MacWilliams Inequalities." *IEEE Trans. Inform. Theory* **23**: 157–66.

V. Pless. 1963. "Power Moment Identities on the Weight Distributions in Error-Correcting Codes." *Inf. Contr.* **6**: 147–62.

M. Plotkin. 1960. "Binary Codes with Specified Minimum Distance." *IRE Trans. Inform. Theory* **6**: 445–50.

N. J. A. Sloane. 1972. "A Survey of Constructive Coding Theory and a Table of Binary Codes of Highest Known Rate." *Discrete Math.* **3**: 265–94.

J. J. Stiffler, ed. 1971. "Special Issue on Error-Correcting Codes." *IEEE Trans. Commun. Technol.* **19,** no. 5, part II.

A. Tietäväinen. 1973. "On the Nonexistence of Perfect Codes over Finite Fields." *SIAM J. Appl. Math.* **24**: 88–96.

R. R. Varshamov. 1957. "Estimate of the Number of Signals in Error-Correcting Codes." *Dokl. Akad. Nauk SSSR* **117**: 739–41.

T. Verhoeff. 1987. "An Updated Table of Minimum Distance Bounds for Binary Linear Codes." *IEEE Trans. on Information Theory* **33**: 665–680.

6

Error Correction II: The Information-Theoretic Viewpoint

6.1. Disruption in the Channel

The director looked worried when Agent 00111 walked into his office. "We have just discovered an enemy agent in the line of communication between you and Agent 11000. He's been changing about 30% of the binary code symbols that are passed through him. I'm afraid that much of the work that you and Agent 11000 have been doing has gone down the tubes!"

"Not so!" replied Agent 00111. "We've known about the 'disruption' in our channel. After all, we do live by the first commandment for good code design!"

Though somewhat relieved, the director was still worried. "But 30% errors inserted at random! To obtain statistically reliable communication, you need long code words. And the Plotkin bound says that d_{min} is at most $n/2$ for nonzero communication rates. You just cannot *guarantee* correcting all those errors!" The director was amazed at himself. He had not realized until now how many of Agent 00111's ideas had become second nature to him.

"You are absolutely correct. 100% reliable communication through a BSC with error rate 0.3 is out of the question at nonzero communication rates." A familiar twinkle lit Agent 00111's eyes as he continued. "But would you settle for 99.9999% reliable communication at a rate of 0.1 bits/symbol?"

The director began to smile as the story came out. "It's all quite simple in theory. Except for perfect or near-perfect codes, it takes a very special error pattern of weight

$$\left\lfloor \frac{d_{min} - 1}{2} \right\rfloor$$

to cause an error in communication. It is highly probable that a randomly selected error pattern even with a few more than

$$\left\lfloor \frac{d_{\min} - 1}{2} \right\rfloor$$

errors will not cause trouble in communication. If we do not shy away from attempting to correct more than

$$\left\lfloor \frac{d_{\min} - 1}{2} \right\rfloor$$

errors, significant communication rates are achievable in BSC's, even when $p > 0.25!$"

The conversation continued for some time; we list its major points:

1. Many noisy channels are characterized by a quantity called channel capacity C. Reliable communication at rates R, $R > C$ is impossible.
2. Reliable though not perfect communication at rates $R < C$ is possible.
3. For binary symmetric channels, the capacity C is greater than the asymptotic Elias bound on R, indicating that a number of errors greater than

$$\left\lfloor \frac{d_{\min} - 1}{2} \right\rfloor$$

   can be corrected in most instances to communicate reliably at rates near C.
4. The cost of communicating reliably at rates approaching C is an ever-increasing complexity in encoding and decoding operations.

Most of Chapter 6 is devoted to developing an understanding of these basic information-theoretic results on communication through noisy channels.

6.2. Data-Processing Theorem and Estimation Problems

Our previous study of mutual information involved only two alphabets; however in situations involving noisy channels, there are many alphabets—the source alphabet **M**, the channel input alphabet **X**, the channel output alphabet **Y**, the decoder output alphabet **M***, and possibly others. For the

Error Correction II

moment, we bypass this complex situation and think about statistical and mutual information relations between three finite alphabets $\mathbf{X}$, $\mathbf{Y}$, $\mathbf{Z}$.

The early work in Section 1.3 indicates that a quantity called the *mutual information between* $\mathbf{X}$ *and* $\mathbf{Y}$, *conditioned on the event z occurring*, can be defined as

$$I(\mathbf{X}; \mathbf{Y}|z) = \sum_{x \in \mathbf{X}} \sum_{y \in \mathbf{Y}} Pr(x, y|z) \log\left[\frac{Pr(x, y|z)}{Pr(x|z)Pr(y|z)}\right] \quad (1)$$

by simply conditioning all probabilities on the fact that z, $z \in \mathbf{Z}$, has occurred. Furthermore, it was shown that $I(\mathbf{X}; \mathbf{Y}|z)$ is nonnegative. Averaging over all possible z in $\mathbf{Z}$, we define

$$I(\mathbf{X}; \mathbf{Y}|\mathbf{Z}) \triangleq \sum_{x \in \mathbf{X}} \sum_{y \in \mathbf{Y}} \sum_{z \in \mathbf{Z}} Pr(x, y, z) \log\left[\frac{Pr(x, y|z)}{Pr(x|z)Pr(y|z)}\right] \quad (2)$$

Since $I(\mathbf{X}; \mathbf{Y}|z)$ is nonnegative, it follows that

$$I(\mathbf{X}; \mathbf{Y}|\mathbf{Z}) \geq 0 \quad (3)$$

with equality in Equation 3 if and only if

$$Pr(x|z)Pr(y|z) = Pr(x, y|z) \quad (4)$$

for all choices of $x \in \mathbf{X}$, $y \in \mathbf{Y}$, $z \in \mathbf{Z}$, for which $Pr(x, y, z)$ is nonzero. That is, the mutual information $I(\mathbf{X}; \mathbf{Y}|\mathbf{Z})$ between the event alphabets $\mathbf{X}$ and $\mathbf{Y}$ is conditionally independent when conditioned on events in $\mathbf{Z}$.

Using the definitions of conditional entropy and mutual information derived in Chapter 1, the reader should also be able to verify mathematically the following relations:

$$I(\mathbf{X}; \mathbf{Y}|\mathbf{Z}) = H(\mathbf{X}|\mathbf{Z}) - H(\mathbf{X}|\mathbf{Y} \times \mathbf{Z}) \quad (5)$$

$$I(\mathbf{X}; \mathbf{Y} \times \mathbf{Z}) = I(\mathbf{X}; \mathbf{Z}) + I(\mathbf{X}; \mathbf{Y}|\mathbf{Z}) \quad (6)$$

Equation 5 yields the usual interpretation of mutual information between $\mathbf{X}$ and $\mathbf{Y}$ as the difference in uncertainties about $\mathbf{X}$ before and after observing $\mathbf{Y}$, all conditioned on $\mathbf{Z}$ having been observed. The right side of Equation 6 is interpretable as the reduction $I(\mathbf{X}; \mathbf{Z})$ in uncertainty about $\mathbf{X}$ when observing $\mathbf{Z}$ and the additional reduction in uncertainty about $\mathbf{X}$ when $\mathbf{Y}$ is observed, given that an event in $\mathbf{Z}$ has already been observed.

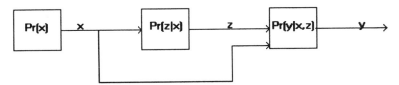

Figure 6.1. A model for the probability of a triplet x, y, z.

The probability of a triplet x, y, z from $\mathbf{X} \times \mathbf{Y} \times \mathbf{Z}$ can be written in many ways, e.g.,

$$Pr(x, y, z) = Pr(x)Pr(z|x)Pr(y|x, z) \qquad (7)$$

If the probabilities in the product are viewed as descriptions of the outcome of testing a black box when the given variables are the inputs, then the right side of Equation 7 describes an interconnection of black boxes of the form shown in Figure 6.1. Of course, Figure 6.1 describes a fictitious model, since $Pr(x, y, z)$ can be decomposed into a product of conditional probabilities in several ways.

What is more interesting is the fact that a physical connection of black boxes whose outputs depend statistically on the external world only through their inputs can imply certain statistical relationships. As an example, consider the diagram shown in Figure 6.2. The statement that y depends statistically on the external world only through z is equivalent to stating that

$$Pr(y|x, z) = Pr(y|z) \qquad (8)$$

Furthermore, when such a serial connection exists, it follows that

$$Pr(x, y|z) = Pr(x|z)Pr(y|x, z) = Pr(x|z)Pr(y|z) \qquad (9)$$

for all possible choices of the triplet x, y, z, and therefore, $\mathbf{X}$ and $\mathbf{Y}$ are conditionally independent alphabets given any element of $\mathbf{Z}$. This implies that $I(\mathbf{X}; \mathbf{Y}|\mathbf{Z})$ is zero, and hence, from Equation 6

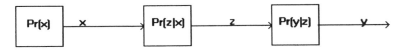

Figure 6.2. A model for the data-processing theorem.

Error Correction II

$$I(X; Z) = I(X; Y \times Z) = I(X; Y) + I(X; Z|Y) \geq I(X; Y) \quad (10)$$

Equation 10 is the mathematical description of Theorem 6.1.

Theorem 6.1 (the Data Processing Theorem). Let **X** and **Z** be any two alphabets with mutual information $I(X; Z)$. If **Y** is an alphabet generated by an operation on **Z** alone, then

$$I(X; Y) \leq I(X; Z)$$

A less formal description of the data processing theorem is that processing data $z \in Z$ about a source **X** to give processed data (a statistic) $y \in Y$ cannot increase the inherent information about **X**.

The condition for equality to hold in the data-processing theorem is that

$$I(X; Z|Y) = 0 \quad (11)$$

or equivalently that there is no reduction in the uncertainty about a source **X** when informed of the raw data z, given that the processed data y is known. In this special case, the mapping from **Z** to **Y** is "information lossless," and y (which is a function of the raw data z) is called a *sufficient statistic for* **X**.

A similar data-processing situation occurs in estimation problems, which may be modeled in the form shown in Figure 6.3. Here, solid lines indicate relations that may physically exist, namely, a source variable x to be estimated, an observation y, and an estimate $\hat{x}$ of x based solely on the observation y. Dashed lines indicate connections that exist only conceptually. We assume that:

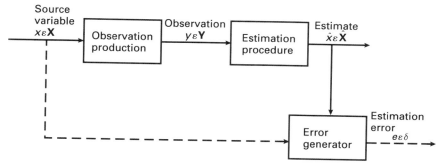

Figure 6.3. A model for estimation problems.

1. $\hat{x}$ is a deterministic function of y.
2. For each value of $\hat{x}$, the mapping from x to e in the error generator is one-to-one.

These assumptions together imply that when y is known, then $\hat{x}$ is known, and hence, x may be determined from e or e from x; thus,

$$Pr(x|y, e) = 0 \text{ or } 1 \qquad \text{for all } x, y, e$$
$$Pr(e|y, x) = 0 \text{ or } 1 \qquad \text{for all } e, y, x$$

As a result,

$$H(\mathbf{X}|\mathbf{Y} \times \mathbf{E}) = 0 = H(\mathbf{E}|\mathbf{Y} \times \mathbf{X}) \qquad (12)$$

Equation 5 in conjunction with Equation 12 gives

$$H(\mathbf{X}|\mathbf{Y}) = I(\mathbf{X}; \mathbf{E}|\mathbf{Y}) = H(\mathbf{E}|\mathbf{Y}) \qquad (13)$$

Using the fact that in general $H(\mathbf{E}|\mathbf{Y})$ is a lower bound on $H(\mathbf{E})$ (see Section 1.3), Equation 13 reduces to

$$H(\mathbf{X}|\mathbf{Y}) \leq H(\mathbf{E}) \qquad (14)$$

That is, regardless of the specifics of the estimation procedure and the error generator (except for Assumptions 1 and 2), the entropy $H(\mathbf{E})$ of the error is lower bounded by $H(\mathbf{X}|\mathbf{Y})$, the uncertainty of the source, given the observation.

If we apply the data-processing theorem to the physical part of the estimation procedure diagram, the result is

$$I(\mathbf{X}; \hat{\mathbf{X}}) \leq I(\mathbf{X}; \mathbf{Y}) \qquad (15)$$

At first, this result seems unexpected, since the object of the estimation procedure is to duplicate x as closely as possible in $\hat{x}$. The $\mathbf{Y}$ alphabet may not look like the $\mathbf{X}$ alphabet, and to the untrained observer, i.e., one unfamiliar with descriptions of the black boxes (observation production and estimation procedure), information about y is useless in estimating x, but $\hat{x}$ is a legitimate estimate in itself. However, information that $\hat{x}$ is an estimate of x and no information about how y is related to either x or $\hat{x}$ is not the basis for the data-processing theorem. The data-processing theorem is based on a probabilistic description of the relation between x and y and between y and $\hat{x}$. For

Error Correction II

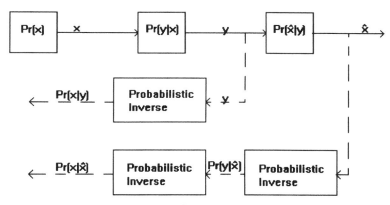

Figure 6.4. A schematic for the data-processing theorem.

an observation y, the *a posteriori* probability of x given y can be computed to provide the most accurate description of the state of knowledge of x; this is shown diagramatically in Figure 6.4.

If we are informed only about $\hat{x}$, then y can be determined only probabilistically, and from this, an *a posteriori* distribution on x can be determined. However, inaccuracies in determining y can lead to only a less accurate probabilistic description of x. This viewpoint is the essence of the data-processing theorem, or as Agent 00111 once said, "Don't throw away your raw data!"

6.3. An Upper Bound on Information Rate for Block-Coding Schemes

Consider the diagram of the communication process shown in Figure 6.5. It is assumed that $\mathbf{X}$, $\mathbf{E}$, and $\hat{\mathbf{X}}$ are identical additive groups so that n-tuple subtraction may be used to determine an estimation error n-tuple. The error indicator outputs a 1 if and only if an error occurs; otherwise, it outputs a zero.

$$d = \begin{cases} 0 & \text{iff } e = 0 \\ 1 & \text{iff } e \neq 0 \end{cases} \quad (16)$$

Hence, $\mathbf{D} = \{0, 1\}$, and we define

probability of decoding error $\triangleq P_E = Pr\{d = 1\}$

probability of correct decoding $= 1 - P_E = Pr\{d = 0\}$ \quad (17)

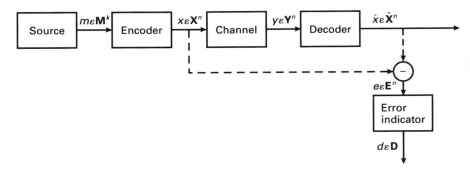

Figure 6.5. A communications process.

The uncertainty about events in **D** is given by

$$H(\mathbf{D}) = H(P_E) = -P_E \log P_E - (1 - P_E) \log(1 - P_E) \qquad (18)$$

Using information-theoretic considerations, we relate the probability of error P_E to source transmission rate and the mutual information between channel inputs and outputs.

The three deterministic mappings in the preceding diagram are

1. The encoder: Since the encoder is a one-to-one mapping from $\mathbf{M}^k$ onto a subset of $\mathbf{X}^n$, it follows

$$H(\mathbf{M}^k) = H(\mathbf{X}^n) \qquad (19)$$

2. The decoder: Since the decoder is deterministic, it follows from Equation 13 in Section 6.2 that

$$H(\mathbf{X}^n | \mathbf{Y}^n) = H(\mathbf{E}^n | \mathbf{Y}^n) \qquad (20)$$

3. The error indicator: Since the error indicator is deterministic, it follows that

$$H(\mathbf{D} | \mathbf{E}^n) = 0 \qquad (21)$$

Considering the mutual information between the channel input and output in the light of Equations 19 and 20 gives

Error Correction II

$$H(\mathbf{M}^k) = H(\mathbf{X}^n) = I(\mathbf{X}^n; \mathbf{Y}^n) + H(\mathbf{X}^n|\mathbf{Y}^n)$$

$$= I(\mathbf{X}^n; \mathbf{Y}^n) + H(\mathbf{E}^n|\mathbf{Y}^n) \quad (22)$$

Therefore, the desired relation between source entropy, channel mutual information, and error probability is available if we relate $H(\mathbf{E}^n|\mathbf{Y}^n)$ to P_E. Property 3 of the error indicator implies that

$$H(\mathbf{D}|\mathbf{E}^n \times \mathbf{Y}^n) = 0 \quad (23)$$

and therefore

$$H(\mathbf{E}^n|\mathbf{Y}^n) = H(\mathbf{D} \times \mathbf{E}^n|\mathbf{Y}^n) = H(\mathbf{D}|\mathbf{Y}^n) + H(\mathbf{E}^n|\mathbf{D} \times \mathbf{Y}^n) \quad (24)$$

The quantity $H(\mathbf{D}|\mathbf{Y}^n)$ is upper bounded by $H(\mathbf{D})$, which is given in terms of P_E in Equation (18). Also

$$H(\mathbf{E}^n|\mathbf{D} \times \mathbf{Y}^n) = Pr\{d=0\}H(\mathbf{E}^n|\mathbf{Y}^n \times \{d=0\})$$

$$+ Pr\{d=1\}H(\mathbf{E}^n|\mathbf{Y}^n \times \{d=1\}) \quad (25)$$

When $d = 0$, then $e = 0$, and there is no uncertainty about $\mathbf{E}^n$. On the other hand, when an error occurs, and y and hence $\hat{x}$ are known. Since there are $|\mathbf{M}|^k - 1$ possible error vectors it follows that

$$H(\mathbf{E}^n|\mathbf{Y}^n \times \{d=1\}) \leq \log(|\mathbf{M}|^k - 1) \quad (26)$$

Using these results in Equation 24 yields

$$H(\mathbf{E}^n|\mathbf{Y}^n) \leq H(P_E) + P_E \log(|\mathbf{M}|^k - 1) \quad (27)$$

and applying Equation 27 to Equation 22 gives the basic result

$$H(\mathbf{M}^k) - I(\mathbf{X}^n; \mathbf{Y}^n) = H(\mathbf{X}^n|\mathbf{Y}^n) \leq H(P_E) + P_E \log(|\mathbf{M}|^k - 1) \quad (28)$$

which is often referred to as *Fano's inequality*. One interpretation of equivocation $H(\mathbf{X}^n|\mathbf{Y}^n)$ in Equation 28 is the decoder's average uncertainty about the transmitted word after observing the received word. This uncertainty can be expressed as a lack of knowledge about (1) whether or not an error was made [i.e., $H(P_E)$] and (2) when an error is made (i.e., P_E of the time)—which of the other $|\mathbf{M}|^k - 1$ signals was actually transmitted. Given an error, the average information for specifying which signal was transmitted is bounded by the information required to specify one of $|\mathbf{M}|^k - 1$ equally likely alternatives, namely, $\log(|\mathbf{M}|^k - 1)$ units of information. Thus, Fano's bound on equivocation has a natural interpretation in information-theoretic terms.

The function $f(P_E)$ of error probability given by

$$f(P_E) = H(P_E) + P_E \log(|\mathbf{M}|^k - 1) \qquad (29)$$

is a monotone increasing function of P_E for P_E in the range

$$\left(0, \frac{|\mathbf{M}|^k - 1}{|\mathbf{M}|^k}\right) \quad \text{where}$$

$$f(0) = 0 \quad \text{and} \qquad (30)$$

$$f\left(\frac{|\mathbf{M}|^k - 1}{|\mathbf{M}|^k}\right) = \log |\mathbf{M}|^k \qquad (31)$$

This function is sketched in Figure 6.6 for several values of $N = |\mathbf{M}|^k$.

When $H(\mathbf{M}^k)$ is less than $I(\mathbf{X}^n; \mathbf{Y}^n)$, nothing can be inferred from Fano's inequality in Equation 28 about the value of P_E. However, when $H(\mathbf{M}^k)$ is greater than $I(\mathbf{X}^n; \mathbf{Y}^n)$, then the left side of Equation 28 is positive, i.e.,

$$0 < H(\mathbf{X}^n | \mathbf{Y}^n) \leq f(P_E) \qquad (32)$$

The monotone increasing property of $f(P_E)$ in this range implies that

$$f^{-1}[H(\mathbf{X}^n | \mathbf{Y}^n)] \leq P_E \qquad (33)$$

where $f^{-1}(\ )$ is the inverse of the transformation $f(\ )$. *Thus, the probability of error is bounded away from zero when the decoder has any residual uncertainty $H(\mathbf{X}^n | \mathbf{Y}^n)$ about the transmitted n-tuple, given the channel output n-tuple.*

We now ask, "If the best possible coding scheme is employed, what is a lower bound on the probability of error?" Of course the bound given in Equation 33 applies to any coding scheme, so the answer to our question is

$$f^{-1}[\min_{\{\text{codes}\}} H(\mathbf{X}^n | \mathbf{Y}^n)] \leq P_{E_{\min}} \qquad (34)$$

Now, using the alternative expression for $H(\mathbf{X}^n | \mathbf{Y}^n)$ given in Equation 28 and noting that $I(\mathbf{X}^n; \mathbf{Y}^n)$ depends on the chosen code only through the probability distribution on $\mathbf{X}^n$ that it imposes, gives

Error Correction II

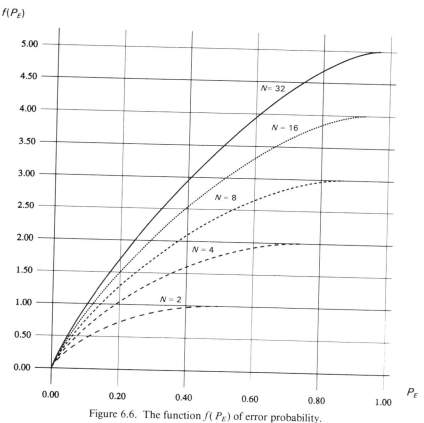

Figure 6.6. The function $f(P_E)$ of error probability.

$$\min_{\{codes\}} H(\mathbf{X}^n | \mathbf{Y}^n) = \min_{\{codes\}} [H(\mathbf{M}^k) - I(\mathbf{X}^n; \mathbf{Y}^n)]$$

$$\geq H(\mathbf{M}^k) - \max_{\{Pr(x)\}} I(\mathbf{X}^n; \mathbf{Y}^n) \quad (35)$$

The inequality in Equation 35 comes from the fact that the set of all probability distributions on $\mathbf{X}^n$ may be larger than those imposed by deterministic encoders operating on the source $\mathbf{M}$. Combining these results yields the conclusion that (1) there is always a residual uncertainty $H(\mathbf{X}^n | \mathbf{Y}^n)$, (2) P_E is bounded away from zero, and (3) *reliable communication cannot be established by a block code of* $\mathbf{M}^k$ *into* $\mathbf{X}^n$ whenever

$$H(\mathbf{M}^k) \geq \max_{\{Pr(x)\}} I(\mathbf{X}^n; \mathbf{Y}^n) \quad (36)$$

i.e., when the source entropy $H(\mathbf{M}^k)$ is greater than the largest possible channel mutual information $I(\mathbf{X}^n; \mathbf{Y}^n)$. It is worth noting that there are no constraints on the source and channel to which this applies; e.g., the source can be Markov, and the channel can have memory.

We assume that the channel symbols cost τ_c units of time and source messages cost τ_s units of time; for continuous operation,

$$k\tau_s = n\tau_c \tag{37}$$

Then for a given channel, another variable n (and simultaneously k by Equation 37) can also be varied in an attempt to minimize the equivocation $H(\mathbf{X}^n|\mathbf{Y}^n)$ and find a value of n where P_E is not bounded away from zero. However, it is easily demonstrated (Problem 1) that for memoryless channels

$$\max_{\{Pr(x)\}} I(\mathbf{X}^n; \mathbf{Y}^n) = n \max_{\{Pr(x)\}} I(\mathbf{X}; \mathbf{Y}) \tag{38}$$

and therefore

$$\min_{\{n,\text{codes}\}} H(\mathbf{Y}^n|\mathbf{X}^n) \geq \min_{\{n\}}[H(\mathbf{M}^k) - n \max_{\{Pr(x)\}} I(\mathbf{X}; \mathbf{Y})] \tag{39}$$

Furthermore, since

$$\frac{1}{k} H(\mathbf{M}^k) \geq H(\mathbf{M}|\mathbf{M}^\infty) \tag{40}$$

for any source (see Chap. 1), Equation 39 can be rewritten as

$$\min_{\{n,\text{codes}\}} H(\mathbf{Y}^n|\mathbf{X}^n) \geq \min_{\{n\}} n\tau_c\left[\frac{H(\mathbf{M}|\mathbf{M}^\infty)}{\tau_s} - \frac{1}{\tau_c} \max_{\{Pr(x)\}} I(\mathbf{X}; \mathbf{Y})\right] \tag{41}$$

which yields the specialized result that for a memoryless channel and any block-coding scheme, there is a residual uncertainty $H(\mathbf{X}^n|\mathbf{Y}^n)$ at the decoder about the transmitted word, and P_E is bounded away from zero when

$$\frac{1}{\tau_s} H(\mathbf{M}|\mathbf{M}^\infty) > \frac{1}{\tau_c} \max_{\{Pr(x)\}} I(\mathbf{X}; \mathbf{Y}) \tag{42}$$

regardless of the block length n. Equivalently, reliable communication through a memoryless channel is impossible when the source entropy rate (or *infor-*

Error Correction II

mation rate) in bits/second is greater than the *channel capacity* C in bits/second, where

$$C = \frac{1}{T_c} \max_{\{Pr(x)\}} I(\mathbf{X}; \mathbf{Y}) \qquad (43)$$

This definition of channel capacity for memoryless channels is somewhat premature. In Section 6.6, we verify Shannon's amazing result that reliable communication is possible when the source information rate is less than channel capacity. This result will complete the justification for calling C the *capacity* of a memoryless channel.

Example 6.1. A binary symmetric memoryless channel that accepts ten symbols per second has error rate 0.1. If we let

$$Pr\{x = 0\} = q, \qquad Pr\{x = 1\} = 1 - q$$

then the channel output probability distribution is given by

$$r \triangleq Pr\{y = 0\} = 0.9q + 0.1(1 - q)$$
$$(1 - r) \triangleq Pr\{y = 1\} = 0.1q + 0.9(1 - q)$$

and the mutual information between the input and output alphabets is given by

$$I(\mathbf{X}; \mathbf{Y}) = 0.9q \log\left(\frac{0.9}{r}\right) + 0.1q \log\left(\frac{0.1}{1-r}\right)$$
$$+ 0.1(1-q) \log\left(\frac{0.1}{r}\right) + 0.9(1-q) \log\left(\frac{0.9}{1-r}\right)$$

Recalling that r is a function of q, the preceding expression can be differentiated with respect to q to determine the maximizing value of q or equivalently, the optimal choice of probability distribution; thus,

$$0 = \frac{\partial}{\partial q} I(\mathbf{X}; \mathbf{Y}) = 0.8 \log\left(\frac{1-r}{r}\right)$$

The preceding equation implies that

$$\frac{1-r}{r} = 1$$

or equivalently that

$$r = 1/2 = q$$

Therefore, the capacity of the channel in bits per channel symbol is

$$C = \max_{\{Pr(x)\}} I(\mathbf{X};\mathbf{Y}) = 1 - H(0.1) \doteq 0.6749 \text{ bits/channel symbol}$$

or since the channel accepts ten symbols per second,

$$C = 10[1 - H(0.1)] \doteq 6.749 \text{ bits/sec}$$

The results of Section 6.3 indicate that it is impossible to communicate any source with entropy greater than 6.749 bits/sec over the channel in this example.

Exercises

1. Consider a stationary memoryless channel with input alphabet $\mathbf{X}$ and output alphabet $\mathbf{Y}$.

    (a) Show that $H(\mathbf{Y}^n|\mathbf{X}^n) = \sum_{i=1}^{n} H(\mathbf{Y}_i|\mathbf{X}_i)$.

    (b) Show that

    $$\max_{\{n, Pr(x)\}} \frac{1}{n} I(\mathbf{X}^n; \mathbf{Y}^n) = \max_{\{Pr(x)\}} I(\mathbf{X}; \mathbf{Y})$$

2. A binary erasure channel with erasure probability p accepts one symbol every 2 seconds. What is the largest source information rate for which it might be possible to establish reliable communication through this channel?

6.4. The Chernoff Bound

Large number laws and the concept of typical sequences play a major role in many coding theorems. In Section 6.4, we explore the law of large numbers using a bound that is both analytically tractable and yields powerful results. This bound is called the Chernoff bound.

Let $r(x)$ be a real-valued random variable defined on $\mathbf{X}$ and let $\phi_A(r)$ be a function that indicates the values of r that are at least as large as A, i.e.,

$$\phi_A(r) = \begin{cases} 1 & r \geq A \\ 0 & \text{otherwise} \end{cases} \quad (44)$$

A probability distribution on $x \in \mathbf{X}$ induces a probability distribution on $r(x) \in \mathbf{R}$ (the real line) and on $\phi_A(r) \in \{0, 1\}$, with the result that

$$Pr\{\phi_A(r) = 1\} = Pr\{r(x) \geq A\}$$
$$Pr\{\phi_A(r) = 0\} = Pr\{r(x) < A\} \quad (45)$$

and hence

$$\mathbb{E}\{\phi_A(r)\} = Pr\{r(x) \geq A\} \quad (46)$$

where $\mathbb{E}\{\ \}$ denotes the expected value operator. Notice that the indicator function $\phi_A(r)$ can be dominated by an exponential function

$$\phi_A(r) \leq e^{s(r-A)} \quad \text{for all } r \in \mathbf{R}, s \geq 0 \quad (47)$$

as shown in Figure 6.7.

Taking the expected value of both sides of Equation 47 and applying Equation 46 gives the basic form of the *Chernoff bound*:

$$Pr\{r(x) \geq A\} \leq e^{-sA}\mathbb{E}\{e^{sr(x)}\} \quad s \geq 0 \quad (48)$$

This result can be made as tight as possible by minimizing the right side of Equation 48 with the proper choice of the nonnegative parameter s.

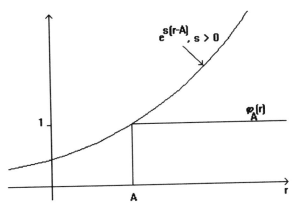

Figure 6.7. The Chernoff bound.

Elementary probability theory tells us that the moment-generating function $g(s)$ of the random variable $r(x)$, given by

$$g(s) \triangleq \mathbb{E}\{e^{sr(x)}\} = \sum_{n=0}^{\infty} \mathbb{E}\{r^n(x)\} \frac{s^n}{n!} \qquad (49)$$

can be used to recover the noncentral moments of $r(x)$:

$$\mathbb{E}\{r^n(x)\} = \left[\frac{d^n}{ds^n} g(s)\right]\bigg|_{s=0} \qquad (50)$$

Furthermore, the semi-invariant generating function $\mu(s)$ of the random variable $r(x)$ can be used to determine the mean and variance of $r(x)$ using the following relations:

$$\mu(s) = \ln g(s) \qquad g(s) = e^{\mu(s)} \qquad (51)$$
$$\mu(0) = 0 \qquad (52)$$
$$\mu'(0) = \mathbb{E}\{r(x)\} \qquad (53)$$
$$\mu''(0) = \text{Var}\{r(x)\} \qquad (54)$$

All of these moment properties involve evaluating derivatives at $s = 0$.

The derivatives of $\mu(s)$ have an interesting interpretation when $s \neq 0$. Notice that

$$\mu(s) = \ln\left[\sum_{x \in \mathbf{X}} Pr(x)e^{sr(x)}\right] \tag{55}$$

$$\mu'(s) = \sum_{x \in \mathbf{X}} r(x)\left[\frac{Pr(x)e^{sr(x)}}{\sum_{x' \in \mathbf{X}} Pr(x')e^{sr(x')}}\right] \tag{56}$$

$$\mu''(s) = \sum_{x \in \mathbf{X}} r^2(x)\left[\frac{Pr(x)e^{sr(x)}}{\sum_{x' \in \mathbf{X}} Pr(x')e^{sr(x')}}\right]$$

$$- \left\{\sum_{x \in \mathbf{X}} r(x)\left[\frac{Pr(x)e^{sr(x)}}{\sum_{x' \in \mathbf{X}} Pr(x')e^{sr(x')}}\right]\right\}^2 \tag{57}$$

If we let the *tilted probability distribution on* $\mathbf{X}$ with respect to the random variable $r(x)$ be defined by

$$T_s(x) = \frac{Pr(x)e^{sr(x)}}{\sum_{x' \in \mathbf{X}} Pr(x')e^{sr(x')}} \tag{58}$$

then

$$\mu'(s) = \mathbb{E}_{T_s}\{r(x)\} \tag{59}$$

$$\mu''(s) = \text{Var}_{T_s}\{r(x)\} \tag{60}$$

where $\mathbb{E}_{T_s}$ and Var_{T_s} denote the expected value operator and variance operator with respect to the tilted probability distribution $T_s(x)$. Since $\mu''(s)$ must be nonnegative, the semi-invariant generating function $\mu(s)$ is convex upward. Using this fact, Equations 52 and 53 allows us to draw rough sketches of $\mu(s)$, as shown in Figure 6.8.

The tightest Chernoff bound takes the form

$$Pr\{r(x) \geq A\} \leq \min_{s>0} e^{-sA+\mu(s)} = \exp\left\{\min_{s>0}\left[-sA + \mu(s)\right]\right\} \tag{61}$$

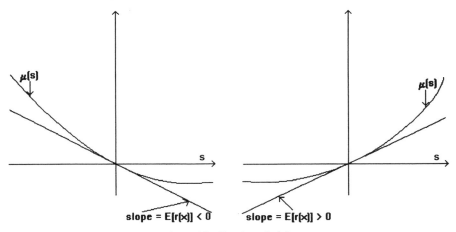

Figure 6.8. Sketches of $\mu(s)$.

Plots of this exponent are shown in Figure 6.9.

The bound in Equation 61 is nontrivial only if the exponent is negative for some positive value of s or equivalently, if A is greater than the mean $E\{r(x)\}$ of the random variable $r(x)$. When this is the case, then the minimizing value of s in Equation 61 satisfies

$$\mu'(s) = A \tag{62}$$

Therefore, if we find a value of s greater than zero for which the $E_{T_s}\{r(x)\}$ is A, this value of s should be used in the Chernoff bound for tightest results.

Laws of large numbers are easily developed from the Chernoff bound. Let $x_1, x_2, \ldots$ be a sequence of independent selections from an alphabet $\mathbf{X}$ and let $r(x)$ be a random variable defined on $\mathbf{X}$. We define the sample mean $\langle r \rangle$ as the average of the n independent random variables $r(x_1), \ldots, r(x_n)$

$$\langle r \rangle = \frac{1}{n} \sum_{i=1}^{n} r(x_i) \tag{63}$$

Of course,

$$E\{\langle r \rangle\} = E\{r\} \tag{64}$$

Since the $r(x_i)$ are independent, the following moment-generating function relation holds:

Error Correction II

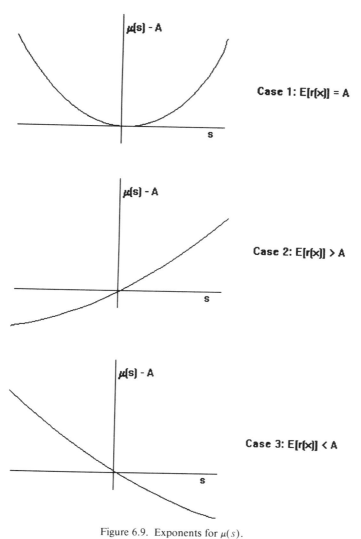

Figure 6.9. Exponents for $\mu(s)$.

$$g_{n\langle r\rangle}(s) = \mathbb{E}\{e^{s\sum_{i=1}^{n} r(x_i)}\} = g_r^n(s) \tag{65}$$

Then applying the Chernoff bound to $n\langle r\rangle$ gives

$$Pr\{\langle r\rangle \geq A\} = Pr\{n\langle r\rangle \geq nA\} \leq e^{-snA}g_{n\langle r\rangle}(s) = e^{n(\mu_r(s)-sA)} \tag{66}$$

The exponent in Equation 66 is negative for some value of s if $\mathbb{E}\{\langle r \rangle\}$ is less than A; i.e., if

$$\mathbb{E}\{r\} + \epsilon = A, \epsilon > 0 \tag{67}$$

Hence, we can state the result that for any $\epsilon > 0$, there exists a $\delta > 0$ such that

$$Pr\{\langle r \rangle - \mathbb{E}\{r\} > \epsilon\} \leq e^{-n\delta} \tag{68}$$

In fact, δ can be determined explicitly from Equation 66 as

$$-\delta = \min_{s>0} \mu_r(s) - As \tag{69}$$

Using the union bound of probability theory, namely,

$$Pr(\mathbf{A} \cup \mathbf{B}) \leq Pr(\mathbf{A}) + Pr(\mathbf{B}) \tag{70}$$

it is easily verified that for any $\epsilon > 0$, there exists a δ' greater than zero such that

$$Pr\{|\langle r \rangle - \mathbb{E}\{r\}| > \epsilon\} \leq 2e^{-n\delta'} \tag{71}$$

Equation 71, a law of large numbers, states that the probability that the sample mean $\langle r \rangle$ deviates from its expected value $\mathbb{E}\{r\}$ by more than a fixed amount ϵ goes to zero exponentially with the number of samples.

Example 6.2. Suppose that $r(x)$ is a binary random variable with

$$Pr\{r(x) = 1\} = p$$
$$Pr\{r(x) = 0\} = 1 - p \triangleq q \tag{72}$$

Then

$$g_r(s) = q + pe^s$$
$$\mu_r(s) = \ln(q + pe^s) \tag{73}$$

Error Correction II

The semi-invariant generating function is shown in Figure 6.10.

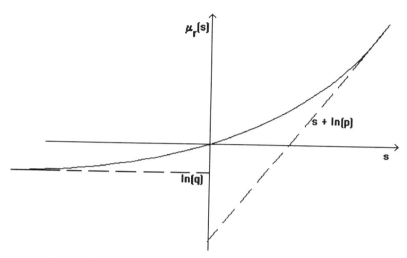

Figure 6.10. The semi-invariant generating function for Example 6.2.

The binomial distribution provides a direct computation of $Pr\{\langle r \rangle \geq A\}$

$$Pr\{\langle r \rangle \geq A\} = \sum_{i=\lceil nA \rceil}^{n} \binom{n}{i} p^i q^{n-i} \qquad (74)$$

On the other hand, the Chernoff bound states that

$$Pr\{\langle r \rangle \geq A\} \leq e^{n[\mu_r(s) - As]} \qquad (75)$$

for all $s > 0$. The optimum choice s_0 of s in the preceding bound is given by the solution to

$$\mu_r'(s_0) = \frac{pe^{s_0}}{q + pe^{s_0}} = A \qquad (76)$$

or

$$s_0 = \ln\left[\frac{A}{(1-A)} \cdot \frac{q}{p}\right] \qquad (77)$$

Then

$$\mu_r(s_0) = \ln\left(\frac{q}{1-A}\right) \tag{78}$$

and after some algebra, the Chernoff bound in Equation 73 reduces to

$$Pr\{\langle r \rangle \geq A\} \leq e^{n\left[\ln\left(\frac{q}{1-A}\right) - A\ln\left(\frac{A}{1-A}\cdot\frac{q}{p}\right)\right]} = e^{n[H(A,1-A)+(1-A)\ln q + A \ln p]} \tag{79}$$

where H is the entropy function in nats. Notice that Equations 74 and 79 combine with $p = 1/2$ to provide the bound

$$2^{-n} \sum_{i=\lceil An \rceil}^{n} \binom{n}{i} \leq e^{n[H(A,1-A)-\ln 2]} \tag{80}$$

or equivalently,

$$\sum_{i=0}^{\lfloor (1-A)n \rfloor} \binom{n}{i} \leq 2^{H(A,1-A)} \tag{81}$$

where in Equation 81 we have converted the units of H into bits. Since in this case the expected value of r is $1/2$, $1 - A$ must be less than $1/2$ for Equation 81 to hold.

Exercises

1. Use the one-sided version of the Chernoff bound and the union bound to develop the law of large numbers in Equation 71 and give an expression for δ'.

2. Use the last term of S,

$$S = \sum_{i=0}^{\alpha n} \binom{n}{i}$$

to develop a lower bound on S involving $H(\alpha, 1-\alpha)$. **Hint:** Use Stirling's approximation:

$$n! \doteq \sqrt{2\pi} n^{n+\frac{1}{2}} \exp\left\{-n + \frac{\theta(n)}{12n}\right\}$$

for n a positive integer and $0 < \theta(n) < 1$.

3. Let $r_j(x), j = 1, 2, \ldots, J$, be J random variables defined on an alphabet **X** and let $Pr(x)$ be a probability distribution defined on a finite alphabet **X**. We say that an n-tuple x from $\mathbf{X}^n$ is ϵ-*typical with respect to the random variable* $r_j(x)$ if

$$|\langle r_j \rangle - \mathbb{E}\{r_j\}| < \epsilon$$

where

$$\langle r_j \rangle = \frac{1}{n} \sum_{i=1}^{n} r_j(x_i)$$

$$\mathbb{E}\{r_j\} = \sum_{x \in \mathbf{X}} Pr(x) r_j(x)$$

Suppose that we now select a sequence x by choosing its elements independently according to the density $Pr(x)$. Show that the probability that x is *simultaneously ϵ-typical with respect to all J random variables* is lower bounded by an expression of the form $1 - Je^{-n\delta}$, where δ is independent of n. Indicate how δ may be determined. **Hint:** Use the union bound.

6.5. Linked Sequences

Consider a channel with finite input alphabet **X** and finite output alphabet **Y** that is completely described by a conditional probability distribution $Pr(y|x)$ for all $y \in \mathbf{Y}^n$, $x \in \mathbf{X}^n$, and all choices of n as shown in Figure 6.11. Assume further that a probability distribution $Pr(x)$ is known that indicates the likelihood that a given x will be transmitted. From this description, it is possible to compute the induced probability distribution on $\mathbf{Y}^n$

$$Pr(y) = \sum_{x \in \mathbf{X}^n} Pr(y|x) Pr(x) \qquad (82)$$

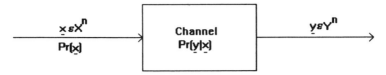

Figure 6.11. A communications channel.

and the *a posteriori* probability distribution on $\mathbf{X}^n$ given an observation of y

$$Pr(x|y) = \frac{Pr(y|x)Pr(x)}{Pr(y)} \qquad (83)$$

On observing y, an observer of the channel output would change his/her statistical description of the transmitted signal from $Pr(x)$ to $Pr(x|y)$. This suggests that candidates for estimates of the transmitted sequence should be vectors x whose ratio $Pr(x|y)/Pr(x)$ is large, or equivalently,

$$i(x, y) \triangleq \frac{1}{n} \ln\left[\frac{Pr(x|y)}{Pr(x)}\right] = \frac{1}{n} \ln\left[\frac{Pr(x, y)}{Pr(x)Pr(y)}\right] \qquad (84)$$

is large. We refer to $i(x, y)$ as the *incremental likelihood of x given y*. It is useful to work with $i(x, y)$ rather than other expressions because for many practical situations, $i(x, y)$ does not go to zero or infinity as n becomes large.

Notice that if x and y are chosen randomly according to the joint probability distribution $Pr(x, y)$, then the expected incremental likelihood is given by

$$\mathbb{E}\{i(x, y)\} = \frac{1}{n} I(\mathbf{X}^n; \mathbf{Y}^n) \qquad (85)$$

Certainly, we might expect that if x is estimated as the transmitted sequence when y is observed, then the incremental likelihood $i(x, y)$ should at least be on the order of its expected value $\mathbb{E}\{i(x, y)\}$. Hence, we define x and y to be ϵ-*linked* if

$$i(x, y) \geq \frac{1}{n} I(\mathbf{X}^n; \mathbf{Y}^n) - \epsilon \qquad (86)$$

The concept of linked sequences can be valuable in understanding the operation of a channel only if there is a high probability that the output sequence

Error Correction II

y will be linked with the input sequence x. We term such a pair of input–channel combinations *information ergodic* if, according to the given statistical descriptions,

$$\lim_{n\to\infty} Pr\{i(x, y) \geq \frac{1}{n} I(\mathbf{X}^n; \mathbf{Y}^n) - \epsilon\} = 1 \qquad (87)$$

when x and y are selected according to their joint probability distribution.

A prime example of an information-ergodic input–channel combination is a stationary memoryless input to a stationary memoryless channel; in this case,

$$Pr(x) = \prod_{i=1}^{n} P(x_i) \qquad (88)$$

where $P(\cdot)$ is a probability distribution on $\mathbf{X}$ and

$$Pr(y|x) = \prod_{i=1}^{n} Q(y_i|x_i) \qquad (89)$$

where $Q(\cdot|x)$ is a probability distribution on $\mathbf{Y}$ conditioned on input x. In this case, the pair of sequences (x, y) can be viewed as a sequence of n independently selected 2-tuples $(x_1, y_1), (x_2, y_2), \ldots, (x_n, y_n)$, each selected according to the probability distribution

$$Pr(x_i, y_i) = Q(y_i|x_i)P(x_i) \qquad (90)$$

This independent pairs property allows us to view the incremental likelihood $i(x, y)$ as a sum of *independent* random variables

$$i(x, y) = -\frac{1}{n} \sum_{j=1}^{n} r(x_j, y_j) = -\langle r \rangle \qquad (91)$$

where the random variable r is given by

$$r(x_j, y_j) = -\ln\left[\frac{Pr(x_j|y_j)}{Pr(x_j)}\right] \qquad (92)$$

The expected value of $r(x_j, y_j)$ with respect to the joint probability distribution $Pr(x_j, y_j)$ is

$$\mathbb{E}\{r(x_j, y_j)\} = -I(\mathbf{X}; \mathbf{Y}) = -\frac{1}{n} I(\mathbf{X}^n; \mathbf{Y}^n) \qquad (93)$$

By inserting an extra minus sign in Equations 91–93, it is now possible to apply the Chernoff bound (as given in Equation 71) directly to prove that the source channel pair under consideration is information ergodic:

$$Pr\{i(x, y) \geq \frac{1}{n} I(\mathbf{X}^n; \mathbf{Y}^n) - \epsilon\} = 1 - Pr\{-\langle r \rangle < I(\mathbf{X}; \mathbf{Y}) - \epsilon\}$$
$$= 1 - Pr\{\langle r \rangle - \mathbb{E}\{r\} > \epsilon\} \geq 1 - e^{-n\delta} \qquad (94)$$

where δ is a positive number that is a function of ϵ, but not of n. Since the right side of Equation 94 goes to 1 as n becomes large, the left side must do the same, and therefore, a stationary memoryless channel paired with a stationary memoryless source is information ergodic.

Now that we have exhibited at least one class of source channel pairs that are information ergodic, we derive the fundamental property of linked sequences. Notice from Equations 84 and 86 that for any pair x and y of linked sequences,

$$\ln \frac{Pr(x, y)}{Pr(x)Pr(y)} \geq n\left[\frac{1}{n} I(\mathbf{X}^n; \mathbf{Y}^n) - \epsilon\right] \qquad (95)$$

or equivalently, that

$$Pr(x)Pr(y) \leq Pr(x, y)e^{-n\left[\frac{1}{n} I(\mathbf{X}^n; \mathbf{Y}^n) - \epsilon\right]} \qquad (96)$$

for all ϵ-linked x and y. Suppose now that an input vector x is selected according to $Pr(x)$ and an output vector y' is selected *independently* according to the marginal density for y, as defined in Equation 82. Then the probability that an x and y selected in this independent manner are ϵ-linked can be bounded using Equation 95

Error Correction II

$$Pr\{\text{independently chosen } x \text{ and } y \text{ are } \epsilon\text{-linked}\}$$

$$= \sum_{(x,y)\in L_\epsilon} Pr(x)Pr(y)$$

$$\leq \sum_{(x,y)\in L_\epsilon} Pr(x,y) e^{-n\left[\frac{1}{n}I(\mathbf{X}^n;\mathbf{Y}^n) - \epsilon\right]} \tag{97}$$

$$\leq e^{-n\left[\frac{1}{n}I(\mathbf{X}^n;\mathbf{Y}^n) - \epsilon\right]}$$

where L_ϵ denotes the set of ϵ-linked pairs. As n goes to infinity, the probability that an independently chosen x and y are linked goes to zero exponentially provided that the asymptotic value of $1/n\ I(\mathbf{X}^n; \mathbf{Y}^n)$ is greater than ϵ.

We summarize the results of Section 6.5 in the following way. Suppose that two input n-tuples x_1 and x_2 are selected independently according to $Pr(x)$. Then x_1 is transmitted over the channel $Pr(y_1|x_1)$ to determine y_1. When the source channel pair is information ergodic,

1. $\lim\limits_{n\to\infty} Pr\{y_1 \text{ is } \epsilon\text{-linked to } x_1\} = 1$

2. $Pr\{y_1 \text{ is } \epsilon\text{-linked to } x_2\} \leq e^{-n\left[\frac{1}{n}I(\mathbf{X}^n;\mathbf{Y}^n) - \epsilon\right]}$

3. $\lim\limits_{n\to\infty} Pr\{y_1 \text{ is } \epsilon\text{-linked to } x_2\} = 0$ provided that

$$\epsilon < \lim_{n\to\infty} \frac{1}{n} I(\mathbf{X}^n; \mathbf{Y}^n)$$

These results are the foundation for the proof of Shannon's coding theorem for noisy channels.

Exercises

1. Give an example of a source that does *not* give an information-ergodic combination when combined with a stationary memoryless channel. Give an example of a channel that does not give an information-ergodic combination when combined with a stationary memoryless source. Prove that the desired properties hold for your examples.

2. Consider the three following random variables defined on $\mathbf{X} \times \mathbf{Y}$:

$$r_1(x, y) = -\ln(Pr(x))$$
$$r_2(x, y) = -\ln(Pr(y))$$
$$r_3(x, y) = \ln\left[\frac{Pr(x, y)}{Pr(x)Pr(y)}\right]$$

Here, $Pr(x, y)$ is a joint distribution on $\mathbf{X} \times \mathbf{Y}$, and $Pr(x)$ and $Pr(y)$ are the corresponding marginal distributions. Let (x, y) represent a sequence of n independently selected symbol pairs from $\mathbf{X} \times \mathbf{Y}$ according to the distribution $Pr(x, y)$.

(a) Determine the expected values of these random variables in information-theoretic terms.

(b) What is the probability that (x, y) is simultaneously ϵ-typical with respect to all three random variables, r_1, r_2, and r_3? (See Problem 3, Section 6.4.)

Only x and y that are simultaneously ϵ-typical with respect to r_1, r_2, and r_3 are considered in the following questions:

(c) If (x, y) is ϵ-typical, are x and y ϵ-linked?

(d) Develop upper and lower bounds on $Pr(x)$, $Pr(y)$, $Pr(x, y)$, $Pr(x|y)$, and $Pr(y|x)$ when (x, y) is ϵ-typical.

(e) Develop upper and lower bounds on

   (i) The number N_x of x that are ϵ-typical.

   (ii) The number N_y of y that are ϵ-typical.

   (iii) The number $N_{x,y}$ of (x, y) that are ϵ-typical.

   (iv) The number $N_{x|y}$ of ϵ-typical x for which (x, y) is ϵ-typical, with y a fixed ϵ-typical sequence.

   (v) The number $N_{y|x}$ of ϵ-typical y for which (x, y) is ϵ-typical, with y a fixed ϵ-typical sequence.

Your answers to (d) and (e) should be in information-theoretic terms.

6.6. Coding Theorem for Noisy Channels

We have just seen that under relatively general conditions, it is possible to pick two input n-tuples at random so that when one is transmitted, the

Error Correction II

channel output n-tuple has a very high probability of not being ϵ-linked to the other randomly selected input n-tuple. If the receiver observing the channel output has knowledge of the randomly selected pair of input sequences, it can then reliably determine which of the two possible input sequences was transmitted by determining which is ϵ-linked to the received output n-tuple. Since the random selection of a code composed of two words works so well, why not add another randomly selected word to the code? And another? In Section 6.6, we see how far we can really go with this *random coding* concept.

A block diagram of a randomly coded communication system is shown in Figure 6.12. The physical line of communication is shown in solid lines, with the message selection process defined by the probability distribution $Pr(m)$ defined on $\mathbf{M}^k$. The message m controls the position i of a switch in the encoder that connects one of $|\mathbf{C}|$ *independent, statistically identical n-tuple sources* to the channel input. Each of the n-tuple sources, called *code word generators*, selects an n-tuple from $\mathbf{X}^n$ according to the probability distribution $P(x)$, with the aggregate of the selections being a randomly selected code. Hence, the random code is a collection of $|\mathbf{C}|$ independent random n-tuples

$$\mathbf{C} = \{x_1, x_2, \ldots, x_{|\mathbf{C}|}\} \tag{98}$$

and the probability of selecting a given code $\mathbf{C}$ is

$$Pr\{\mathbf{C}\} = \prod_{i=1}^{|\mathbf{C}|} P(x_i) \tag{99}$$

Notice that it is possible for two selected code words to be identical, although this is an extremely unlikely occurrence when n is large.

When the switch is in position i, the randomly selected code word is transmitted over the channel, which is described by $Q(y|x)$. Since the codeword generators are statistically identical, the channel input x_i and output y are selected according to the distribution $P(x_i)Q(y|x_i)$ regardless of the value of i. To use the concept of ϵ-linked sequences in the decoding process effectively, we assume that $P(x)$ and $Q(y|x)$ are an information-ergodic source channel pair.

The error-correcting portion of the decoder is asked to estimate the position i of the switch in the encoder from knowledge of the channel output y and the randomly selected code. The *decoding* algorithm follows: $\hat{\imath}$ is the encoder switch position if and only if $x_{\hat{\imath}}$ is the only n-tuple in $\mathbf{C}$ that is ϵ-linked to the received vector y, based on the joint probability distribution $P(x)Q(y|x)$.

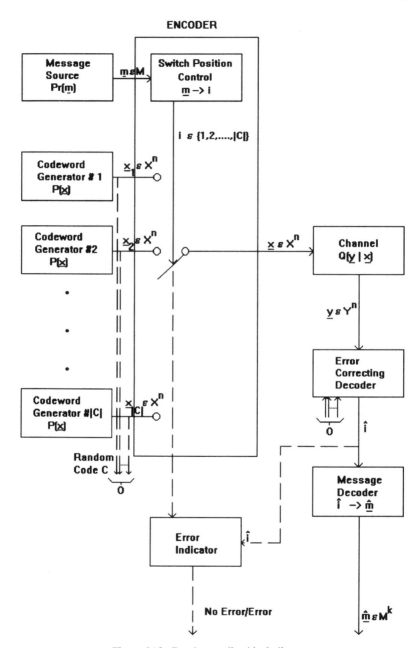

Figure 6.12. Random coding block diagram.

Error Correction II

The event ERROR occurs whenever the switch position estimate $\hat{i}$ is not equal to the true switch position i. The *design* of such a communication system can be viewed as follows:

Given

1. a *channel* $Q(y|x)$ defined on $\mathbf{X}^n \times \mathbf{Y}^n$ for all n, and
2. a *source* $Pr(m)$ defined on $\mathbf{M}^k$ for all k,

design the system by selecting

1. a word length n
2. a linking parameter ϵ, $\epsilon > 0$
3. a code size $|\mathbf{C}|$
4. a code-word generator probability distribution $P(x)$ such that $P(x)Q(y|x)$ is information ergodic,

all to be employed in the preceding system with the designated decoding algorithm.

The design procedure can be evaluated rather easily by fixing the switch position i in the encoder and then computing the probability of error when the following sequence of events occurs:

1. $x_1, \ldots, x_{|\mathbf{C}|}$ are randomly selected according to $P(x)$ (see Equation 99).
2. x_i is transmitted over the channel, and y is generated randomly according to $Q(y|x_i)$.
3. The ϵ-linked decoding algorithm previously discussed is carried out.

The decoding algorithm makes an error when either

1. x_i is not ϵ-linked to y.
2. x_j is ϵ-linked to y for some $j \neq i$.

Hence, if $\mathbf{L}_j$ is defined as the event x_j that is ϵ-linked to y, then when the true switch position is i, the error event is given by

$$\text{error} = \mathbf{L}_i^c \cup \left(\bigcup_{j \neq i} \mathbf{L}_j \right) \tag{100}$$

where $(\ )^c$ denotes the complement of the event. Then, using the union bound,

$$Pr(\text{error}|i) = Pr\left[\mathbf{L}_i^c \cup \left(\bigcup_{j \neq i} \mathbf{L}_i | i\right)\right] \leq Pr(\mathbf{L}_i^c|i) + \sum_{j \neq i} Pr(\mathbf{L}_j|i) \quad (101)$$

Any nontransmitted code generator output x_j and the received n-tuple y are independent; i.e.,

$$Pr(x_j, y) = P(x_j)P'(y) \quad (102)$$

where $P'(y)$ is the marginal distribution

$$P'(y) = \sum_{x \in \mathbf{X}^n} P(x)Q(y|x) \quad (103)$$

The probability that an independently selected pair is ϵ-linked under the distribution $P(x)Q(y|x)$ was bounded in the previous section, and hence:

$$Pr(\mathbf{L}_j|i) \leq e^{-n\left[\frac{1}{n}I(\mathbf{X}^n;\mathbf{Y}^n) - \epsilon\right]} \quad j \neq i \quad (104)$$

In addition, since all code-word generators are identical, we can write

$$Pr(\mathbf{L}_i^c|i) = f(n) \quad (105)$$

where $f(n)$ is independent of the choice of switch position i. Furthermore, since the code-word generators were designed so that the generator channel combination is information ergodic, the probability that an input–output pair (x_i, y) selected according to $P(x_i)Q(y|x_i)$ is ϵ-linked goes to 1 as n increases. Hence,

$$\lim_{n \to \infty} Pr(\mathbf{L}_i^c|i) = \lim_{n \to \infty} f(n) = 0 \quad (106)$$

Substituting Equations 104 and 106 into Equation 101 and averaging over all switch positions gives

$$Pr(\text{error}) = \sum_{i=1}^{|\mathbf{C}|} Pr(\text{error}|i) Pr(i) < f(n) + (|\mathbf{C}| - 1)e^{-n\left[\frac{1}{n}I(\mathbf{X}^n;\mathbf{Y}^n) - \epsilon\right]} \quad (107)$$

Hence, for a fixed code size $|\mathbf{C}|$, the probability of error goes to zero as n increases provided that ϵ is chosen to satisfy

Error Correction II

$$\lim_{n\to\infty} \frac{1}{n} I(\mathbf{X}^n; \mathbf{Y}^n) > \epsilon > 0 \tag{108}$$

This requirement can be satisfied for virtually all information-ergodic source channel combinations of interest.

The fact that $Pr(\text{error})$ goes to zero for a fixed value of $|\mathbf{C}|$ as n increases is not really an astounding result, since the transmission rate R given by

$$R = \frac{1}{n} \ln |\mathbf{C}| \text{ nats/channel symbol} \tag{109}$$

also goes to zero as n increases. The interesting point is that if a sequence of system designs is considered with n increasing and R effectively held constant, i.e., $|\mathbf{C}|$ is increased exponentially with n,

$$|\mathbf{C}| = e^{nR} \tag{110}$$

then

$$Pr(\text{error}) < f(n) + e^{-n\left[\frac{1}{n}I(\mathbf{X}^n;\mathbf{Y}^n) - R - \epsilon\right]} \tag{111}$$

and the probability of error tends to zero as n increases provided that

$$\lim_{n\to\infty} \frac{1}{n} I(\mathbf{X}^n; \mathbf{Y}^n) > R + \epsilon \tag{112}$$

Since n must become large to achieve a small probability of error, the only design parameter that can be used to increase the upper bound on R in Equation 112 is the code-word generator's probability distribution. We define the *per symbol capacity* C of a channel as

$$C = \max_{\mathbf{P}_a} \left[\lim_{n\to\infty} \frac{1}{n} I(\mathbf{X}^n; \mathbf{Y}^n) \right] \text{ nats/symbol} \tag{113}$$

where $\mathbf{P}_a$ represents the collection of all sources that combine with the given channel to form an information-ergodic source channel combination. In other words, *the per symbol capacity C is the largest possible limiting value of mutual information that can be achieved by good code-word generator design.* Then using the optimal code-word generator, for any $\epsilon' > 0$,

$$\frac{1}{n} I(\mathbf{X}^n; \mathbf{Y}^n) > C - \epsilon' \tag{114}$$

when n is sufficiently large. Applying Equation 114 to Equation 111 means that there exists a code-word generator such that for any $\epsilon, \epsilon' > 0$,

$$Pr(\text{error}) < f(n) + e^{-n[C-\epsilon'-R-\epsilon]} \tag{115}$$

and hence a random coding system exists for which

$$R < C \Rightarrow \lim_{n \to \infty} Pr(\text{error}) = 0 \tag{116}$$

This is basically Shannon's famous result.

Of course, we would really like to pick the code $\mathbf{C}$ in advance, build it into the encoder and decoder, and then compute the probability $Pr(\text{error}|\mathbf{C})$ of the error event when a switch position i is chosen at random, x_i is transmitted, and y received and decoded. This would correspond to evaluating performance for a given code $\mathbf{C}$. Unfortunately, this is quite difficult, but if we average over $\mathbf{C}$ using the distribution described in Equation 99, then

$$Pr(\text{error}) = \sum_{\{\mathbf{C}\}} Pr(\text{error}|\mathbf{C}) Pr(\mathbf{C}) \tag{117}$$

is exactly the same probability of error computed in Equation 111. Since the right side of Equation 117 is an average, there must exist at least one $\mathbf{C}$, say, $\mathbf{C}_{\text{opt}}$, for which

$$Pr(\text{error}) \geq Pr(\text{error}|\mathbf{C}_{\text{opt}}) \tag{118}$$

Hence, it is not necessary to build the random coding system shown in Figure 6.12 to attain low error probability for large n; we need only find $\mathbf{C}_{\text{opt}}$ and built it into our coding system. Shannon's result can now be rephrased: For any source and channel with $R < C$ there exists a sequence of codes $\mathbf{C}_n$, $n = 1, 2, \ldots$, such that

$$\lim_{n \to \infty} Pr(\text{error}|\mathbf{C}_n) = 0 \tag{119}$$

This result, though deceptively simple, is difficult to implement for two reasons:

1. The optimum sequence of codes $\mathbf{C}_n$ is not specified; however when n is large, the derivation indicates that we can do quite well on the average by selecting the code at random.
2. To make $Pr(\text{error}|\mathbf{C}_n)$ small, we must increase word length n.

Error Correction II

The first of these difficulties is not insurmountable, since when n is large, our analysis indicates that we can do quite well on the average by simply randomly picking a code. The real crux of the problem is that n must be large, requiring the decoder to determine which of e^{nR} code words is ϵ-linked to the received n-tuple. This exponential growth rate of $|\mathbf{C}|$ with n makes both encoding and decoding formidable tasks.

Actually, the accuracy of the switch position estimate is not of primary concern but rather the accuracy of the message estimate $\hat{m}$, which occurs at the output of the message decoder. Since we now know how to create a channel from i to $\hat{i}$ that is virtually noiseless, it follows that the results of our study of coding for the noiseless channel can be adapted to the design of the message encoder (i.e., switch position control) and message decoder. More specifically, we dichotomize $\mathbf{M}^k$ into two sets $\mathbf{T}_{\epsilon''}$ and its complement, where $\mathbf{T}_{\epsilon''}$ is the set of all message k-tuples that are ϵ''-typical (see Section 1.6). We know that for any $\epsilon'' > 0$ and k large enough,

$$|\mathbf{T}_{\epsilon''}| < e^{k[H(\mathbf{M}|\mathbf{M}^\infty)+\epsilon'']} \qquad (120)$$

and furthermore, defining

$$Pr(m \notin \mathbf{T}_{\epsilon''}) = g(k) \qquad (121)$$

then

$$\lim_{k \to \infty} g(k) = 0 \qquad (122)$$

The switch control is now assumed to map all ϵ''-typical source sequences into distinct switch positions, leaving the last switch position to represent all nontypical sequences from $\mathbf{T}_{\epsilon''}^c$. Certainly, the message decoder can correctly decode if and only if the message produced by the source is ϵ''-typical. Hence, for a good code $\mathbf{C}_{opt}$, two events can cause errors:

1. The error-correcting decoder makes a mistake.
2. The source output is not ϵ-typical.

Using the union bound,

$$Pr(\text{message error}|\mathbf{C}_{opt}) \leq Pr(\text{error}|\mathbf{C}_{opt}) + Pr(m \notin \mathbf{T}_{\epsilon''}) \qquad (123)$$

Noting that $|\mathbf{C}_{opt}|$ is determined by Equation 120 and assuming a fixed-rate encoder for which

$$k\tau_s = n\tau_c \tag{124}$$

we can combine Equations 107, 115, 118, 120–124 to give the bound

$$Pr(\text{message error} | \mathbf{C}_{\text{opt}}) \leq f(n) + g\left(\frac{\tau_c}{\tau_s} n\right) + e^{-n\left[C - \epsilon' - \epsilon \frac{\tau_c}{\tau_s}(H(\mathbf{M}|\mathbf{M}^\infty) + \epsilon'')\right]} \tag{125}$$

Since f and g go to zero as n increases and since ϵ, ϵ', and ϵ'' can be arbitrarily small numbers, we have another version of Shannon's result:
When

$$\frac{C}{\tau_c} > \frac{H(\mathbf{M}|\mathbf{M}^\infty)}{\tau_s}$$

then it is possible to find a sequence $\mathbf{C}_n$, $n = 1, 2, \cdots$ of optimum codes with increasing word length such that

$$\lim_{n \to \infty} Pr(\text{message error} | \mathbf{C}_n) = 0$$

Notice that since C is the channel capacity in nats/channel symbol and since one channel symbol takes τ_c seconds, C/τ_c may be interpreted simply as the channel capacity in nats/second. Similarly

$$\frac{H(\mathbf{M}|\mathbf{M}^\infty)}{\tau_s}$$

can be viewed as source entropy in nats/sec. Conversion by the factors τ_s and τ_c can be viewed simply as a units adjustment that must be made before comparisons can be drawn.

Exercises

1. Consider the collection of sequence pairs (x, y) that are simultaneously ϵ-typical with respect to the three random variables in Problem 2 in Section 6.5. Let $\mathbf{T}_X$ be the set of n-tuples from x that appear in some ϵ-typical pair (x, y). Pick $|\mathbf{C}|$ n-tuples independently from $\mathbf{T}_X$ using a uniform probability distribution on $\mathbf{T}_\infty$ and use the selected code in a random coding scheme of the type discussed in Chapter 6. Using the results of Problem 2 in Section 6.5, prove Shannon's theorem.

2. Suppose that our random coding scheme yields

$$Pr(\text{error}) < P_0$$

Show that if we select a code at random according to $Pr(\mathbf{C})$, then

$$Pr[\{\mathbf{C}: Pr(\text{error}|\mathbf{C}) > kP_0\}] < \frac{1}{k}$$

and hence the probability that a randomly selected code performs well is quite high.

3. Simplify the ϵ-linking decoding algorithm to the binary symmetric channel and give an interpretation of the algorithm in terms of Hamming distance. Based on this interpretation, develop a specialized derivation of Shannon's theorem for the BSC similar to the proof given in Section 6.6. State all pertinent probabilities in terms of the BSC error rate p, code-word length n, and appropriate Hamming distances. What is the optimum code generator design?

4. Consider a first-order binary Markov source with the state diagram shown in Figure 6.13.

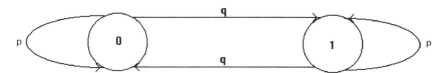

Figure 6.13. A first-order binary Markov source.

The source makes one transition every 5 seconds: A binary symmetric channel with channel matrix

$$\begin{bmatrix} 1-p & p \\ p & 1-p \end{bmatrix}$$

transmits one input symbol every $1/2$ second. Determine the range of values for the pair (p, q) for which reliable communication is possible and sketch this range in the p, q plane.

5. This problem exemplifies the random-coding arguments applied to an r

× r channel matrix, each row and column of the matrix having exactly k nonzero equal entries and $r - k$ zero entries.

(a) How many input sequences of length n are there?

(b) How many different input n-tuples could cause a specific output n-tuple?

(c) If we choose 2^{nR} input n-tuples at random for use in the code, what is the probability that a given input n-tuple is selected for the code?

(d) Given an input n-tuple and its resultant output n-tuple, find a bound on the error probability (i.e., compute a bound on the probability that another input n-tuple in the code could have caused the same output n-tuple).

(e) Show that the error probability goes to zero for rates less than $\log(r/k)$.

6.7. The Situation for Reliable Communication

We pause to consider what we have learned thus far about conditions for reliable communication. For any stationary Markov source with entropy $H(\mathbf{M}|\mathbf{M}^\infty)$ and any stationary channel, Fano's inequality can be used to show that

$$\left[\frac{1}{\tau_s} H(\mathbf{M}|\mathbf{M}^\infty) > \frac{1}{\tau_c} \max_{P(x) \in \mathbf{P}_n} \frac{I(\mathbf{X}^n; \mathbf{Y}^n)}{n} \right] \Rightarrow$$

Reliable communication impossible with a block code of length n (126)

Here, $\mathbf{P}_n$ represents the set of all possible probability distributions on $\mathbf{X}^n$. Furthermore, the coding theorem just derived states that

$$\left[\frac{1}{\tau_s} H(\mathbf{M}|\mathbf{M}^\infty) < \frac{1}{\tau_c} \max_{\mathbf{X}^\infty \in \mathbf{S}_Q} \left(\lim_{n \to \infty} \frac{I(\mathbf{X}^n; \mathbf{Y}^n)}{n} \right) \right] \Rightarrow$$

Reliable communication possible with block codes for large n (127)

The set $\mathbf{S}_Q$ represents the collection of all possible code-word generators that produce an information-ergodic source channel pair when combined with the given channel. If the bracketed terms in Equations 126 and 127 were identical, we would have the ingredients necessary to define a quantity called

Error Correction II

channel capacity, which specifies the rates at which reliable communication is possible.

For the moment, we restrict the channels under consideration to stationary memoryless channels. Then it is easily shown (Problem 1, Section 6.3) Equation 126 can be simplified by observing that

$$\max_{P(x) \in \mathbf{P}_n} \frac{I(\mathbf{X}^n; \mathbf{Y}^n)}{n} = \max_{Pr(x) \in \mathbf{P}_1} I(\mathbf{X}; \mathbf{Y}) \qquad (128)$$

by demonstrating that the maximizing distribution $P(x)$ should correspond to a stationary memoryless n-tuple source $\mathbf{X}^n$. Since stationary memoryless code-word generators combine with stationary memoryless channels to be information ergodic, $\mathbf{S}_Q$ contains all stationary memoryless code-word generators, and hence,

$$\max_{\mathbf{X}^\infty \in \mathbf{S}_Q} \left(\lim_{n \to \infty} \frac{I(\mathbf{X}^n; \mathbf{Y}^n)}{n} \right) \geq \max_{Pr(x) \in \mathbf{P}_1} I(\mathbf{X}; \mathbf{Y}) \qquad (129)$$

However, results from Equations 126 and 127 imply that Equation 129 must hold with equality or else there would exist source information rates for which reliable communication is both possible and impossible—a contradiction. If we define the stationary Markov source information rate as

$$R = \frac{1}{\tau_s} H(\mathbf{M} | \mathbf{M}^\infty) \text{ nats/sec} \qquad (130)$$

and the capacity C of the *stationary memoryless channel* as

$$C = \frac{1}{\tau_c} \max_{Pr(x) \in \mathbf{P}_1} I(\mathbf{X}; \mathbf{Y}) \text{ nats/sec} \qquad (131)$$

then reliable communication of a source with rate R nats/sec through a stationary memoryless channel with capacity C nats/sec is possible if $R < C$, and it is impossible if $R > C$. Of course, reliable communication is guaranteed by the previous results only when the block length n of the code is large. Furthermore, in the unusual case when $R = C$, the previous results are not sensitive enough to answer questions about reliable communication.

When the channel contains memory, Equations 126 and 127 are not easily reconciled (if at all) to yield a capacity C that specifies precisely the set of rates for which reliable communication is possible.

6.8. Convex Functions and Mutual Information Maximization

A *space* **R** is said to be *convex* if for every choice of points r_1 and r_2 in **R**, all the points on the straight line $\alpha r_1 + (1 - \alpha)r_2$, $\alpha \in [0, 1]$, connecting r_1 with r_2 are also in the space.

Examples 6.3.

1. Let **R** be the space of all n-dimensional vectors p whose components are nonnegative; that is, **R** is the positive quadrant in an n-dimensional Euclidean space. Then **R** is a convex space.

2. Hyperplanes in Euclidean n-space are convex; that is, for any fixed vector c and scalar d, the set of vectors r for which

$$r^t c = d$$

   forms a convex set.

3. Linear transformations of convex spaces are convex spaces; hence, if **R** is convex, then the set $\mathbf{R}_Q$ defined by

$$\mathbf{R}_Q = \{s : s = Qr, r \in \mathbf{R}\}$$

   is also convex.

4. The intersection of two convex sets is convex.

The reader is left to verify that the preceding statements are correct.

A real-valued function $f(r)$ defined on a convex space **R** is a *convex function* if for every choice of r_1 and r_2 in **R**, the chord $\alpha f(r_1) + (1 - \alpha)f(r_2)$ lies below or in the surface $f(\alpha r_1 + (1 - \alpha)r_2)$ of the function, i.e.

$$\alpha f(r_1) + (1 - \alpha)f(r_2) \leq f(\alpha r_1 + (1 - \alpha)r_2) \text{ for all } r_1, r_2 \in \mathbf{R} \quad (132)$$

Error Correction II

A plot of both sides of this equation as α goes from 0 to 1 yields Figure 6.14.

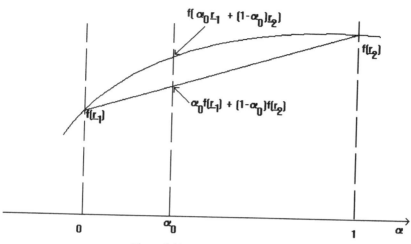

Figure 6.14. A convex function.

Example 6.4 and a few relations yield some interesting results.

Examples 6.4.

1. For any fixed s, the linear function consisting of an inner product with a known vector s,

$$f(r) = s^t r \tag{133}$$

is convex and satisfies Equation 132 with equality for all r_1 and r_2 in Euclidean n-space, because

$$\alpha s^t r_1 + (1 - \alpha) s^t r_2 = s^t[\alpha r_1 + (1 - \alpha) r_2] \tag{134}$$

For example, for any specified channel $P(y|x), x \in \mathbf{X}, y \in \mathbf{Y}$, the negative of the conditional entropy (equivocation)

$$-H(\mathbf{Y}|\mathbf{X}) = \sum_{x \in \mathbf{X}} Pr(x) \sum_{y \in \mathbf{Y}} Pr(y|x) \ln Pr(y|x)$$

is a convex function of the vector r whose components are $Pr(x), x \in \mathbf{X}$.

2. A function $f(p)$ of a one-dimensional vector p is convex on an interval if

$$\frac{d^2}{dp^2} f(p) \leq 0$$

for all p in the interval; hence for example,

$$f(p) = -p \ln p$$
$$\frac{d}{dp} f(p) = -1 - \ln p$$
$$\frac{d^2}{dp^2} f(p) = -\frac{1}{p} \tag{135}$$

and $f(p)$ is convex on the positive real line.

3. Positive linear combinations of convex functions are convex functions; that is, if $f_i(r)$, $i = 1, \ldots, n$, are convex on $\mathbf{P}$, then for any choice of $c_i \geq 0$, $i = 1, \ldots, n$, the function

$$f(r) = \sum_{i=1}^{n} c_i f_i(r) \tag{136}$$

is convex on $\mathbf{P}$. For the special case where $f_i(r) = -r_i \ln r_i$, the results in Equations (135) and (136) imply that the entropy function

$$H(r) = -\sum_{i=1}^{n} r_i \ln r_i \tag{137}$$

is convex everywhere in the positive "quadrant" of Euclidean n-space.

4. If $\mathbf{S}$ is a convex space generated from the convex space $\mathbf{R}$ by the linear relation

$$s = Qr \tag{138}$$

where $r \in \mathbf{R}$, $s \in \mathbf{S}$, and if $f(s)$ is a convex function on $\mathbf{S}$, then $f(Qr)$ is a convex function on $\mathbf{R}$; equivalently,

Error Correction II

$$\alpha f(Qr_1) + (1 - \alpha)f(Qr_2) \le f(\alpha Qr_1 + (1 - \alpha)Qr_2)$$
$$= f(Q(\alpha r_1 + (1 - \alpha)r_2)) \quad (139)$$

Hence, if p is a vector of channel input probabilities and Q is a channel matrix, then the entropy of the output probability distribution, namely,

$$H(\mathbf{Y}) = H(Qp) \quad (140)$$

is a convex function on the space of input probability vectors p, and in fact, is convex for all p in the positive "quadrant."

Finally, we think of the mutual information $I(\mathbf{X};\mathbf{Y})$ between the input and output alphabets, $\mathbf{X}$ and $\mathbf{Y}$, respectively,

$$\mathbf{X} = \{x_i, i = 1, \ldots, |\mathbf{X}|\} \quad (141)$$
$$\mathbf{Y} = \{y_j, j = 1, \ldots, |\mathbf{Y}|\} \quad (142)$$

as a function of the input probability vector

$$p = [Pr(x_1), \ldots, Pr(x_{|\mathbf{X}|})]^t \quad (143)$$

and a $|\mathbf{Y}| \times |\mathbf{X}|$ channel matrix,

$$Q = [Pr(y_i | x_j)] \quad (144)$$

We can make this notationally explicit by defining

$$I(p, Q) \triangleq I(\mathbf{X};\mathbf{Y}) = H(\mathbf{Y}) - H(\mathbf{Y} | \mathbf{X}) = H(Qp) - p^t h \quad (145)$$

where the ith element of h is given by

$$h_i = -\sum_{j=1}^{|\mathbf{Y}|} Pr(y_j | x_i) \ln Pr(y_j | x_i) \quad (146)$$

Notice that both $H(Qp)$ and $-p^t h$ are convex functions defined on the positive "quadrant", which includes the convex space of all vectors. Therefore by Case 3 of Example 6.4, $I(p, Q)$ *is a convex function of p for all p in the convex space of probability vectors* and in fact, for all p in the positive "quadrant."

Figure 6.15 shows $I(p, Q)$ for the channel matrix

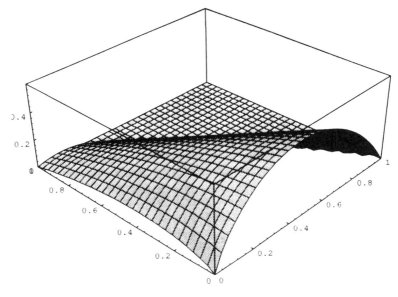

Figure 6.15. The function $I(p, Q)$ on the unit square.

$$Q = \begin{bmatrix} 0.9 & 0.3 \\ 0.1 & 0.7 \end{bmatrix} \qquad (147)$$

and for p in a section of the positive quadrant. Certainly, Figure 6.15 indicates that the maximum of $I(p, Q)$ is at a distant point in the quadrant, and not in the convex subspace of probability vectors. An attempt to solve the equations

$$\frac{\partial}{\partial p_i} I(p, Q) = 0 \qquad i = 1, 2, p_i > 0 \qquad (148)$$

for the location p of the maximum does not yield a point in the subspace of probability vectors. Hence, a Lagrange multiplier λ must be used to add the constraint

$$p_1 + p_2 = 1 \qquad (149)$$

The maximum of the function

$$I(p, Q) - \lambda(p_1 + p_2) \qquad (150)$$

Error Correction II

can then be located for any λ by differentiating Equation 150 with respect to each element of p and solving the resulting equations

$$\frac{\partial}{\partial p_i} I(p, Q) = \lambda \qquad i = 1, 2, p_i > 0 \qquad (151)$$

for p. A unique solution for the maximum value of $I(p, Q)$ can be expected using this technique, since $I(p, Q)$ is convex. Figure 6.16 shows the result of applying the constraint in Equation 149 to this channel. In Figure 6.16, the maximum is 0.206 ... nats when $p(x_1)$ is 0.528 ....

When the dimension of p is greater than 2, it is conceivable that the maximum of the mutual information function $I(p, Q)$ lies on the boundary

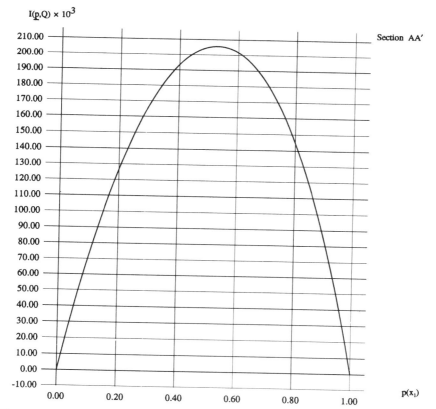

Figure 6.16. Section AA' from Figure 6.15 indicating $I(p, Q)$ under the constraint $p(x_1) + p(x_2) = 1$.

of the positive "quadrant," with, for example, $p_j = 0$. In this case, equations for locating the maximum in the "quadrant" must be modified for the jth coordinate to

$$\frac{\partial}{\partial p_j} I(p, Q) \leq 0 \qquad \text{for all } j \text{ such that } p_j = 0 \qquad (152)$$

allowing the value of $I(p, Q)$ to decrease in the p_j direction at the boundary of the "quadrant." Similarly, the maximum locating equations in probability vector space must be similarly modified:

$$\frac{\partial}{\partial p_j} I(p, Q) \leq \lambda \qquad \text{for all } j \text{ such that } p_j = 0 \qquad (153)$$

Rather than develop the above approach rigorously, we prove the result directly for convex functions.

Theorem 6.2. Let $f(p)$ be a convex function defined on the space of probability vectors p, and assume that

$$\frac{\partial f(p)}{\partial p_i}$$

are continuous for all i everywhere within the space except possibly at the boundaries of the positive "quadrant" where

$$\lim_{p_j \to 0} \frac{\partial f(p)}{\partial p_j} = +\infty$$

Then a necessary and sufficient condition for p to maximize $f(p)$ in the space of probability vectors is that there exist a constant λ for which

$$\frac{\partial f(p)}{\partial p_i} = \lambda \qquad \text{for all } i \text{ such that } p_i > 0 \qquad (154a)$$

$$\frac{\partial f(p)}{\partial p_i} \leq \lambda \qquad \text{for all } i \text{ such that } p_i = 0 \qquad (154b)$$

Proof: We first assume Equation 154 is satisfied for p and λ and show that for all other q in the probability vector space, $f(q)$ is less than $f(p)$. By the convexity of $f(\cdot)$,

Error Correction II

$$\alpha f(q) + (1-\alpha)f(p) \le f[\alpha q + (1-\alpha)p] \qquad 0 < \alpha < 1$$

where

$$f(q) - f(p) \le \frac{f[\alpha q + (1-\alpha)p] - f(p)}{\alpha} \qquad \text{for } 0 < \alpha < 1$$

As α approaches 0, the right side of the preceding equation approaches a scale factor times the derivative of $f(p)$ in the direction of q. More precisely, using the continuity of the partial derivatives of $f(\cdot)$,

$$\lim_{\alpha \to 0} \frac{f[\alpha q + (1-\alpha)p] - f(p)}{\alpha} = \left. \frac{\partial f[\alpha q + (1-\alpha)p]}{\partial \alpha} \right|_{(\alpha=0)} = \sum_{i=1}^{|X|} \frac{\partial f(z)}{\partial z_i} \cdot \frac{\partial z_i}{\partial \alpha}$$

where

$$z = \alpha q + (1-\alpha)p$$

Substituting these results into the preceding equations and applying the assumption in Equation 154 gives

$$f(q) - f(p) \le \sum_{i=1}^{|X|} \frac{\partial f(p)}{\partial p_i}(q_i - p_i) \le \lambda \sum_{i=1}^{|X|} (q_i - p_i) = 0$$

Hence, $f(p)$ is the location of the maximum. QED (sufficiency)

Conversely, now suppose that p is a location of the maximum of $f(\cdot)$ on the space of probability vectors. We now show that Equation 154 must be satisfied for some λ. For any probability vector q,

$$\frac{f[\alpha q + (1-\alpha)p] - f(p)}{\alpha} \le 0 \qquad 0 < \alpha < 1$$

Then by the continuity of $f(\cdot)$, taking the limit as α approaches zero yields

$$\sum_i \frac{\partial f(p)}{\partial p_i}(q_i - p_i) \le 0$$

Suppose that $p_j > 0$ and hence ϵ can be chosen so that $p_j > \epsilon > 0$. Then q can be chosen so that for any particular k,

$$q_k = p_k + \epsilon$$

$$q_j = p_j - \epsilon$$

$$q_i = p_i \qquad i \neq j \qquad i \neq k$$

Then it follows that

$$\frac{\partial}{\partial p_j} f(p) \geq \frac{\partial}{\partial p_k} f(p) \qquad \text{when } p_j > 0$$

Of course, if $p_k > 0$, the roles of j and k may be reversed and

$$\frac{\partial}{\partial p_j} f(p) = \frac{\partial}{\partial p_k} f(p) \qquad \text{when } p_j > 0 \qquad p_k > 0$$

and since j and k are arbitrary, the result follows. QED (necessity)

An application of Theorem 6.2 to the maximization of $I(p, Q)$ requires careful differentiation of Equation 145 with respect to each of the elements in p. If we let

$$r = Qp \tag{155}$$

where the elements of r are $Pr(y_1), Pr(y_2), \ldots, Pr(y_{|Y|})$, then

$$\frac{\partial I(p, Q)}{\partial p_i} = \sum_{j=1}^{|Y|} \frac{\partial H(r)}{\partial r_j} \cdot \frac{\partial r_j}{\partial p_i} - h_i$$

$$= \sum_{j=1}^{|Y|} Pr(y_j | x_i)[-1 - \ln r_i + \ln Pr(y_j | x_i)]$$

$$= -1 + \sum_{j=1}^{|Y|} Pr(y_j | x_i) \ln \frac{Pr(y_j | x_i)}{Pr(y_j)} \tag{156}$$

where for convenience in differentiating, all entropies have been evaluated in nats. Remembering that $Pr(y_j)$ is a function of p, Theorem 6.2 states that p maximizes $I(p, Q)$ if there exists a constant λ such that

Error Correction II

$$I(x_i; \mathbf{Y}) \triangleq \sum_{j=1}^{|\mathbf{Y}|} Pr(y_j | x_i) \ln \frac{Pr(y_j | x_i)}{\sum_{k=1}^{|\mathbf{X}|} Pr(y_j | x_k) Pr(x_k)} \begin{cases} = \lambda + 1 & Pr(x_i) > 0 \\ \leq \lambda + 1 & Pr(x_i) = 0 \end{cases}$$

(157)

Notice that the expected value of $I(x_i; \mathbf{Y})$ with respect to p is the value of the mutual information $I(p, Q)$ [or $I(\mathbf{X}; \mathbf{Y})$] at its maximum, and from Equation 157, this must equal $\lambda + 1$. To summarize, a necessary and sufficient condition for $Pr(x_1), Pr(x_2), \ldots, Pr(x_{|\mathbf{X}|})$ to maximize $I(\mathbf{X}; \mathbf{Y})$ is the existence of a constant C, the capacity of the memoryless channel, for which

$$I(x_i; \mathbf{Y}) = C \quad \text{when } Pr(x_i) > 0$$
$$I(x_i; \mathbf{Y}) \leq C \quad \text{when } Pr(x_i) = 0 \tag{158}$$

where $I(x_i; \mathbf{Y})$ is defined in Equation 157. This result indicates that to operate a memoryless channel at capacity, the contribution $I(x; \mathbf{Y})$ when x is transmitted to the mutual information $I(\mathbf{X}; \mathbf{Y})$ should be independent of which x is used. Equivalently, if $I(x_1; \mathbf{Y}) > I(x_2; \mathbf{Y})$ and both x_1 and x_2 are being used with probability greater than zero, then the situation can be improved by increasing $Pr(x_1)$ and decreasing $Pr(x_2)$ until $I(x_1; \mathbf{Y}) = I(x_2; \mathbf{Y})$. Section 6.9 shows that the condition in Equation 158 is the key to a general approach to computing the capacity C.

Exercises

1. Verify that Cases 1–4 of convex spaces in Example 6.4 are indeed convex.

2. Give an example of a function $f(\cdot)$ defined on a space $\mathbf{R}$ such that $f(\cdot)$ satisfies the convexity requirement in Example 132 but to which Theorem 6.2 does not apply. *Hint:* $\mathbf{R}$ cannot be a convex space.

3. Gallagher (1968) offers a proof of the convexity of $I(p; Q)$ based on the following model. Q describes a memoryless channel mapping the $\mathbf{X}$ alphabet into the $\mathbf{Y}$ alphabet. Consider an additional mapping from an alphabet $\mathbf{Z}$ to $\mathbf{X}$ using the channel matrix

$$P = [p_1 | p_2]$$

Obviously, $\mathbf{Z}$ has only two symbols, with symbol 1 creating distribution p_1 on the $\mathbf{X}$ alphabet and symbol 2 creating distribution p_2 on the

X alphabet. Let

$$Pr(z = 1) = \alpha$$

(a) Write expressions for $I(\mathbf{X}; \mathbf{Y} | \mathbf{Z})$ and $I(\mathbf{X}; \mathbf{Y})$ in terms of $I(p; Q)$.

(b) State the condition for $I(p; Q)$ to be convex in terms of $I(\mathbf{X}; \mathbf{Y})$ and $I(\mathbf{X}; \mathbf{Y} | \mathbf{Z})$.

(c) You can now apply techniques similar to those used to develop Theorem 6.1, the data processing theorem, to show that condition (b) is satisfied, and thus, $I(p; Q)$ is convex.

6.9. Memoryless Channel Capacity Computations

In certain special cases, no difficult computations are required to determine the channel capacity of memoryless channels. For example, when the channel is input uniform, then the set of numbers $\{Pr(y|x): y \in \mathbf{Y}\}$ is not a function of x. Denoting these numbers by $q_1, q_2, \ldots, q_{|\mathbf{Y}|}$, we have

$$H(\mathbf{Y}|\mathbf{X}) = \sum_j Pr(x_j) \left[\sum_i Pr(y_i|x_j) \log \frac{1}{Pr(y_i|x_j)} \right] = \sum_i q_i \log \frac{1}{q_i} \quad (159)$$

Hence, the capacity of the input uniform channel reduces to

$$C = \max_p H(\mathbf{Y}) - H(\mathbf{Y}|\mathbf{X}) = \max_p H(Qp) - \left(\sum_i q_i \log \frac{1}{q_i} \right) \quad (160)$$

Then only the maximum of the output alphabet entropy has to be determined. If in addition there exists a probability vector p such that

$$Qp = \frac{1}{|\mathbf{Y}|} \underline{j} \quad (161)$$

where $\underline{j}$ is the all 1s vector, then it is possible to choose p to make the output symbols equally likely, and in this case,

$$C = \log |\mathbf{Y}| - \sum_i q_i \log \frac{1}{q_i} \quad (162)$$

Error Correction II

One constraint on Q that guarantees this result is that the channel must be output uniform, i.e., $\{Pr(y|x): x \in X\}$ is independent of y. Hence, Equation 162 is the capacity of input-output uniform memoryless channels.

Example 6.5. The binary symmetric channel with error probability p is input output uniform, since both rows and columns of the channel matrix are simply permutations of the numbers p and $(1-p)$. Therefore, by Equation 162, the channel capacity of the BSC is given by

$$C = 1 - H(p) \text{ bits/symbol}$$

where $H(p)$ is the entropy function of p. The source that achieves C bits per symbol of mutual information is zero memory with two equiprobable symbols. The capacity of the channel is zero if and only if $p = 1/2$, in which case the output symbols are independent of the input symbols.

When the input-output uniform symmetries are not present, effectively we must solve Equations 158 for C and the maximizing probabilities. Assuming that components of the maximizing input probability vector are all greater than zero, we must solve the equations

$$\sum_{i=1}^{|Y|} Pr(y_i|x_j) \log\left[\frac{Pr(y_i|x_i)}{Pr(y_i)}\right] = C \quad j = 1, 2, \ldots, |X| \quad (163)$$

with the aid of the auxiliary equations

$$Pr(y_i) = \sum_{j=1}^{|X|} Pr(y_i|x_j) Pr(x_j) \quad i = 1, 2, \ldots, |Y| \quad (164)$$

$$\sum_{j=1}^{|X|} Pr(x_j) = 1 \quad (165)$$

where the left side of Equation 163 is $I(x_j; Y)$. There are in fact $|X| + |Y| + 1$ equations and an equal number of unknowns, namely, the $|X|$ input and $|Y|$ output probabilities and the capacity C.

Due to the transcendental nature of Equation 163, solutions to Equations 163–165 are quite difficult. However, it often happens that the channel matrix Q, whose entries are $Pr(y_i|x_j)$, is square and nonsingular. In this case, a solution can be found as follows. We manipulate Equation 163 into the form

$$\sum_{i=1}^{|\mathbf{Y}|} Pr(y_i|x_j)\left[\log \frac{1}{Pr(y_i)} - C\right] = H(\mathbf{Y}|x_j) \qquad (166)$$

and since Equation 166 holds for $j = 1, \ldots, |\mathbf{X}|$, it can be rewritten in vector form as

$$(L + Cj)^t Q = -h^t \qquad (167)$$

Here, h is the vector of conditional entropies $H(\mathbf{Y}|x_i)$ that also appeared in Section 6.8, and L is given by

$$L^t = [\log Pr(y_1), \log Pr(y_2), \ldots, \log Pr(y_{|\mathbf{Y}|})] \qquad (168)$$

Equation 167 is easily solved for L in terms of C. Denoting the ith element of the vector $h^t Q^{-1}$ by $(h^t Q^{-1})_i$ and assuming logs to the base e, we then have

$$Pr(y_i) = e^{-[C+(h^t Q^{-1})_i]} \qquad i = 1, \ldots, |\mathbf{Y}| \qquad (169)$$

Summing both sides over i then yields the result

$$e^C = \sum_{i=1}^{|\mathbf{Y}|} e^{-(h^t Q^{-1})_i} \qquad (170)$$

or

$$C = \ln\left[\sum_{i=1}^{|\mathbf{Y}|} e^{-(h^t Q^{-1})_i}\right] \qquad (171)$$

Equation 171 yields the capacity *only* if we have found a legitimate solution for p, that is, one where $Pr(x_j) \geq 0$ for all j. Hence, before using Equation 171, we should substitute the results of Equation 169 into Equation 164 to verify that the input probabilities are nonnegative.

Search techniques performed by computer can be used to find the capacity of an arbitrary memoryless channel, and these techniques will probably work well, since $I(p, Q)$ is a convex function of p. One such search technique that takes advantage of the properties of mutual information is based on the following enlargement of the capacity computation problem.

Error Correction II

Theorem 6.3. Let Q be a $|\mathbf{Y}| \times |\mathbf{X}|$ channel matrix $[Pr(y_i|x_j)]$ and let P be any $|\mathbf{X}| \times |\mathbf{Y}|$ transition probability matrix. Define

$$J(p, Q, P) = \sum_i \sum_j Pr(x_j) Pr(y_i|x_j) \ln\left[\frac{P_{j|i}}{Pr(x_j)}\right] \qquad (172)$$

where $P_{j|i}$ is the entry in the jth row and i^{th} column of P. Then,

1. $C = \max_{p,P} J(p, Q, P)$

2. For a fixed p,

$$\max_P J(p, Q, P) = I(p, Q) \qquad (173)$$

where the entries in the maximizing value of P are given by

$$P_{j|i} = \frac{Pr(x_j) Pr(y_i|x_j)}{\sum_k Pr(x_k) Pr(y_i|x_k)} \qquad (174)$$

3. For a fixed P,

$$\max_p J(p, Q, P) = \ln \sum_j \exp\left[\sum_i Pr(y_i|x_j) \ln P_{j|i}\right] \qquad (175)$$

where entries in the maximizing vector p are given by

$$P(x_j) = \frac{\exp\left[\sum_i Pr(y_i|x_j) \ln P_{j|i}\right]}{\sum_k \exp\left[\sum_i Pr(y_i|x_k) \ln P_{k|i}\right]} \qquad (176)$$

Proof of Theorem 6.3 is relatively simple and left as an exercise (see Problem 3).

The algorithm for determining C based on Theorem 6.3 follows. Let $p^{(k)}$ be the kth estimate of the probability vector that maximizes $I(p, Q)$. Then we compute the $(k+1)^{st}$ estimate using Parts 2 and 3 in Theorem 6.3 and Equations 174 and 176.

$$P_{j|i}^{(k+1)} = \frac{Pr^{(k)}(x_j) Pr(y_i | x_j)}{\sum_j Pr^{(k)}(x_j) Pr(y_i | x_j)} \qquad (177)$$

and

$$Pr^{(k+1)}(x_j) = \frac{\exp\left[\sum_i Pr(y_i | x_j) \ln P_{j|i}^{(k+1)}\right]}{\sum_j \left\{\exp\left[\sum_i Pr(y_i | x_j) \ln P_{j|i}^{(k+1)}\right]\right\}} \qquad (178)$$

From Parts 2 and 3 in Theorem 6.3, it is clear that

$$J[p^{(k)}, Q, P^{(k)}] \le J[p^{(k)}, Q, P^{(k+1)}]$$
$$\le J[p^{(k+1)}, Q, P^{(k+1)}] \le I(p^*, Q) = C \qquad (179)$$

for all values of k, where the entries of $p^{(k)}$ are determined by Equation 178 and p^* denotes an input probability vector that achieves capacity. The sequence $J(p^{(k)}, Q, P^{(k)}), k = 1, 2, \ldots$, must converge, since it is monotone and bounded. The question is simply, does the generated sequence of values of $J[p^{(k)}, Q, P^{(k)}]$ converge to C or a number less than C?

As we now see, the preceding algorithm yields a sequence converging to C provided it is started properly. We define a quantity f_k as follows:

$$f_k = \sum_j Pr^*(x_j) \ln\left[\frac{Pr^{(k+1)}(x_j)}{Pr^{(k)}(x_j)}\right] \qquad (180)$$

We note that

$$f_k = \sum_j Pr^*(x_j) \left\{ \sum_i Pr(y_i | x_j) \ln\left[\frac{P_{j|i}^{(k+1)}}{Pr^{(k)}(x_j)}\right] - \max_p J[p, Q, P^{(k+1)}] \right\} \qquad (181)$$

This follows from substituting Equation 177 into f_k and Part 3 of Theorem 6.3. Recognizing that the maximum in this expression is simply $J[p^{(k+1)}, Q, P^{(k+1)}]$ and applying Equation 177 to the argument of the logarithm further simplifies Equation 181 to

$$f_k = \sum_j Pr^*(x_j) \sum_i Pr(y_i | x_j) \ln\left[\frac{Pr(y_i | x_j)}{Pr^{(k)}(y_i)}\right] - J[p^{(k+1)}, Q, P^{(k+1)}] \qquad (182)$$

Error Correction II

where

$$Pr^{(k)}(y_i) = \sum_j Pr(y_i | x_j) Pr^{(k)}(x_j) \quad (183)$$

Hence,

$$f_k = \sum_j Pr^*(x_j) \sum_i Pr(y_i | x_j) \left\{ \ln\left[\frac{Pr(y_i|x_j)}{Pr^*(y_i)}\right] + \ln\left[\frac{Pr^*(y_i)}{Pr^{(k)}(y_i)}\right] \right\}$$
$$- J[p^{(k+1)}, Q, P^{(k+1)}]$$
$$= [C - J(p^{(k+1)}, Q, P^{(k+1)})] + \left\{ \sum_i Pr^*(y_i) \ln\left[\frac{Pr^*(y_i)}{Pr^{(k)}(y_i)}\right] \right\} \quad (184)$$

Both terms on the right side of Equation 184 are nonnegative, and hence, f_k is nonnegative and $\sum_{k=1}^{K} f_k$ is a monotone nondecreasing function of K. Now,

$$\sum_j Pr^*(x_j) \ln\left[\frac{Pr^{(K+1)}(x_j)}{Pr^{(1)}(x_j)}\right] = \sum_{k=1}^{K} f_k = \sum_{k=1}^{K+1} \{C - J[p^{(k+1)}, Q, P^{(k+1)}]\}$$
$$+ \sum_{k=1}^{K+1} \left\{ \sum_i Pr^*(y_i) \ln\left[\frac{Pr^*(y_i)}{Pr^{(k)}(y_i)}\right] \right\} \quad (185)$$

The quantity on the left side of Equation 185 is bounded away from $+\infty$ provided $Pr^{(1)}(x_j) \neq 0$ for any j (the proper starting condition), and hence, as k increases, the terms in both sums on the right side must go to zero. Thus,

$$\lim_{k \to \infty} J[p^{(k+1)}, Q, P^{(k+1)}] = C \quad (186)$$

$$\lim_{k \to \infty} Pr^{(k)}(y_i) = Pr^*(y_i) \quad (187)$$

Hence, Equation 186 states that the computational algorithm yields a quantity converging to capacity, and Equation 187 implies that the maximizing output probability distribution is unique.

Example 6.6. Suppose

$$Q = \begin{bmatrix} 0.9 & 0.3 \\ 0.1 & 0.7 \end{bmatrix}$$

Using the direct approach from Equations 169 and 171,

$$h^t = [H(0.9, 0.1), H(0.3, 0.7)] = (0.325, 0.611) \text{ nats}$$

and

$$Q^{-1} = \begin{bmatrix} 7/6 & -1/2 \\ -1/6 & 3/2 \end{bmatrix}$$

$$h^t Q^{-1} = (0.277, 0.754)$$

Hence, from Equation 171,

$$C \doteq 0.206 \text{ nats}$$

and from Equation 169,

$$Pr(y_1) \doteq 0.617 \qquad Pr(y_2) \doteq 0.383$$

Inverting Equations 164, we have

$$\begin{bmatrix} Pr(x_1) \\ Pr(x_2) \end{bmatrix} = Q^{-1} \begin{bmatrix} Pr(y_1) \\ Pr(y_2) \end{bmatrix} \doteq \begin{bmatrix} 7/6 & -1/2 \\ -1/6 & 3/2 \end{bmatrix} \begin{bmatrix} 0.617 \\ 0.383 \end{bmatrix} \doteq \begin{bmatrix} 0.528 \\ 0.472 \end{bmatrix}$$

Since both input probabilities are positive, the quantity C is the capacity of the channel.

As a check, we test the iterative algorithm to see what answer is produced using Equations 177 and 178 with an initial input probability vector $p^0 = (0.5, 0.5)$ (see Table 6.1).

Error Correction II

Table 6.1. Algorithm Results

k	$P_{1\mid 1}^{(k)}$	$P_{1\mid 2}^{(k)}$	$P_{2\mid 1}^{(k)}$	$P_{2\mid 2}^{(k)}$	$Pr^{(k)}(x_1)$	$Pr^{(k)}(x_2)$	$J(p^{(k)}, Q, P^{(k+1)})$
1	0.75000	0.12500	0.25000	0.87500	0.51062	0.48938	0.20526
2	0.75788	0.12972	0.24212	0.87028	0.51725	0.48275	0.20549
3	0.76272	0.13275	0.23728	0.86725	0.52137	0.47863	0.20558
4	0.76569	0.13466	0.23431	0.86534	0.52393	0.47607	0.20562
5	0.76753	0.13586	0.23247	0.86414	0.52552	0.47448	0.20563
6	0.76867	0.13661	0.23133	0.86339	0.52651	0.47349	0.20563
7	0.76937	0.13708	0.23063	0.86292	0.52712	0.47288	0.20564
8	0.76980	0.13737	0.23020	0.86263	0.52750	0.47250	0.20564
9	0.77007	0.13755	0.22993	0.86245	0.52774	0.47226	0.20564
10	0.77024	0.13766	0.22976	0.86234	0.52788	0.47212	0.20564
11	0.77035	0.13773	0.22965	0.86227	0.52798	0.47202	0.20564
12	0.77041	0.13778	0.22959	0.86222	0.52803	0.47197	0.20564
13	0.77045	0.13780	0.22955	0.86220	0.52807	0.47193	0.20564
14	0.77048	0.13782	0.22952	0.86218	0.52809	0.47191	0.20564
15	0.77049	0.13783	0.22951	0.86217	0.52810	0.47190	0.20564
16	0.77050	0.13784	0.22950	0.86216	0.52811	0.47189	0.20564
17	0.77051	0.13784	0.22949	0.86216	0.52812	0.47188	0.20564
18	0.77051	0.13784	0.22949	0.86216	0.52812	0.47188	0.20564

Clearly the algorithm converges.

Exercises

1. Determine the capacities of the memoryless channels with the following channel matrices:

    (a) A binary erasure channel:
    $$\begin{bmatrix} 1-p & 0 \\ p & p \\ 0 & 1-p \end{bmatrix}$$

    (b) The m-ary symmetric channel:
    $$\begin{bmatrix} q & p & p & \cdots & p \\ p & q & p & \cdots & p \\ p & p & q & \cdots & p \\ \vdots & & & & \vdots \\ p & p & p & \cdots & q \end{bmatrix}$$

(c) An unbalanced ternary channel

$$\begin{bmatrix} 0.9 & 0.1 & 0.05 \\ 0.07 & 0.3 & 0.95 \\ 0.03 & 0.6 & 0 \end{bmatrix}$$

(d) A general binary channel

$$\begin{bmatrix} 1-p_1 & 1-p_2 \\ p_1 & p_2 \end{bmatrix}$$

2. Let $C_1, C_2, \ldots, C_M$ be the capacities in nats of M memoryless channels having channel matrices $Q_1, Q_2, \ldots, Q_M$, respectively. Show that the memoryless channel with channel matrix given by

$$\begin{bmatrix} Q_1 & O & \cdots & O \\ O & Q_2 & \cdots & O \\ \vdots & \vdots & & \vdots \\ O & O & \cdots & Q_M \end{bmatrix}$$

where O denotes a matrix of zeros, has capacity C given by

$$C = \ln \left(\sum_{m=1}^{M} e^{C_m} \right)$$

3. Prove Theorem 6.3. *Hint:* The divergence inequality and the convexity results of Section 6.8 are useful.

4. Show that the capacity C of a memoryless channel is upper bounded as follows:

$$C \leq \max_{j} \sum_{i} Pr(y_i | x_j) \log \frac{Pr(y_i | x_j)}{\sum_{k} Pr(y_i | x_k) Pr(x_k)}$$

for any choice of the input probabilities. This relation is helpful in determining when to stop algorithms designed to search for the maximum of $I(p, Q)$.

5. Consider the memoryless channel whose transition probabilities are described in Figure 6.17.

Error Correction II

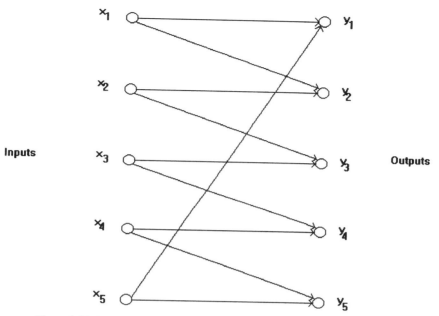

Figure 6.17. Transition probabilities for the memoryless channel in Problem 5.

All lines indicate transitions occurring with probability $1/2$.

(a) What is the capacity of the channel?

(b) How much information can be transmitted through the channel by exactly 1 symbol transmission with exactly zero error probability?

(c) Repeat (b) for the case of 2 symbol transmissions (a code of length 2).

(d) Under what conditions on the channel matrix of an arbitrary memoryless channel do you think that information rates greater than zero are possible with error probability zero?

6.10. Notes

Shannon (1948) deserves credit for the random-coding concept and its application to the proof of coding theorems. His first 1948 paper gives the basic results for discrete memoryless channels, and the second handles the more difficult case of continuous channels. This work has been generalized in many ways by many researchers over several decades. Feinstein (1958) and Wolfowitz (1961) have written books exclusively on the coding theorems for information theory. Gallagher (1968) contains an excellent presentation of coding theorems, channel capacity computations, and exponential bounds on error probability. Beyond these textbook treatments, the reader will find a comprehensive introduction to the literature in Slepian (1974).

The lower bound on error probability (often called the converse to the coding theorem) based on Fano's inequality (see Section 6.3) was developed over a period of time beginning in the available literature in Fano (1961), and followed by Gallagher (1964) and Reiffen (1966). The treatment of the Data Processing Theorem and the Chernoff bound given here follows Forney (1972). The conditions of Theorem 6.2 in Section 6.8 are often referred to as Kuhn–Tucker conditions after Kuhn and Tucker (1951), who stated a more general convex programming result. The method of computing capacity for nonsingular channel matrices was given in Fano (1961), and the iterative computation of channel capacity follows Arimoto (1972). The subject of zero error capacity touched on in Problem 5c is covered by Shannon (1956).

References

S. Arimoto. 1972. "An Algorithm for Computing the Capacity of Arbitrary Discrete Memoryless Channels." *IEEE Trans. Inform. Theory* **18**: 14–20.
A. Feinstein. 1958. *Foundations of Information Theory.* McGraw-Hill, New York.
G. D. Forney, Jr. *Information Theory,* Class Notes, Stanford University, 1972.
R. G. Gallagher. 1964. "Information Theory." In *The Mathematics of Physics and Chemistry,* edited by H. Margenau and G. Murphy. Van Nostrand, Princeton.
———. 1968. *Information Theory and Reliable Communication.* Wiley., New York.
H. W. Kuhn and A. W. Tucker. 1951. "Nonlinear Programming." In *Proc. 2d Berkeley Symposium on Mathematics, Statistics, and Probability.* University of California Press, Berkeley, CA. Pp. 481–92.
R. M. Fano. 1961. *Transmission of Information.* MIT Press and Wiley, New York–London.
B. Reiffen. 1966. "A Per-Letter Converse to the Channel Coding Theorem." *IEEE Trans. Inform. Theory* **12**: 475–80.
C. E. Shannon. 1948. "A Mathematical Theory of Communication." *Bell Sys. Tech. J.* **27**: 379–423; **27**: 623–56.
———. 1956. "The Zero Error Capacity of a Noisy Channel." *IRE Trans. Inform. Theory* **2**: 8–19.
D. Slepian, ed. 1974. *Key Papers in the Development of Information Theory.* IEEE Press, New York.
J. Wolfowitz. 1961. *Coding Theorems of Information Theory.* Springer-Verlag and Prentice-Hall, Englewood Cliffs, NY.

7

Practical Aspects of Coding

7.1. Agent 00111 Is Not Concerned

Feeling at an ebb after so much theory, Agent 00111 glanced at the pile of academic books and theoretical papers on coding that his scientists tried to persuade him were essential reading. They were not. He had long ago come to the conclusion that the number of practical error correction codes he needed to use on his missions were limited to a few techniques. It had taken him a long time, innumerable briefings and any number of overly long meetings before most of the scientists had conceded the point. There were still practical problems that he wanted solved; however, many problems had been solved since he had first felt the need for error correction, and it had been many years since he was hampered by the lack of coding technology.

7.2. Types of Practical Codes

Chapter 7 offers a perspective on the practical side of coding from an information-theoretic viewpoint. It has been stressed in previous chapters that there are many different types of codes and the optimal code for a given set of channel conditions may not resemble the optimal code for another set of channel conditions. It is also probably true that starting with any given code, we could devise a set of channel conditions for which that code is optimal. Agent 00111's complacency is justified because many devised channel conditions are extremely unlikely in practice. If attention is restricted to the channels that are most often encountered, the set of both suitable and implementable codes is correspondingly smaller. In fact, there are probably only five overlapping codes classes in general use:

1. Convolutional codes
2. Short binary block codes
3. Reed–Solomon codes
4. Concatenated codes
5. Codesigned coding and modulation schemes

Section 7.2 outlines some of the more practical properties of these code classes. A *concatenated code,* derived from the other types, consists of placing codes in series. For example, a received encoded signal is first processed by a convolutional decoder whose output is subsequently processed by a Reed–Solomon decoder. *Codesigned coding and modulation* represents the most recent addition to the technology. The principles on which much of this subject is based (involving some ring theory, combinatorial sphere packing, and finite field code theory) are relatively less well-known. This code class is discussed in Section 7.3 and in more depth in Appendix 7A.

7.2.1. Convolutional Codes

Convolutional codes can be characterized by their suitability for analog engineering and their high performance whenever they can be decoded by the Viterbi algorithm (VA). They are widely used in the satellite industry as an economical way of improving power budgets. More recently, convolutional codes used in conjunction with sophisticated modulation schemes (e.g., trellis-coded modulation) have been central in increasing the throughput of commercial voiceband data modems.

Convolutional decoders are excellent at combating additive white Gaussian noise (AWGN). The key to their success is the VA's ability to process "soft-decision" data (the best real-valued estimates of received binary data) or reliability information into maximum-likelihood estimates of the transmitted sequence. A hard-decision demodulator provides only a binary data stream as an output. Internally, the demodulator might have much more information about the relative reliabilities of entries in the binary stream, often in the form of a real number. If these reliabilities are output in addition to data, the combined output is called soft-decision data. In a typical example, a demodulator supplies 4 bits of data for every received encoded bit. The 4 bits indicate not only the most likely bit of information but also the confidence level that should be attached to that bit. The VA processes the soft-decision data into a global estimate of the code word most likely to have been transmitted.

Practical Aspects of Coding

For any given code with equally likely code words, maximum likelihood decoding is the best that a decoder can achieve, and in AWGN, implementations of the VA approach this bound; however, this statement has to be considered carefully. It is possible (and quite common) for a stronger code with a weaker decoding algorithm to outperform a weaker code with a maximum-likelihood algorithm. In fact, the complexity of the VA limits implementable codes to a small class. (A 64-state code is about maximal in most commercial satellite applications. NASA has an advanced project that implements a 16,384-state decoder for deep-space missions.)

An example of a widely used convolutional encoder is shown in Figure 7.1. (This code was discovered by Odenwalder.) The device accepts 1 bit of information data and outputs 2 bits of encoded data; i.e., it is a (2,1) code. Encoding is a continuous process whereby a pair of encoded output bits depends on the present information bit and the values of the 6 most recent information bits held in the shift register. These last 6 bits have a value referred to as the encoder state. There are 64 states for the code in Figure 7.1. Note that the encoding may be considered as spreading the influence of any one information bit over at least seven transmitted pairs of encoded bits. The degree of spreading is referred to as the constraint length K of the code, explaining the (2,1) $K = 7$ notation. When used to combat AWGN and supply a user with a bit error rate of about 10^{-6}, this code gives about 5 dB of gain over uncoded modulation. The (2,1) $K = 7$ code is often referred to as a *de facto* standard for commercial satellite applications. It is certainly true that Very Large Scale Integration (VLSI) encoder-decoder implementations are readily available from suppliers.

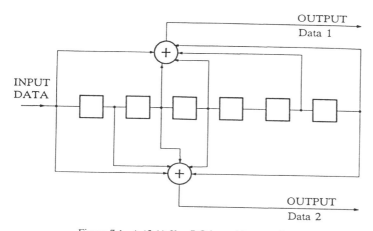

Figure 7.1. A (2,1) $K = 7$ Odenwalder encoder.

Although the interested reader is referred to any of the numerous books on coding for details, the *state notation* is the key to understanding the VA. The encoder can be regarded as a time-invariant finite-state machine and the encoded sequence as a function of the sequence of states through the machine. The decoder's task is to find the sequence of states whose corresponding output most closely resembles the received signal.

Convolutional coding seems to integrate better into an analog context than into a purely digital context. Taking the two points separately, convolutional codes mesh well with both modulation and equalization. Increasingly, the line between demodulation and convolutional decoding is disappearing, or at least it is recognized as artificial. The VA can be adapted to help in clock recovery, intersymbol interference, and other analog problems, with only the complexity issue preventing further integration. For example, maximum-likelihood sequence estimation over a known multipath channel can theoretically be achieved by the VA; in practice, however, there can be too many states for an unmodified application of the VA. Future generations of radios and/or modems are expected to have integrated convolutional codes and VA implementations, the logical technique for processing soft-decision data. In the hard-decision digital context, the VA still achieves maximum-likelihood decoding for the class of codes to which it is applicable; however, these codes have to be compared with classes of inherently stronger codes for which digital algebraic decoding algorithms are practical. The point is illustrated in Figure 7.2, which shows an output (decoded) Bit Error Rate (BER) versus normalized power curve for AWGN and coherent binary phase-shift keying modulation. Two of the curves are for a decoder of the $(2,1)$ $K = 7$ code just described. One curve, the better of the two, shows the performance when soft-decision data are processed; the second curve shows the performance on hard-decision data. There is a loss of 2–3 dB; on a fading channel, the loss can be 4–6 dB. The third curve shows the performance of a block code described in Section 7.2.2. This code is decoded by an algebraic-decoding algorithm that is not maximum likelihood even for the hard-decision case. It is apparent that any verdict on which code is better depends on factors external to Figure 7.2.

The performance of convolutional codes encountering non-AWGN conditions deserves mention. If appropriate modifications are made to the soft-decision metrics, excellent performance can be obtained with an uncorrelated Rayleigh or Rician fading channel. However, modifications are intimately involved with the internal operation of the receiver (the relative power levels of the received signal has to be factored into the metrics), reinforcing the conclusion that integrating of convolutional codes with radio technology is desirable and inevitable. With correlated fading channels, convolutional codes suffer from their short constraint length unless interleaving is used (see Section 7.2.5). Without sufficient interleaving, a deep fade can "hit" all of the trans-

Practical Aspects of Coding

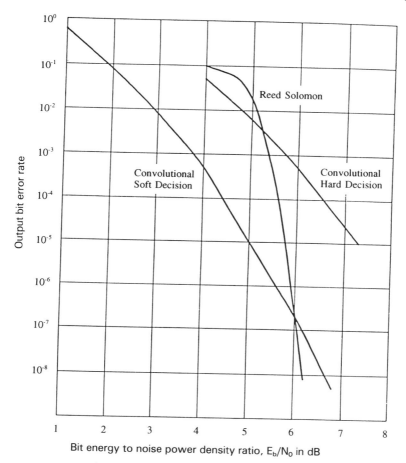

Figure 7.2. Gaussian noise performance curves.

mitted information influenced by a data bit; however, the worst type of noise for a convolutional code is probably high-power burst interference, e.g., periodic tone jamming. Unless specifically prevented, it is possible for the soft-decision metrics to be counterproductive, with the interference-afflicted signal samples regarded as higher quality than the unafflicted signals. (In defense of convolutional codes, this possibility exists for any decoder using soft-decision data.)

Finally, note that alternative decoding algorithms to the VA exist; these alternatives, known as sequential decoding algorithms, attempt to decode convolutional codes whose constraint length is too large for the VA to be applicable. The inherent weakness of sequential decoding algorithms is that they necessarily exhibit a large variation in the number of operations per decoded bit. Theoretically, the standard deviation is infinite (see Jacobs and Berlekamp). In practice, this means that a sequential decoder can "get stuck" at one place; i.e., the decoder may spend an unacceptably long time trying to decode 1 bit. Historically, sequential decoders existed before the invention of the VA, and they are still used in certain applications.

7.2.2. Block Codes: A General Overview

To review, a *block code* over A is a subset of A^N, where A is a preset alphabet and N is an integer referred to as the length of the code. If C denotes the subset of A^N ($C \subseteq A^N$), the elements of C are referred to as code words. If A has the structure of a field and C is a vector space of dimension K over A, C is called a linear (N,K) code over A. Obviously, if $|A| = q$, an (N,K) code has q^K code words. A code over A has minimum Hamming *symbol* distance $d_{\min}$ if every pair of distinct code words differs in at least $d_{\min}$ locations and at least two distinct code words differ in exactly $d_{\min}$ symbols of A. By far the most used and studied codes have A equal to $GF(2^m)$, the field with 2^m elements, for some m. If A is the binary field $GF(2)$, the code C is called a binary code. If A is a field with more than two elements, the code is called a symbol code. Note that if $A = GF(2^m)$, the elements of A can be labeled with m-bit sequences. It follows that a code over $GF(2^m)$ can give rise to many different binary codes of length Nm. (In fact, a different labeling may be chosen for each code coordinate.) If a basis for $GF(2^m)$ over $GF(2)$ is chosen, linear binary codes are obtained from linear symbol codes over $GF(2^m)$. However, the difference between a binary code and a symbol code is more than an internal matter for code designers. Good symbol codes do not necessarily give good binary codes, or vice versa. There are additional reasons to distinguish the categories:

1. If errors tend to cluster in short bursts, the symbol error rate can be lower than the bit error rate, indicating that a symbol code with a good minimum distance may perform better than a binary code with a good minimum distance.

2. A symbol code with a well-defined decoding algorithm may no longer have a well-defined decoding algorithm when regarded as a binary code, regardless of the binary minimum distance. (*Justesen codes* are good binary codes derived from good symbol codes; however, a binary decoding algorithm is not apparent from their derivation. The interested reader is referred to MacWilliams and Sloane.)

Practical Aspects of Coding

It is interesting to examine how the parameters N, K, and d_{min} reflect on code performance and the choice of a decoding algorithm. For a fixed N, symbol size b, and minimum distance d_{min}, putting decoding complexity aside, it is desirable to maximize the amount of data that a code transmits, i.e., maximize the value of K relative to a fixed N and d_{min}. Tables of the best known values of K for binary codes can be found in the literature (note that such tables may not give codes that are practical to decode). It is important to distinguish between types of decoding algorithms; the three major types are

1. Soft-decision decoding. A received sequence of samples, quantized to give relative reliabilities, is processed to estimate the most likely code word transmitted, assuming equally likely code words.

2. Minimum-distance decoding: A sequence a from A^N is compared to the code words; the sequence is decoded to a code word c provided the Hamming distance from a to c is strictly less than the distance of a to any other code word.

3. Bounded-distance decoding: If the code has minimum Hamming distance d_{min}, the bounded-distance decoding algorithm decodes a to a code word c provided $2t < d_{min}$, where t is the number of symbols received in error. The behavior of the algorithm is, in general, not predictable if $2t \geq d_{min}$.

The decoding algorithms are listed in terms of decreasing complexity and with regard to AWGN conditions, decreasing performance. There are many other decoding algorithms that fall somewhere between the preceding algorithms. Forney introduced generalized minimum-distance decoding to obtain some fraction of full soft-decision performance from minimum-distance and bounded-distance decoding techniques. Chase decoding obtains some fraction of full soft-decision performance by using multiple applications of a bounded-distance decoder. (Chase's idea is to "dither" the least reliable bits of a in all possible ways; decode each pattern; then score the answers. If g bits are dithered, 2^g bounded-distance decodings are necessary.) Other examples exist where even bounded-distance decoding may be difficult. For example, the (23,11,7) Golay code is three-error correcting, while two errors can be corrected using a Bose–Chaudhuri–Hocquenghem (BCH) decoding algorithm; patterns of three errors do not fall so readily within the BCH approach.

As the value of N increases, the codes tend to behave less like a perfect code; i.e., the bounded-distance spheres exclude increasing numbers of vectors. Although this might appear to indicate that bounded-distance algorithms are less useful as N increases, such is not necessarily the case: Increasing N can improve code performance to the extent that the degradation of bounded-distance decoding is insignificant. This can be illustrated by fixing K/N and d_{min}/N and allowing N to increase.

Figure 7.3 shows the results for a class of codes discussed more fully in Section 7.2.3. The channel is taken to be a binary symmetric channel where the x-axis is the channel Bit Error Rate (BER); i.e., it is assumed that an identical and independently distributed error process is applied to the transmitted bits. The four codes have binary length $b\,2^b$, where $b = 4, 5, 6, 7,$ and 8. The redundancy is 7/8 for each of the codes, and all values of $d_{\min}/N$ are approximately 1/8. (Note that since the codes are symbol codes with different values of b, the symbol error rates are different for the codes.) The codes are all decoded with a bounded-distance decoder, as discussed in Section 7.2.3. It is evident that in most regions of practical interest, performance increases with block length.

This effect, often referred to as noise averaging, results from the "Law of Large Numbers". Noise conditions become more predictable when viewed

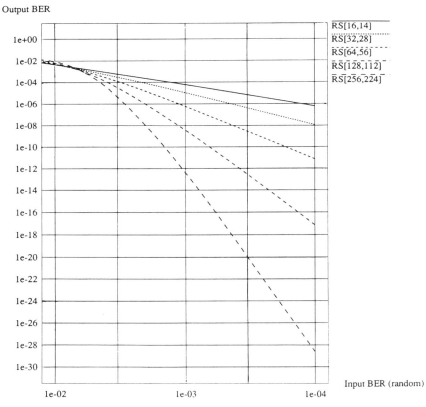

Figure 7.3. Cross-comparison of different rate 7/8 RS codes.

Practical Aspects of Coding

on a long-term basis, and being predictable, a given level of code protection either copes, with high probability, or does not cope, with high probability; i.e., as block length increases, a relatively sharp descending error probability curve results. (Of course, in the extreme, this is nothing more than a paraphrase of Shannon's capacity theorem.) It is a code designer's task to balance three competing factors:

1. Keeping an implemented code functioning on an acceptable part of the performance curve
2. Keeping code redundancy to a minimum
3. Keeping decoder complexity to a minimum

Since channels may be poorly understood in practice, safety margins may have to be added. As pointed out, the superiority or otherwise of a sharply descending curve over a more gracefully degrading curve, e.g., Figure 7.2, is a function of what the user wishes to achieve. Bounded-distance algebraic decoding complexity normally increases with increasing redundancy and block length. A traditional and fundamental trade-off in practical coding concerns the point at which a designer, balancing the three preceding factors, should select a code type, a redundancy, and a decoding strategy. At the present time, readily available convolutional and block decoders supply the user with a range of possible code options all mechanized in a single coding system. It has been suggested that matching code to channel could be automated as an adaptive real-time decision (see Peile and Scholtz). This technology could have saved Agent 00111 from some lengthy meetings.

7.2.3. Reed–Solomon Codes

Reed–Solomon (RS) codes are a class of symbol codes for which encoding and bounded-distance decoding algorithms are practical to implement. Every owner of a reputable compact disc player owns an RS decoder. Pictures of distant planets are received using concatenated techniques that include RS coding.

These algorithms employ sophisticated finite algebraic concepts. Although this makes RS coding a rather specialized area, finite algebra is well suited to digital implementation, allowing for high-speed operation. (Of course, this does not imply that every digital implementation is equally efficient.)

The parameters of RS codes follow. Let b be the symbol size in bits. An $RS(N, K)$ symbol code over $GF(2^b)$ exists whenever:

$$0 < K < N \leq 2^b + 1$$

The minimum Hamming distance of an $RS(N,K)$ code is

$$d_{\min} = N - K + 1$$

It should be noted that the parameters N, K, and b do not fully specify an RS code, and many engineering trade-offs and incompatibilities exist within these parameters. There are many ways of viewing the encoding process, ranging from the conceptually pleasant to the hardware efficient. The original paper of Reed and Solomon presents the easiest conceptual explanation, and this viewpoint is outlined for the case of $N = 2^b - 1$.

1. Let $a_0, a_1, \ldots, a_{K-1}$ be elements in $GF(2^b)$ representing information to be encoded. Form a polynomial $a(x)$ of degree at most $K - 1$,

$$a(x) = \sum_{i=0}^{K-1} a_i x^i$$

2. Evaluate the polynomial at the nonzero elements of $GF(2^b)$ to obtain $2^b - 1$ elements that by definition form the code word c. With respect to the minimum distance, note that the sum of two polynomials of degree at most $K - 1$ is also a polynomial of degree at most $K - 1$, i.e., the code is linear.

3. The minimum distance property then follows from what has been called the fundamental myth of modern algebra: A polynomial of degree t has at most t roots. (This result is true for all polynomials with coefficients in a field.)

It is worth noting that RS codes meet the Singleton bound of $d = N - K + 1$ (MacWilliams and Sloane). Codes that meet this bound are called maximal distance separable (MDS) and have several fascinating properties. Although it is not true that an MDS code is necessarily an RS code, it is true that all known MDS codes, with one class of exceptions, have the same parameters as RS codes. From this point of view, RS codes are an important class of optimal codes.

The correction power of RS codes follows. Let R equal the number of redundant symbols. Suppose that a transmitted code word was received with t symbol errors (symbols believed by the decoder to be correct but which are incorrect) and s symbol erasures (symbols believed by the decoder to be unreliable). Then a bounded-distance decoder is guaranteed to replicate the original code word provided that

$$2t + s \leq R$$

With regard to symbol erasures, some external agent (Agent 11000?) has to tell the decoder which symbols are trustworthy. (This can also be regarded as a simple form of soft-decision information.) It is a design decision whether the decoder should process the erasure information. In the absence of any erasure processing, the conventional $2t \leq R = d - 1$ formula for bounded-distance decoding applies. In the absence of errors (erasures only), implications of the MDS property become evident. It is possible to decode provided any $B \geq K$ symbols are correctly received and it is known which B symbols are received. It should be apparent that it is impossible to correct errors unambiguously if fewer than K symbols are received; i.e., this is the best possible result. The property that we can decode the original K symbols from a knowledge of any K symbols seems a reasonable and desirable stipulation for a code. In fact, this is exactly the MDS property, and most codes are not MDS. It has been conjectured (see MacWilliams and Sloane) that except for two known counterexamples, $N = q + 1$ is the largest possible block length for an MDS symbol code over $GF(q)$. This has been proved for many cases; e.g., if q is odd. The known counterexamples occur for even values of m: There exist ($N = 2^m + 2$, $K = 3$, $d = 2^m$) and ($N = 2^m + 2$, $K = 2^m - 1$, $d = 4$) MDS codes over $GF(2^m)$ (see MacWilliams and Sloane).

Exercise

1. Research the definition of a generator matrix.

2. Show that the MDS property states that every square submatrix of a generator matrix is nonsingular.

The search for good (N,K) codes over $GF(q)$ with $N > q + 1$ and distance $d = N - K + 1 - g$ has generalized RS codes into areas of algebraic geometry. Note that g measures the shortfall of a code from being MDS (i.e. for a MDS code $g = 0$); the object is to minimize the value of g. The algebraic geometry codes retain some flavor of the RS definition. Data are viewed as multinomials of bounded degree with coefficients from a finite field. The multinomial is evaluated at preset points on a preset algebraic curve to give the code word. Finally, a classic algebraic result [the Riemann–Roch theorem with a curve of genus g (see van Lint and van der Geer)] is evoked to guarantee the minimum distance of the code. The interested reader is referred to the references.

To return to RS codes, Figure 7.4 shows the performance of $RS(64,k)$ codes for all values of k divisible by 4. Performance involves error decoding (no erasures) of transmissions over a binary symmetric channel with error probability p. The x-axis shows the channel BER, p. The y-axis shows the

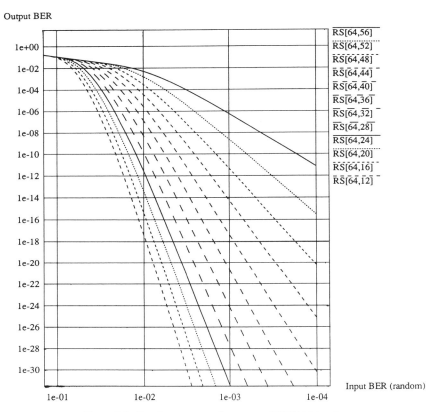

Figure 7.4. Random error performance of RS(64,k) codes.

probability of a bit being in error after the decoder has finished processing the received sequence. (Output errors need not have independent distributions, since most decoders give bursty error statistics.) It is important to note that $RS(N,K)$ decoders do not necessarily give the same output BER at the same operating point even when N and K are fixed. The decoding formula guarantees a probability of correct decoding; it does not guarantee how many bit errors are made if the decoder cannot correct the sequence. At one extreme, a *transform-based* decoder (see Blahut) will give an output BER of about 1/2 the block error rate; i.e., when the decoder decodes incorrectly, data are completely scrambled. At the other extreme, a *systematic* decoder (see Berlekamp) may produce a much lower output BER than the block error rate. In more detail, the preceding definition of encoding data by evaluating a polynomial over the field elements corresponds to the transform approach. Note that the

Practical Aspects of Coding

data will, in general, appear changed in the code word. The systematic approach encodes by leaving the K data symbols unchanged and appends $N - K$ redundant symbols to obtain a code word.

Exercise

1. Show that any linear block code can be implemented by a systematic encoder.

Figure 7.4 shows the performance of a systematic decoder that uses the following decoding strategy. A bounded-distance decoder corrects patterns that meet the $2t \leq R$ bound. If the decoder is aware that a pattern is too corrupt for correction, parity checks are removed and the remaining symbols are output. Since RS codes are reasonably effective at detecting uncorrectable patterns, the result is to make output data for such sequences no worse than what was received; in other words, the presence of a systematic RS decoder can, with very high probability, only help. (Analysis of the output BER becomes more difficult, since it is necessary to consider (1) the statistics of detecting or not detecting an undecodable received word, (2) the statistical split between symbol errors in the parity checks and data checks, and (3) the number of bit errors per incorrectly received symbol.) Although RS codes are more difficult to analyze, they are effective against combined random and burst errors. The symbol structure of RS codes implies that a symbol may as well be completely wrong as wrong in 1 bit. The major drawback of RS codes in the analog arena is that the present generation of decoders does not fully use bit- or symbol-based soft-decision information. Intermediate levels of confidence, other than erasure information, can not, to date, be processed by a single algebraic algorithm. Conversely, the long block length of RS codes allows high performance to be achieved even in the absence of soft-decision data. Some of the other advantages of RS codes include:

1. RS codes are more suited to high-speed implementation than comparable binary codes. For example, the $RS(255,223)$ code takes 223 bytes of data and encodes them into 255 bytes. The decoder and encoder logic can be implemented to work on byte-based arithmetic throughout. The decoder processes 255 objects of data rather than 255×8 items of binary data, which reduces the complexity of the logic as compared to a true binary code of the same length. Moreover, data transfers can be made on a byte-wide basis; a 25-MHz RS decoder translates into an 8 times 25 = 200-Mbit/s binary decoder.

2. Low-redundancy RS codes are less complex to implement than high-redundancy RS codes, which has proven an especially attractive feature for mass memory applications where the channel error rate is low and the need for integrity high; e.g., it may be required to reduce an error rate from 10^{-4} to 10^{-10} and to keep redundancy below 10%.

3. RS codes have a low rate of decoding errors. If the number of symbol errors exceeds the amount that the code can correct, the decoder will primarily recognize the impossibility of decoding and flag the data as corrupt. The error rate of data that are both undecodable and *not* recognized as such by the decoder may typically be as low as five orders of magnitude below the overall symbol error rate. (The rate of decoding errors can be severely degraded if erasure information is processed; for example, if R erasures are present, no residual detection is possible.) The ability to flag decoding errors can be valuable in intermachine data communications, channel assessment, automatic power control, and related areas.

4. Systematic RS codes can often be easily accommodated within a multiplexed traffic stream. This allows for a multiplexed frame where the majority of the channels are dedicated to user data, frame synchronization, or control channels, and a minority of the channels are dedicated to code check characters. Note that the code delay is divided equally among the individual channels.

7.2.4. Interleaving

In interleaving (see Section 5.13) several code words are transmitted in an intermingled fashion. Interleaving prevents a long noise burst from corrupting any one code word too excessively for correct decoding.

Sections 7.2.1 and 7.2.3 showed that convolutional codes are weak when it comes to burst noise, but Reed–Solomon codes are superior. A common objection to this classification is that codes can be interleaved so that errors in uninterleaved received data are near random and therefore more or less optimal for the convolutional decoder; after all, Reed–Solomon codes are often interleaved. The argument is appealing but fallacious. In a very precise information-theoretic sense, the worst type of noise is produced by independent and identically distributed noise processes. If the noise process has additional structure, this structure can be exploited. More simply stated, the effect of short interference bursts on packetized data transmission can be less damaging than a constant level of random bit errors, given that the total number of errors is comparable. The reason is not profound: Given trouble, it is better to confine the trouble to one packet. In an interleaved code, the deinterleaver carefully tries to convert a less damaging type of noise (i.e., bursts) into a more damaging type of noise (i.e., random).

Given that interleaving is wrong on a philosophical level but practically necessary in many applications, a useful design methodology starts with a code that has some burst error compression (i.e., a symbol code) and interleaves to a lesser extent. In more detail, system delay constraints may preclude interleaving to a depth so great as to make the deinterleaved channel appear memoryless. There is often a nonnegligible probability that a channel noise burst will have a length that is a significant fraction of the system delay constraint. In such cases, the proper comparison is not between idealized coder interleaver systems with infinite interleaving depth but rather between specific schemes that meet the delay constraint. When this delay constraint is hundreds of bits and bursts approach such lengths, long RS codes with well-designed interleavers enjoy an enormous advantage over interleaved convolutional codes. Conversely, when burst length and delay constraints are both only tens or hundreds of bits, long RS codes are precluded. In this case, interleaved convolutional codes may be the only viable option.

7.2.5. Concatenated Codes

Concatenated codes use two or more decoders in series; a typical schematic is shown in Figure 7.5. The *inner code* is typically the code nearest the channel; i.e., the signal is first processed by the inner decoder. The *outer decoder,* which is nearly always an RS decoder in practice, is the last decoder to process the signal.

Concatenated codes drive a neat wedge between the practical nature of coding and the information-theoretic side of coding. In his monograph *Concatenated Codes* Forney (1966) proved that the Shannon capacity could be approached using longer and longer concatenated codes but the rate of ap-

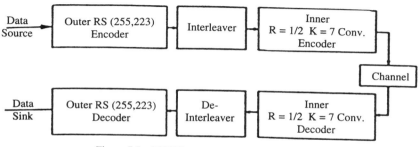

Figure 7.5. CCSDS concatenated coding scheme.

proach was inferior to that of using single-stage decoders of the same lengths. On the other hand, a practical engineer has every right to an initial scepticism about concatenated codes. If coding technologists cannot make one code work, what makes two so good? In practice, a concatenated code may have a combined block length far longer than any *implementable* single code, implying that we can build concatenated code systems that perform closer to capacity.

Another answer is that two codes can split the task at hand. For example, when faced with AWGN, an inner convolutional code is often used with an outer RS code. The inner convolutional code is responsible for processing all the available soft-decision data and making a maximum-likelihood decision; the VA can implement this strategy effectively. However, the limited constraint length of the convolutional code restricts improvement from a convolutional decoder alone. Output from the convolutional decoder may have a BER of, say, 10^{-4}; moreover, statistics of the convolutional decoder output will exhibit bursts of errors. Prior to RS decoding, output from the convolutional code is normally deinterleaved, where interleaving is on a symbol rather than a bit basis. Deinterleaving separates the bursts so that the RS decoder operates on symbols subject to an error process that is less correlated than that leaving the convolutional decoder. In other words, the convolutional decoder plus symbol deinterleaver provides an output well suited to the input of the Reed–Solomon code. The RS code, typically using no more than 6–10% redundancy if it is an 8-bit symbol code, is responsible for reducing the symbol error rate to an output BER below 10^{-10}. This exposition points out a valuable partnership: The convolutional decoder specializes in reducing a highly adverse channel to a medium-quality channel. The RS decoder specializes in raising a medium-quality channel to an extremely high-integrity channel using a minimum of redundancy. The whole is greater than the parts used separately.

Concatenated codes are used with both block codes as the inner code and convolutional codes as the inner code. Convolutional inner codes are most frequently associated with combating AWGN on power-limited channels and in particular, with the successes NASA has had in retransmitting compressed video from distant planets. Inner block codes are most often associated with applications of concatenated RS codes to frequency-hopping radios faced with a jamming threat or to other harsh links where soft-decision data are not readily available or unreliable.

Concatenation of the $(2,1)$ $K = 7$ convolutional inner code with an outer $RS(255,223)$ code on 8-bit symbols deserves mention. This scheme, devised by NASA's Jet Propulsion laboratory, has been adopted by the Consultative Committee on Space Development Services (CCSDS) as a standard for all deep-space telemetry data. Perhaps more importantly, the code has proved to work well. Figure 7.6 shows several performance curves on an output BER versus signal-to-noise power ratio curve. On the far right is the no-coding

Practical Aspects of Coding

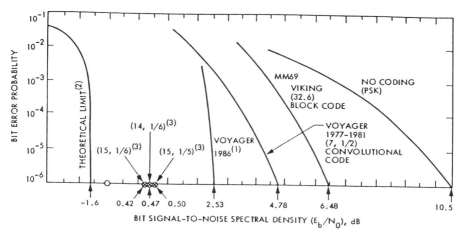

Figure 7.6. Deep-space telemetry performance. [1](7, 1/2) Convolutional code (Viterbi decoding) concatenated with a (255,223) Reed–Solomon outer code. [2]Infinite bandwidth expansion. [3](14,1/6) Concatenated with a (1023,959) Reed–Solomon code. ○, Theoretical limit for rate 1/5 codes is −1.01 dB, and for rate 1/6 codes it is −1.10 dB. (Included by permission of the Communications Systems Research Section at the Jet Propulsion Laboratory.)

curve for Phase Shift Key (PSK) modulation. On the next right-hand curve is the curve for the (32,6) binary Reed–Muller code that was used in an early probe on Mars. Note that the power saving, relative to uncoded PSK and a delivered BER of 10^{-6}, was about 4.0 dB (10.5−6.48).

From 1977 to 1981, Voyager communications used the (2,1) $K = 7$ convolutional decoder without concatenation; this improved the gain to about 5.25 dB. Finally, the 1986 Voyager mission used the (2,1) $K = 7$ code concatenated with the $RS(255,223)$ 8-bit code, improving the gain to nearly 8 dB. Note that Shannon's capacity indicates that a maximum gain of about 11 dB is achievable. Note also from Figure 7.6 that more advanced concatenations are planned to provide another 2 dB gain over the 1986 Voyager concatenation. The plan is to use a much stronger (4,1) $K = 15$ convolutional code with an $RS(1023,959)$ code on 10-bit symbols.

Figure 7.7 shows the performance of these codes in a different setting. The x-axis is the code rate, and the y-axis displays the code's shortfall in dB from meeting Shannon's infinite bandwidth expansion capacity lower bound. The curve exposes the large gain that occurs in going from a convolutional code alone to a convolutional plus RS concatenated system. All of the concatenated schemes use symbol interleaving.

The RS codes can be concatenated with block codes and used, for example, on a frequency-hopping radio. Motivation for such schemes is that a

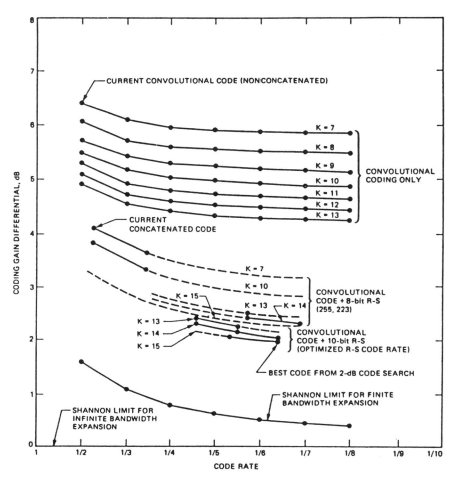

Figure 7.7. Coding gain relative to the Shannon limit for infinite bandwidth expansion at a bit error rate of 10^{-6}. ———, Actual performance of good codes found during code search; ----, extrapolation based on matching slope of Shannon limit curve. (Included by permission of the Communications Systems Research Section at the Jet Propulsion Laboratory.)

hop may be accidentally or deliberately afflicted with high-power interference. Since it is undesirable to spread this damage from one hop to another, it is more appropriate to confine the inner code words to one hop rather than to have a convolutional code acting across hops. It is possible to see another partnership: Inner code words trying to clean up noise within a hop or failing that, diagnosing the damage as an erasure. The RS code words, interleaved

Practical Aspects of Coding 387

across hops, try to mend the damage on jammed hops by reference to the hops successfully cleaned up by the inner decoder.

There are a variety of ways that an inner decoder can act in cooperation with the receiver to diagnose an entire hop of data as having been jammed by interference. The result of such processing allows the RS decoder to have primarily good symbols interspersed with occasional erased symbols. As one example, a slow hopper may have about ten inner code words on a hop. If the hop is randomized, the decoder will with high probability refuse to decode for some fraction of the corrupted code words. One possible strategy is to erase the entire hop of data if the number of refusals exceeds a threshold T. To give some perspective, a concatenation of a $(15,5,7)$ binary code with an $RS(31,15)$ code on 5-bit symbols can combat a high level of memoryless binary errors. Figure 7.8 shows the performance of this code. A channel error rate of 1 in 10 is reduced to below an output BER of 10^{-8}. When adapted for a slow frequency-hopping radio, the same pair of codes can deliver block

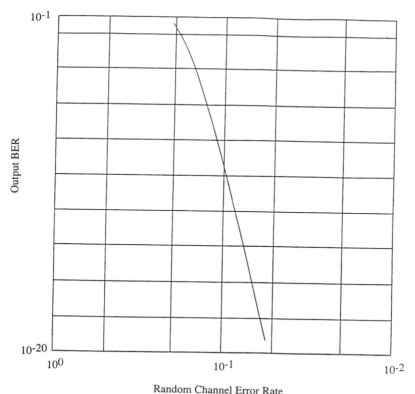

Figure 7.8. Performance of a concatenated code.

messages with high probability even when faced with 10–15% blocked channels and an error rate of 1 in 10 on a good channel. Historians of concatenated RS codes should also research the Joint Tactical Information Distribution System (JTIDS) and the origins of spread-spectrum communications (see Scholtz *et al.*).

7.3. Coding and Modulation

In 1948, Shannon introduced his formula for the channel capacity C in bits per second of an ideal band-limited Gaussian channel

$$C = W \log_2(1 + SNR) \text{ bits/sec}$$

where W is the channel bandwidth in Hz and SNR is the channel signal-to-noise ratio. If the SNR is high compared to the bandwidth, i.e., the channel is band limited, Shannon's results indicate that an upper limit of about 9 dB of gain can be expected from coding. In 1982, Ungerboeck published his work on trellis-coded modulation (TCM) which showed that easily implemented schemes could achieve 3–4 dB of coding gain. The primary application has been to commercial voice-band modems; for example, it is now possible to purchase modems that communicate at 19.2 kbps over a 3-kHz bandwidth, whereas previous generations of equipment communicated at 2.4 kbps or less. (The Shannon capacity of these channels has been estimated at about 23–30 kbps.) The rapid acceptance and implementation of TCM for the band-limited voice-band telephone channel has been one of the more spectacular success stories of coding technology.

Most TCM schemes known to the authors operate on the same principles; there are five major steps:

1. Selecting a block lattice.

2. Selecting a finite number of points from the lattice as a set of modulation points; this is sometimes called the *constellation*.

3. Labeling selected points to ensure that the Euclidean distance between distinct points is related to the difference in the labels.

4. Selecting a convolutional encoder that generates a succession of point labels; in other words, a sequence of constellation points is transmitted that includes code structure.

5. Decoding and demodulation via the VA.

Appendix 7A concentrates on Step 1, i.e., selecting a block lattice.

Practical Aspects of Coding

Coding and modulation technology has come full circle since Shannon's original 1948 and 1949 papers. These papers explicitly stated that a set of possible signal sequences can be regarded as a set of points in a multidimensional space. By considering subsets of multidimensional points in real space, Shannon exposed the concept of channel capacity in AWGN for an average power-limited signal. His results can be paraphrased as follows: It is possible to find a set of points in real N-dimensional space that has the following properties if the AWGN is below a certain power:

1. The set has enough points to allow a large amount of data to be communicated.

2. The squared Euclidean distance between any pair of distinct points is large enough to give an acceptably small average error probability.

3. The squared Euclidean distances of the points from the origin obey an average power constraint.

Mathematically, this formulation points toward the study of *sphere-packings* (see Conway and Sloane) where a sphere-packing refers to a countable subset of points in multidimensional real space. The name comes from centering equal nonoverlapping (but possibly touching) spheres at the points in the subset. Efficient sphere packing encloses a high percentage of the vector space; i.e., there is little gap between the spheres. Appendix 7A examines the theory of certain structured sphere packings and derives some of the simpler examples.

The restatement of Shannon's results does not appear to involve codes over a finite alphabet with minimum Hamming symbol distance. However, in the search for good codes, much 1960s and 1970s research focused on codes defined over finite fields. Although it may appear that the resulting codes have little application to the multidimensional sphere-packing problem, there are close connections between codes over finite fields and sphere packings, particularly if we confine the study to modulation schemes with implementable decoders and demodulators. The connection is found in ring theory and in particular, the theory of modules over principal ideal domains (PIDs); this theory is outlined in Appendix 7A.

7.4. Hybrid Forward Error Correction (FEC) and Retransmission (ARQ) Schemes

Section 7.4.1 makes a few general statements comparing the properties of ARQ and FEC when used in isolation as error control techniques. Section

7.4.2 describes some of the major classifications of ARQ strategies. Sections 7.4.3–7.4.6 describe various possibilities of codesigning FEC and ARQ into hybrid error control strategies. Most ARQ techniques make up a small part of computer protocols of a far greater utility. (We do not attempt to comment on the protocol aspects.) It should be noted that the basic philosophy of FEC differs from ARQ: FEC attempts to "get it right the first time," while ARQ attempts to "get it right next time if it is wrong this time or maybe the time after that if it is still wrong or maybe the time after that."

7.4.1. General Comments

An ARQ strategy is a transmission strategy that relies on retransmitting data that was received with errors. Error correction typically requires a much higher proportion of check symbols than does error detection; i.e., for a given block size, FEC normally requires more redundant symbols than does ARQ, leaving fewer symbols for information and control. Error correction requires much more computation, and decoder hardware is consequently more complex and expensive, although these costs continue to drop. The ARQ error detection technology is trivial in comparison.

We now consider some situations where using FEC is advantageous or unavoidable. Link performance is often measured in terms of average throughput (information delivered per unit of time) assuming a constant backlog of messages to be sent (or equivalently, one long message). At high channel error rates, ARQ is inefficient because of the large number of retransmissions, but using FEC, or a hybrid of FEC with ARQ, can often increase throughput substantially. In many communication systems, delays for short messages may be critical. An important property of FEC is fixed delay. Quality of the delivered data may vary with link quality, but the time taken to deliver a message using FEC does not depend on the channel BER. With ARQ, data quality is relatively independent of channel conditions but delay may be subject to large variations with the channel error rate; i.e., the delay distribution has a long "tail." Finally, in some communications scenarios, no feedback information is available, ARQ is impossible, and FEC is the only technique available to help improve data integrity. Agent 00111 often faced such situations when transmitting from behind enemy lines.

7.4.2. Some Basic ARQ Strategies

As an alternative to FEC, retransmission schemes are common. In such schemes, messages are split into convenient size units, and along with control data, redundancy bits are attached to detect the presence of errors; if errors are detected, the message is sent again. Within this framework, there are many different types of ARQ, including "stop-and-wait," "selective-repeat," and "go-back N"; in addition, there is a logical split between positive and negative acknowledgment systems.

In a positive acknowledgment (ACK) system, the data source expects confirmation from the data sink on the correct receipt of every block; for example, an ACK may contain the sequence number of the next packet awaited, indicating receipt of all packets with lower sequence numbers. In a negative acknowledgment (NACK) system, the data source retransmits data only on specific request. This type of response may be appropriate in a multiple-access environment with a low error probability and a shortage of bandwidth. The majority of ARQ systems use positive acknowledgments.

In some data transmission environments, blocks of data must be delivered to the data sink (receiving host) in proper numerical sequence. One simple method of achieving this is called stop-and-wait ARQ, where the protocol waits for an ACK for each packet before sending the next.

For links with long round-trip delay relative to packet duration (such as satellite channels), stop-and-wait is clearly inefficient. If the receiver has the ability to buffer packets, it can accept out-of-sequence packets and rearrange the sequence at a later date. ARQ protocols that manage such buffers are often called sliding window protocols. The window refers to the amount of memory, i.e., the range of packet sequence numbers that the receiver can accept and rearrange in proper order.

In a go-back-N system, when the receiving protocol requests retransmission (negative acknowledgment) or the transmitter times out after waiting for a positive acknowledgment, the transmitter retransmits a total of N packets. Provided that round-trip delays are within preset bounds, this strategy ensures that a missing packet is included among the repeated data. Although simple, go-back-N schemes are inefficient, since they retransmit more data than necessary. The selective repeat protocols allow specific packets to be repeated on request. Note that a selective repeat system can be made into a positive acknowledgment system by having the sender retransmits one packet when a time-out occurs and resetting the time-outs for the other packets. A selective repeat strategy works best when multiple packet failures are rare.

7.4.3. Type 1 Hybrid FEC/ARQ

One common FEC/ARQ system is called Type 1 hybrid FEC/ARQ. In such systems, error correction is used to reduce the probability of packet failure. If the FEC fails, error detection is used to trigger retransmission. At one extreme, FEC and ARQ mechanisms can be kept totally distinct (even at different levels of the network architecture). At another extreme, the FEC decoder's residual error detection can be used to trigger ARQ requests.

It is illustrative to compare the performance of a pure ARQ system to that of a Type 1 hybrid FEC/ARQ system. In the following, two (idealized) systems are postulated. System 1 uses ARQ alone. Each packet has 14 bytes (112 bits) of control data that include error detection redundancy. The total packet length is chosen to be optimal for the prevailing binary symmetric channel random BER. In general, a system cannot know the exact error rate and may suffer some performance degradation due to a suboptimal choice of packet length. In System 2, Type 1 FEC/ARQ is used. The data, with the same 14 bytes of overhead, are protected by an $RS(64,48)$ code on 6-bit symbols. The length of the packet is again optimized to give the maximum throughput (with the additional constraint that a packet should contain an integer number of code words). Figure 7.9 shows the optimal length versus the prevailing error rate. Figure 7.10 shows the comparative throughput of

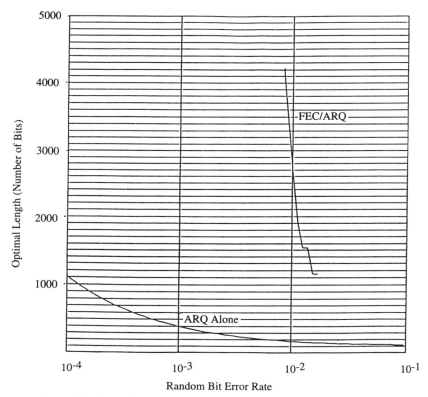

Figure 7.9. Optimal length vs. error rate. FEC: RS (64,48) on 6-bit symbols.

Practical Aspects of Coding

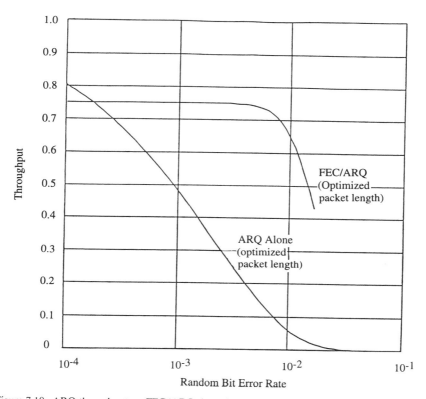

Figure 7.10. ARQ throughput vs. FEC/ARQ throughput. FEC: (64,48) RS code on 6-bit symbols.

the two systems. System 2, the Type 1 system, is clearly superior in the worse channel rates. For extremely benign channels, the coding overhead is not necessary to achieve throughput, and System 1, ARQ alone, is superior. Such results are typical of these comparisons. Note that the optimal packet length for the Type 1 system is much longer than for ARQ alone. In practice, packets are often kept short for practical reasons and/or to ameliorate the effect of sporadic interference or fading. Figure 7.9 may be interpreted as stating that there is little reason to tie the FEC block-coding boundaries to the packet boundaries in a Type 1 system.

7.4.4. Type 2 Hybrid ARQ/FEC

One objection to Type 1 systems is that the FEC redundancy is present even when the noise is not. This, in addition to other reasons, has led to the design of Type 2 ARQ/FEC systems. There are many different kinds of Type 2 ARQ systems with the characteristic of using ARQ first and then FEC; for example in the following schemes the transmitter responds on receiving an ARQ request by sending FEC parity symbols for the original data rather than the original data. (For this reason, *repetitive parity* is a synonym for Type 2 ARQ/FEC.) The method is best explained by example. Data are initially sent without FEC protection but with redundancy attached for error detection and ARQ purposes. Suppose that k bytes of data are sent in a packet. If the packet is received correctly, there is no problem; if errors are detected, a request is sent back to the data source. The data source does not repeat the packet; instead the source uses a $(k + r, k)$ systematic code to obtain r redundant symbols. These parity symbols are sent to the data sink; typically r may be much less than k (examples of r being about 10% of k have been proposed). At the receiver, the two blocks are sent to a $(k + r, r)$ error-correcting decoder that attempts to extract the correct data. If this succeeds, the process is over; if not, another request is sent to the data source, and one of several possible back-up strategies is activated; for example, the original packet is repeated. Alternatively, a $(k + r + s, k + r)$ systematic code is used to generate s further redundant bytes that are sent back to the data sink. There are many variations of similar themes, and although individual variants have been analyzed, a general theory does not yet seem to have emerged.

Given the following conditions:

1. The average noise is mild enough to give the ARQ a reasonable throughput.

2. The worst-case noise is not routinely severe enough to exceed the protection of the underlying FEC code [e.g., the $(k + r, k)$ code in the preceding example].

3. The noise level changes slowly relative to the block length.

then the throughput of a Type 2 system may be close to the theoretical channel capacity. In practice, some of these conditions may be hard to ensure in advance of an application.

Type 2 systems have been suggested as ingredients in further hybrids that try to use the power of concatenated codes. This involves using a code with a fairly high throughput that is also easy to decode; an example is a rate 3/4 convolutional code processing soft-decision data, where the soft-decision data are generally reliable. If a block of data encounters a noise burst or fade that swamps the convolutional decoder, an ARQ request for an afflicted block of data is sent. However, the request generates parity symbols from a systematic encoder at the transmitter, which are sent over the channel. After receiving both the corrupted data packet and the (possibly corrupted) parity checks and applying the Type 2 processing, the received packet has the full strength

Practical Aspects of Coding

of a concatenated RS/convolutional code. Thus the strength of a concatenated code is deployed but only on those blocks erroneous enough to swamp the inner decoder. (Note that the method relies on the infrequent use of the RS decoder.) This strategy has certain merits and demerits: One advantage is that if the RS code is used infrequently, the RS decoding can be performed in non-real-time decoding. One disadvantage is that the RS code may not be very redundant, so it may be simpler to have the RS code permanently installed on the link.

7.4.5. Chase Code Combining

Chase code combining (Chase, 1985) is a technique for combining retransmissions in which errors have been detected. In essence, the idea is simply not to throw away partial information; for example, if a packet originally contained the data 00111, and three data transmissions are received that are all known to be wrong from error detection techniques. Suppose the three received data packets contain 10111, 00011, and 00110. Then simple majority voting on the three copies of the data will extract the true message. Code combining contains more details than this above example exposes, since packet repeats are accumulated and processed until data are extracted. If there are k bits of data in a packet, n encoded bits per packet, and L packets have been received, the process is treated as a decoding problem for an (nL,k)-code. Note that code combining may be technically regarded as a special case of a repetitive parity Type 2 scheme.

7.4.6. Coding and Future Networks

It is traditional to treat both FEC and ARQ as link-level techniques that improve packet reception. The process of generating, reconstituting, and routing packets is treated as part of a higher level discipline beyond error control techniques, but two factors indicate that error control may have a role at these higher levels. First, many networks have moved out of benign office environments into much harsher operational environments; second, extremely high-speed optical technologies threaten to tear away the memory-intensive protocols at a conventional switch, and there is a lack of static mass memory operating at these speeds. In both cases, despite the best efforts of link-level error control, the major problem facing a message recipient could be missing packets rather than erroneous packets; i.e., switch buffers drop data and/or links are jammed completely. Given present technology, such situations lead to unacceptable message delivery delays. As Section 7.2.2 shows, this situation can be readily improved.

We now turn to an example from Agent 00111's colorful career. Consider a large military network that must communicate a vital message from a data source to a data sink in the shortest possible time despite massive electronic and conventional warfare attacks. Before the attacks, there were many possible routes connecting the source and destination; now, however, many or most of the long-distance cables have been cut, and switches are being redeployed, so radio is the principal mode of contact. Directional jammers are sweeping high-energy interference across the network, attempting to trick adaptive-routing protocols into closing down all routes. In reaction, adaptive-routing protocols have been disabled, but due to the attacks, routes have become unpredictable with wildly fluctuating data qualities and time delays despite the best efforts of link-level error control. Route quality is impossible to predict in advance. We examine how two systems react to this scenario.

System 1, a Conventional Approach

A message is split into K packets at the data source. Each packet has an identification number, error detection redundancy, and other control data attached to it. Due to the extreme urgency of the message, we assume the data source continuously retransmits the packets in parallel, assigning packets to routes in random fashion with a different route for each packet, stopping only when the destination positively acknowledges the message. At the destination, we assume packets are either received correctly or rejected. (The presence of link-level error control, HDLC framing data, and cryptographic equipment all support this assumption.) The destination has set up a list of packets that have arrived; only when all K out of K packets have arrived can the message be assembled. The essential problem in System 1 is that message delivery time equals the time it takes to deliver the *most* delayed packet.

System 2, an Unconventional Approach

The message is split into K packets of user data as before. A systematic (N, K) MDS code (see Section 7.2.2) is used in conjunction with interleaving to produce N packets of encoded data. Interleaving is performed so that each packet contains precisely 1 symbol from each code word. Each of the N packets has an identification number, error detection redundancy, and other control data attached to it, as in System 1. As in System 1, the data source continuously retransmits packets in parallel, randomly assigning the N packets to routes, with a different route for each packet, and stopping only when the destination positively acknowledges the message. At the destination, we assume, as with System 1, that packets are either received correctly or rejected and data sink has a list of packets that have arrived. However, the MDS interpacket–coding

Practical Aspects of Coding

property means that when any K out of N packets have arrived, the original message can be decoded. Note that it is easier to obtain any K out of N packets than to obtain all K out of K. Message delivery time equals the time it takes to deliver the K *least* delayed packets.

Figure 7.11 shows an analysis of the delivery time for the System 2 scenario. Delivery time is normalized so that in the absence of any noise, it takes unit time to deliver K parallel packets. The x-axis shows the percentage of links between the source and sink that are jammed at any one time. The y-axis shows the average time required to deliver the complete message. Packet coding makes delivery time relatively resilient to jamming.

Research Problem

1. Explore and analyze how the preceding packet-coding concept can be combined with Chase code combining.

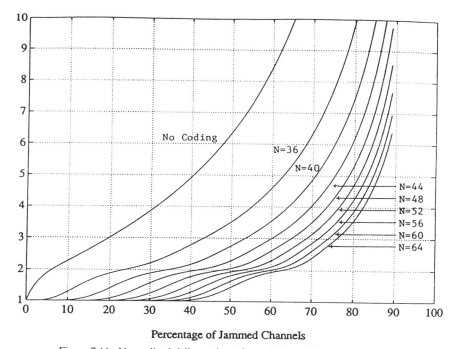

Figure 7.11. Normalized delivery times for coded $K = 32$ packet messages.

7.A. Appendix: Lattices and Rings

Section 7.3 stated that sphere packing relies on the theory of modules defined over PIDs, an aspect of ring theory. This appendix outlines this relationship, concentrating on the simpler examples of how to construct interesting sphere packings from finite codes. We confine our attention to a subclass of sphere packings known as *lattices*. A *lattice* L is a countable set of points in a real vector space having the property that the sum of any two points in the lattice is also in the lattice:

$$x \in L \text{ and } y \in L \to x + y \in L$$

In ring theory terminology, L is a *finitely generated Z-module.*

The starting point for our constructions is a low-dimensional lattice L that possesses additional algebraic structure. In practice, the three most important examples are

1. L equal to Z, the set of integers

2. L equal to the Gaussian integers G; i.e.,

$$G = \{a + ib \mid a, b \in Z, i = \sqrt{-1}\}$$

3. L equal to the Eisenstein integers E; i.e.,

$$E = \left\{\frac{a}{2} + \theta\frac{b}{2} \mid a, b \in Z, a \equiv b \,(mod\, 2), \theta = \sqrt{-3}\right\}$$

Note that if ω is defined as

$$(1/2)(1 + \theta) = (1/2)(1 + \sqrt{3}i)$$

ω is a primitive sixth root of unity; that is

$$\omega = e^{2\pi i/6}$$

where ω is an element of E, and an equivalent definition of E is

$$E = \{c + d\omega \mid c, d \in Z\}$$

Less formally, G is a two-dimensional square grid of points (i.e., graph paper; see Figure 7.12), and E is a two-dimensional hexagonal grid of points (see Figure 7.18). If we take a large number of equally sized coins and place as many as possible on a large table, the coin centers would lie on a hexagonal grid similar to E; E is the optimal sphere packing in two dimensions. Note that Z, G, and E are closed under addition and multiplication and the squared Euclidean distance of any point from the origin is an integer.

Z, G, and E all have considerable algebraic structure beyond the lattice requirement: They are Euclidean domains with respect to squared Euclidean distance and *a fortiori*, PIDs and unique factorization domains (UFDs) (see Hartley and Hawkes). It follows that prime numbers and prime decomposition can be defined for each choice of L; however, some formal definitions are needed.

Definitions

1. R refers to a commutative ring with an identity element and no zero divisors; that is, R is a set with two operations known as *addition* and *multiplication* ($+ = +_R$ and $\cdot = \cdot_R$). $(R,+)$ is an additive abelian group;

Figure 7.12. Some points in G.

multiplication is commutative, associative, and distributive over addition. There exists an element 1_R with the property that $1_R \cdot x = x$ for all $x \in R$; and if $a \cdot b = 0$, either $a = 0$ or $b = 0$. Denote the nonzero elements of R by R^*.

2. An *R-module* is an abelian group closed under multiplication by elements of R.

3. A *cyclic* R-module C is an R-module whose elements are multiples of a single element $x \in C$ by the elements of R; i.e., $C = \{y | y = rx, r \in R\}$.

4. An *ideal* of R is an R-module contained in R.

5. A PID is a ring R with the additional property that every ideal is cyclic; in the following definitions, R is always a PID.

6. An element $u \in R$ is a *unit* if it has a multiplicative inverse; i.e., $uv = 1$ for some $v \in R$. Note that there are two units in Z: $(+1, -1)$, four in G: $(1, i, -1, \text{ and } -i)$, and six in E: $(1, \omega, \omega^2, -1, -\omega, -\omega^2)$, where $\omega = (1/2)(1 + \sqrt{3}i)$ is a primitive sixth root of unity.

7. An element $p \in R$ is *prime* if and only if

   a. p is not zero or a unit.

   b. p divides ab implies p divides a or p divides b (a divides b is written $a|b$).

8. An element $p \in R$ is said to be *irreducible* if $a|p$ implies that a is either a unit or p times a unit. Note that in a PID, an element is a prime if and only if it is irreducible; in more general structures, this is not always the case.

9. A ring R is a *UFD* if:

   a. Every nonzero element is equal to a unit times a product of primes.

   b. Up to reordering, the primes in the product are uniquely determined by the element.

10. For an R-module S, define $o(S)$ to be the ideal of R equal to

    $$\{e | e \in R \text{ and } em = 0 \text{ for all } m \in S\}$$

    $o(S)$ is called the *order ideal* of S.

11. An R-module S *has order* d, $d \in R$, if $d \cdot m = 0$ for all $m \in M$ and d generates $o(S)$; more precisely, since R is a PID, $o(S)$ is generated by a

single element $d \in R$, where d is well defined up to unit multiplication, and d is set equal to the order of S.

12. A *finitely generated* R-module is equal to a finite sum (not necessarily direct) of cyclic R-modules.

13. A *Euclidean domain* is a ring R (in the sense of Definition 1) and a function $f: R^* \to Z_{\geq 0}$ such that

   a. a divides b implies that $f(a) \leq f(b)$.

   b. Given $a \in R$ and $b \in R^*$, there exist $q, r \in R$ such that $a = qr + b$ with either $r = 0$ or $f(r) < f(b)$.

14. If I is an ideal in R, R/I is defined to be the set $\{I + r \mid r \in R\}$. R/I is also a ring where $(I + r) + (I + s) = I + r + s$ and $(I + r) \cdot (I + s) = I + rs$. In fact, if I is generated by a prime element and R/I is finite, R/I is a field. An element of R/I, $I + r$, say, is referred to as a *coset* of I in R.

Exercises

1. Find a prime that is not irreducible in $Z(\sqrt{-5})$. (*Hint:* factor 6 in two different ways.)

2. Show that in a UFD, a prime element is an irreducible element, and vice versa.

3. Prove that $o(S)$ is an ideal of R.

4. Show that a Euclidean domain is a PID.

5. Show that a PID is a UFD.

6. Show that G and E are Euclidean domains with respect to squared Euclidean distance.

7. Show that if I is an ideal generated by a prime element and R/I is finite, then R/I is a field. *Hint:* show that R/I does not have any zero divisors. By definition, this makes R/I a finite *integral domain* and invokes the result that a finite integral domain is a field.

   Let p equal 2 for Z, $p = \phi = 1 + i$ for G, and $p = \theta$ for E. In each case, p is a prime number of the smallest squared Euclidean distance, i.e., with distance equal to 4, 2, 3 for Z, G, and E, respectively. Note that finite fields are involved, since L/pL equals a finite field F_r with r elements. In more detail,

$$\frac{Z}{2Z} = F_2$$

$$\frac{G}{\phi G} = F_2$$

$$\frac{E}{\theta E} = F_3.$$

Dividing L into cosets of pL has a geometrical interpretation essential to coded modulation. Consider a pair of points in L that are a minimum squared Euclidean distance apart; i.e., they are adjacent in the lattice. The points are in different cosets of pL in L. In communication terms, the most error-prone pairs of points are labeled with different least significant bits; as examples, Z can be split into $0 + 2Z$ and $1 + 2Z$; G into $0 + \phi G$ and $1 + \phi G$; and E into $0 + \theta E$, $1 + \theta E$, and $\omega + \theta E = 2 + \phi E$. This process of looking at the cosets of pL in L is repeated, since it is possible to look at the cosets of $p^2 L$ in pL and the cosets of $p^{i+1} L$ in $p^i L$. The result can be viewed geometrically as a partial labeling of the lattice or algebraically as a primary decomposition of the points.

Example 7.A.1.

1. In Z, any integer can be taken modulo 16 and labeled with four binary bits (a_0, a_1, a_2, a_3):

$$z = \sum_{i=0}^{3} a_i 2^i + 16Z$$

where $a_i \in F_2$.

2. Any Gaussian integer $g \in G$ can be taken modulo ϕ^4 and labeled with four binary bits (a_0, a_1, a_2, a_3):

$$g = \sum_{i=0}^{3} a_i \phi^i + \phi^4 G$$

where $a_i \in F_2$.

3. Any Eisenstein integer $e \in E$ can be taken modulo θ^4 and labeled with four ternary digits (b_0, b_1, b_2, b_3):

$$e = \sum_{i=0}^{3} b_i \theta^i + \theta^4 E$$

where $b_i \in F_3$.

In all cases, the ith coordinate refers to the coefficient of the ith power of the appropriate prime. It is important to note in these labelings that the value of the squared Euclidean distance is partially determined by the labels. Let q equal the squared Euclidean distance of p from the origin. (Recall that $q = 4$, 2, and 3 for Z, G, and E, respectively.) Consider a point in L that has its ith coordinate equal to zero for all $i < j$. Then from the preceding discussion, it is known that the squared Euclidean distance of this point from the origin is equal to 0 modulo q^j. This algebraic result has a geometric interpretation.

Example 7.A.2 (Gaussian Integers G, $p = \phi$). The geometry of prime factorization of G with respect to ϕ is illustrated in Figures 7.12–7.17 for G. Figure 7.12 shows part of the G lattice or an unlabeled rectangular grid. In

+1	+0	+1	+0	+1	+0	+1	+0
+0	+1	+0	+1	+0	+1	+0	+1
+1	+0	+1	+0	+1	+0	+1	+0
+0	+1	+0	+1	+0	+1	+0	+1
+1	+0	+1	+0	+1	+0	+1	+0
+0	+1	+0	+1	+0	+1	+0	+1
+1	+0	+1	+0	+1	+0	+1	+0
+0	+1	+0	+1	+0	+1	+0	+1

Figure 7.13. Some labeled points in G.

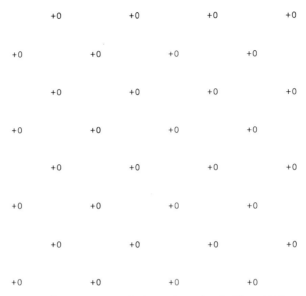

Figure 7.14. Identically labeled points in Figure 7.13.

Figure 7.13, the grid is labeled so that the closest points have different labels. Figure 7.14 selects the sublattice consisting of the points of Figure 7.13 ending in zero. Note that these points can be described algebraically by ϕG; i.e., G is rotated 45 degrees and magnified by $\sqrt{2}$. The minimum squared Euclidean distance in Figure 7.14 is twice that in Figure 7.12. In Figure 7.15, the points in Figure 7.14 are labeled so that the closest (Figure 7.13) points have different labels. Figure 7.16 shows the effect of isolating the points in Figure 7.15 that end in 10. The set is identical to $\phi^2 G$. The minimum squared Euclidean distance of the points in Figure 7.15 is 4. It is possible to continue splitting points by their labels until all 64 points in Figure 7.12 have distinct labels; this is shown in Figure 7.17. The label corresponds to the decomposition of G modulo $\phi^6 G$.

Example 7.A.3 (Eisenstein Integers and $p = \theta$). The geometry of prime factorization in E with respect to θ is illustrated in Figures 7.18–7.20. Figure 7.18 shows some of the points in E; these points are labeled 0, 1, or 2, so that the closest points have distinct labels. Figure 7.19 shows the labels. If points ending in 2 are extracted, the sublattice of points shown in Figure 7.20 results. This sublattice is essentially an offset, rotated, and magnified version of E. The minimum squared Euclidean distance in Figure 7.20 is 3.

Practical Aspects of Coding

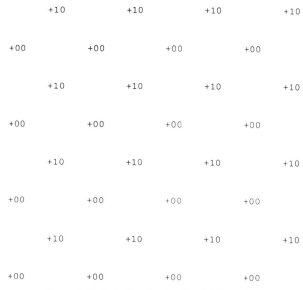

Figure 7.15. Labeling the 0 points in Figure 7.14.

Given a low-dimensional lattice, a primary decomposition and/or a labeling that reflects Euclidean distance, it is now possible to apply coding to obtain higher dimensional and superior lattices.

All coded modulation schemes control the squared Euclidean distance by restricting the sequences of labels. Since the labels equal coset representatives, this process has been called *coset coding* (see Calderbank and Sloane). Using block codes leads to block lattices, while convolutional codes lead to trellis-coded modulations. In the following discussion, we focus on block codes.

A lattice Λ is constructed that is contained in L^N; i.e., a lattice point of Λ is a sequence of N points selected from L. The following definitions are helpful:

1. $M_i = (p^i L)^N$
2. $\Lambda_i = M_i \cap \Lambda$
3. A depth-j lattice of length N is a lattice Λ for which

$$\Lambda_j = M_j$$

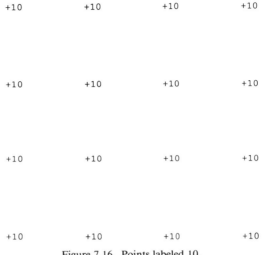

Figure 7.16. Points labeled 10.

Note that an element of Λ_i is an element of Λ with all N coordinates having squared Euclidean distance equal to 0 modulo q^i. In general, Λ_i is contained in M_i, but it is not necessarily equal to M_i.

The classification of depth-j lattices is identical to defining a set of nested codes of length N where the codes are defined over $F_q = L/pL$. Moreover, the minimum squared Euclidean distance of a point in Λ can be lower bounded by an expression involving the minimum distance of the nested codes; the result is stated in Theorem 7.A.2. This classification arises from a more general and fundamental result in ring theory, i.e., the classification of finitely generated modules defined over a PID; see Theorem 7.A.1. (Theorem 8.2 in (Hartley and Hawkes)). The following notation is used in Theorem 7.A.1:

1. R refers to a commutative ring with an identity element and no zero divisors.

2. An R-module is an abelian group closed under multiplication by elements of R.

3. A cyclic R-module C is an R-module whose elements are the multiples of a single element $x \in C$ by the elements of R; i.e., $C = \{y | y = rx, r \in R\}$.

4. A finitely generated R-module is equal to a finite sum (not necessarily direct) of cyclic R-modules.

+010101+010110+011001+011010+100101+100110+101001+101010

+010100+010111+011000+011011+100100+101111+101000+101011

+010001+010010+011101+011110+100001+100010+101101+101110

+010000+010011+011100+011111+100000+100011+101100+101111

+000101+000110+001001+001010+110101+110110+111001+111010

+000100+000111+001000+001011+110100+110111+111000+111011

+000001+000010+001101+001110+110001+110010+111101+111110

+000000+000011+001100+001111+110000+110011+111100+111111

Figure 7.17. Labeling the 64 points in Figure 7.12.

Theorem 7.A.1. Let R be a PID and S a finitely generated R-module. Then S can be expressed as a direct sum

$$S = S_1 + S_2 + \cdots + S_s \qquad (s \geq 1)$$

where

1. S_i is a nontrivial cyclic submodule of S of order d_i.
2. $d_1 | d_2 | \cdots | d_s$.

Proof: Theorem 7.A.1 is proved by several different methods in (Hartley and Hawkes). The reader is urged to look at the proof for the special case of Euclidean domains (p. 111). Note that in Hartley and Hawkes, the cyclic modules can be divided between those where d_i equals 0 and those where d_i is nonzero. A module is called a torsion-free or a torsion module depending on whether its order is zero or nonzero, respectively. It is a matter of definition that $r|0$ for all $r \in R$ and Part 2 of Theorem 7.A.1 covers both torsion and torsion-free submodules.

Figure 7.18. Some points in E.

It is now possible to state Theorem 7.A.2; in the following presentation, a *linear code* is a synonym for any vector space over F_r.

Theorem 7.A.2. If Λ is a depth-j lattice, then there exist linear codes C_i with parameters (N, k_i, d_i), $0 \le i \le j$, for which:

1. C_i is a code over F_r and C_j is the trivial $(N, N, 1)$ code.

2. $C_i \subseteq C_{i+1}$

3. $(\Lambda_i / \Lambda_{i+1})$ is isomorphic to C_i.

4. The minimum squared Euclidean distance of a nonzero point from the origin is $d_{\min}^2$ where:

$$d_{\min}^2 \ge \min_{0 \le i \le j} (q^i d_i)$$

and the product and minimization are over the real numbers.

In particular, if $d_i = q^{j-i}$ for all i: $0 \le i \le j$, then

Practical Aspects of Coding

	+2	+0	+1	+2	+0	+1	+2	+0	+1	+2	
+0	+1	+2	+0	+1	+2	+0	+1	+2	+0	+1
	+2	+0	+1	+2	+0	+1	+2	+0	+1	+2
+0	+1	+2	+0	+1	+2	+0	+1	+2	+0	+1
	+2	+0	+1	+2	+0	+1	+2	+0	+1	+2
+0	+1	+2	+0	+1	+2	+0	+1	+2	+0	+1
	+2	+0	+1	+2	+0	+1	+2	+0	+1	+2
+0	+1	+2	+0	+1	+2	+0	+1	+2	+0	+1
	+2	+0	+1	+2	+0	+1	+2	+0	+1	+2
+0	+1	+2	+0	+1	+2	+0	+1	+2	+0	+1
	+2	+0	+1	+2	+0	+1	+2	+0	+1	+2
+0	+1	+2	+0	+1	+2	+0	+1	+2	+0	+1

Figure 7.19. Labeling the points in Figure 7.18.

$$N = q^j \quad \text{and} \quad d^2_{\min} \geq q^j.$$

Proof: Apply Theorem 7.A.1 to the L-module $S = \Lambda/M_j$. Since S is finitely generated, S is a sum of cyclic modules. Since $p^j S = 0$ in S, all cyclic modules are torsion modules. Furthermore, it is a consequence of unique factorization that the order of each submodule has order that is a divisor of p^j; i.e., the orders are p^i, $i \leq j$. If the direct sum of all the cyclic modules with order equal to p^i is taken, the result is a module defined to be H_i. It follows that

$$\frac{\Lambda}{M_j} = \sum_{i=0}^{j-1} H_i$$

where H_i is contained in Λ_{j-i}, but not in Λ_{j-i+1}, since Λ_{j-i} consists of elements $x \in \Lambda$, where $p^k x = M_j$ for $k \geq i$. Note that pH_i is contained in H_{i-1}, since pH_i has order $i - 1$. It follows that

$$\Lambda_i = \Lambda_{i+1} + H_{j-i}$$

```
        +2                  +2                  +2                  +2
                +2                  +2                  +2
        +2              +2                  +2                  +2
                +2                  +2                  +2
        +2              +2                  +2                  +2
                +2                  +2                  +2
        +2              +2                  +2                  +2
                +2                  +2                  +2
        +2              +2                  +2                  +2
                +2                  +2                  +2
        +2              +2                  +2                  +2
                +2                  +2                  +2
```

Figure 7.20. Points in Figure 7.19 ending in 2.

and

$$\Lambda_{i+1} = \Lambda_i \cap M_{i+1}$$

To finish the structural part of Theorem 7.A.2, note the following chain of isomorphisms:

$$\frac{\Lambda_i}{\Lambda_{i+1}} = \frac{\Lambda_i}{\Lambda_i \cap M_{i+1}} \equiv \frac{\Lambda_i + M_{i+1}}{M_{i+1}} \subseteq \frac{M_i}{M_{i+1}} \equiv \frac{M_0}{M_1} = V_N(F_r)$$

This chain of isomorphisms shows that

$$\frac{\Lambda_i}{\Lambda_{i+1}}$$

is a subspace of $V_N(F_r)$; i.e., it is a code C_i. Note that the dimension of C_i is equal to the number of cyclic modules in H_{j-i}. The nested property of codes follows immediately from $p\Lambda_i \subseteq \Lambda_{i+1}$.

Practical Aspects of Coding

The Euclidean distance portion of Theorem 7.A.2 follows almost immediately. Let $Q \in \Lambda$ be a nonzero point in the lattice and let k be the smallest value of i for which $Q \in \Lambda_i$. Then every nonzero entry in the coordinates of Q mod M_{i+1} contributes a quantity to the squared Euclidean distance of Q that is nonzero but 0 modulo q^i. By definition, there are at least d_i such nonzero coordinates, and the squared Euclidean distance of Q is d_Q^2, where

$$d_Q^2 \geq d_i q^i$$

Minimizing this last expression over all the possible points in Λ gives Result 4 in Theorem 7.A.2.

Theorem 7.A.2 can be restated in terms of generator matrices and/or coordinate arrays. Since these interpretations tend to be more intuitive, they are developed in the following definitions.

Definitions

1. A generator matrix G for a length N, depth D lattice Λ over L is a D by N matrix ($D \leq N$) with entries from L. Any element v of L may be expressed by the matrix equation

$$v = xG$$

where x is a 1 by D vector with elements selected from L.

2. Let $a = (a_0, a_1, \ldots, a_{N-1})$ be any element of $M = L^N$. Let

$$a_k = \sum_{l=0}^{j-1} a_{l,k} \cdot p^l$$

be the primary expansion of a_k modulo p^j; i.e., $a_{l,k} \in F_r$ for all $k: k < N$ and for all $l < j$. Put $A(a) = (a_{l,k})$ equal to the j by N matrix over F_r with the $a_{l,k}$ as entries. $A(a)$ is referred to as the (depth-j) coordinate array for a.

For a depth-j lattice Λ, the generator matrix and coordinate arrays of elements of Λ are closely related to the results in Theorem 7.A.2, and these are recorded in Theorem 7.A.3. The notation and assumptions are the same as in Theorems 7.A.1 and 7.A.2.

Theorem 7.A.3

1. Let a be an element of Λ. Suppose that i is the first nonzero row in $A(a)$; i.e., a belongs to Λ_i, but not to Λ_{i-1}. Then the ith row of $A(a)$ is a code word in C_i. Conversely, given a code word in C_i, it is possible to find an element in Λ where the first nonzero row of the coordinate array of that element occurs at i and equals that code word.

2. A lower triangular matrix T can be found that is a generator matrix for Λ. T has the following properties. Define $n_i = k_i - k_{i-1}$ for $0 \leq i \leq j - 1$, where k_{-1} is 0 and k_i is the dimension of C_i. Then T is a

$$\sum_{i=0}^{j-1} n_i \text{ by } N$$

matrix, and every entry on the tth row of T is a multiple of p^l, where l is determined as the integer for which

$$\sum_{i=0}^{l-1} n_i < t \leq \sum_{i=0}^{l} n_i$$

Furthermore, the tth row of T has an entry of p^l as its diagonal entry.

Proof: The solution follows immediately from the proof of Theorem 7.A.2.

It remains to give some examples of lattices that can be constructed from Theorem 7.A.2. Note that Theorems 7.A.2 and 7.A.3 do not state that a nested sequence of codes uniquely determines a lattice; nonisomorphic lattices can be found with the same sequence of nested codes in the primary decomposition. Theorem 7.A.3 guarantees that given a nested sequence of codes, a lattice Λ can be found with a minimum squared Euclidean distance at least as large as the bound in Theorem 7.A.2. For any particular nested sequence of codes, it may be possible to manipulate the exact entries of the generator matrix to obtain a minimum squared Euclidean distance that is larger than the bound in Theorem 7.A.2.

Example 7.A.4

1. D_n: These depth-1 Z lattices are obtained by setting C_0 to be the $(n, n-1, 2)$ even parity code, i.e., the code consisting of all binary sequences of length n with an even number of 1 entries. Equivalently, a lattice point consists of n integers where an even number of the integers are odd. The D_n lattices

do not give high densities for large n; however, D_4 is the densest known lattice in four dimensions, and it has many special properties. For example, multiplication may be defined on the elements of D_4 that identifies D_4 with a unique algebraic object known as the Hurwitz quaternion integers H. Although noncommutative, H is a PID and a theorem similar to Theorem 7.2 can be proved by taking a nested sequence of codes defined over F_4 to a lattice that is an H-module.

2. Gosset lattice E_8: This is often called the second most interesting lattice. The lattice has eight real dimensions and a variety of equivalent constructions. Define E_8 as a depth-2 complex binary lattice; i.e., $L = G$ and $p = 1 + i$. The binary codes in question are $C_0 = (4,1,4)$ and $C_1 = (4,3,2)$. The Gosset lattice can also be defined using $L = Z$ and $p = 2$. In this interpretation, E_8 is a depth-1 lattice with $C_0 = (8,4,4)$, the extended Hamming code. E_8 has been shown to be the densest possible lattice in eight real dimensions; it remains undecided if E_8 is the densest possible sphere packing in eight dimensions, although this is suspected to be true. It is possible to define a multiplication on E_8 to give it the structure of an *alternative division ring,* known as the Cayley–Dickson integers. With this approach, it is possible to prove a theorem similar to Theorem 7.A.2 that goes from a series of codes defined over E_8 to a lattice.

3. Barnes–Wall lattices: These are complex binary lattices; i.e., the construction uses L equal to the Gaussian integers G and $p = 1 + i$. The construction uses the binary Reed–Muller codes on m variables. There is a Barnes–Wall (BW) lattice for every positive integer m; BW(m) has dimension $N = 2^m$ over the complex numbers and dimension 2^{m+1} over the real numbers. The BW (m) is a depth-m lattice, and C_k equals $RM(m,k)$, the kth order Reed–Muller code of length 2^m and minimum Hamming distance 2^{m-k}, $k: 0 \leq k < m$. From Theorem 7.2, the BW lattice has minimum squared Euclidean distance of at least 2^m. As m becomes large, the density of the BW lattices is suboptimal. However, BW(1), BW(2), and BW(3) are optimal lattices in their respective dimensions. Note that BW(1) is D_4, and BW(2) is E_8. The BW(3), a depth-3 complex lattice, is the optimal lattice known in 16 real dimensions. The codes for BW(3) are

$C_0 = (8,1,8)$, the repetition code

$C_1 = (8,4,4)$, the extended Hamming code

$C_2 = (8,7,2)$, the even-parity code

The BW(m), $m \leq 5$, are examples of a *decomposable* lattice. Consider the coordinate array of any point Q in a lattice. If the lattice is decomposable, then the ith row of the array is a code word in C_i for all possible i. (The general definition ensures only that the ith row is a code word if the earlier rows are the all-zero code word.) The condition of being decomposable can also be translated into a binary code property, although the details are not presented here. Contrary to popular myth, BW(m) is not a decomposable lattice for $m > 5$, and BW(4) is not the densest lattice of real dimension 32. The BW lattices can also be constructed as Z-lattices.

4. Half-Leech and Leech lattice: The densest possible lattice in 24 dimensions is the Leech lattice. This lattice has the most remarkable structure of all known lattices; in fact, the details of this lattice fill most of a large volume by two of the twentieth-century's leading mathematicians [see (Conway and Sloane)]. The starting point is to define the half-Leech lattice H_{24}, which is described as a decomposable depth-2 Z-lattice with $p = 2$. Note that C_0 is equal to the binary (24,12,8) code, and C_1 is equal to the binary (23,12,2) even-parity code. The minimum squared Euclidean distance of H_{24} equals 8. Define the full-Leech lattice Λ_{24} as the union of two cosets of $2H_{24}$ in Z^{24}, where

$$2H_{24} = \{2y | y \in H_{24}\}$$

a scaled version of H_{24} with minimum squared Euclidean distance equal to 32. Λ_{24} is equal to the disjoint union of two cosets:

The even-parity coset (0,0,0,0,0,0,0,0,0,0,0,0,0,0,0,0,0,0,0,0,0,0,0,0) + $2H_{24}$

The odd-parity coset (−3,1,1,1,1,1,1,1,1,1,1,1,1,1,1,1,1,1,1,1,1,1,1,1) + $2H_{24}$.

From the coordinate array point of view, it is useful to note that −3 in the preceding vector can be changed to 5 without changing the coset, since 5 is equivalent to −3 modulo 8.

In nested code terms, the result is a depth-3 Z-lattice with $C_0 = (24,1,24)$, $C_1 = (24,12,8)$, and $C_2 = (24,23,12)$. The coordinate array approach is also revealing. Let $z = (z_i | 0 \leq i \leq 24)$ be a sequence from Z^{24}. Now look at the coordinate array for z modulo 8. If z belongs to Λ_{24}, the following conditions must hold:

a. The first row of the coordinate array (z_i mod 2) is δ^{24}, where δ is 0 or 1.

b. The next row, the coefficients of 2 in the binary expansion of z_i, forms a binary Golay code word.

c. The next row, the coefficients of 4 in the binary expansion of z_i, depends on the value of δ in condition a. (This is behavior characteristic of a nondecomposable lattice.) If $\delta = 0$, the row is an even-parity (24,23,2) code word. If $\delta = 1$, the row has odd parity; i.e., there are an odd number of 1s in the row of the coordinate array.

If the codes used in the definition of the Leech lattice are examined, note that Theorem 7.2 guarantees that d_{min}^2 is 24. However, by using the exact coset vectors previously defined, the lattice exhibits a minimum squared Euclidean distance of 32. Note that the lattice vector (−3,1,1,1,1,1,1,1,1,1, 1,1,1,1,1,1,1,1,1,1,1,1,1,1) has squared distance equal to 32.) It is this remarkable excess of distance that makes the Leech lattice a very dense lattice. It is the densest known lattice in 24 dimensions, and in light of the unusual closeness of its density to the theoretical lower bounds, Λ_{24} is almost certainly the densest possible sphere packing in 24 dimensions, although this is not proved. As an aside, Λ_{24} is conventionally scaled by a factor of $1/\sqrt{8}$ [See Conway and Sloane (p. 131, equations 135–36) or Sloane (p. 146)]. This reduces the squared distance from 32 to 4, but it does not of course affect the density. The construction has taken two cosets of $2H_{24}$, each of minimum squared distance equal to 32, and packed them together to obtain a lattice that also has a minimum squared distance of 32!

The Leech lattice has an alternative descriptions as a G lattice and an E lattice (and others). As a ternary lattice, it is depth-3 with C_0 equal to a ternary (12,1,12) repetition code, C_1 equal to the (12,6,6) ternary Golay code, and C_2 equal to the (12,11,2) ternary parity check code.

Exercises

1. Prove that the minimum squared Euclidean distance of the Leech lattice as just constructed is 32.

No figure of merit has been defined for the lattices described in Example 7.4. There are several contenders. Forney (1988) introduces perhaps the most communications-driven measure of the normalized gain of a lattice as compared to using uncoded modulations; the following gains use his terminology:

D_n give 3.01 dB of gain under high-power conditions as n increases. D_4 gives 1.51 dB of gain.

E_8 gives 3.01 dB of gain over uncoded modulation.

BW(m) gives

$$10 \log_{10}(2^{\binom{m}{2}}) = 1.55m \text{ dB}$$

of gain over uncoded modulation; BW(3) gives 4.52 dB of gain; BW(4) gives 6.02 dB of gain.

The Leech lattice Λ_{24} gives 6.02 dB of gain, and the half-Leech lattice H_{24} gives 5.77 dB.

Conway and Sloane present more extensive definitions of density and other figures of merit. They also use a normalized dB assessment that differs from the preceding assessment.

It is worth mentioning that elliptic curves and algebraic geometry have recently been used by Elkies *et al.* to unify many of the densest known lattices, including improvements over the previous densest known lattices in higher dimensions, and to expose an area for new exploration.

References

General Coding for Communication Systems

G. C. Clark and J. B. Cain. 1981. *Error Correcting Codes for Digital Communications.* New York: Plenum Press.

S. Lin and D. J. Costello, Jr. 1983. *Error Control Coding: Fundamentals and Applications.* Englewood Cliffs, NJ: Prentice-Hall.

W. W. Wu, D. Haccoun, R. E. Peile, and Y. Hirata. 1987. "Coding for Satellite Communication." *IEEE J. Sel. Areas Commun.* **SAC-5**, No. 4, pp. 724–748.

Convolutional Codes and the Viterbi Algorithm

H. Bustamante, I. Kang, C. Nguyen, and R. E. Peile. 1989. "STEL Design of a VLSI Convolutional Decoder." MILCOM '89, Boston: Coherence Proceedings.

S. Lin and D. J. Costello, Jr. 1983. *Error Control Coding: Fundamentals and Applications.* Englewood Cliffs, NJ: Prentice-Hall.

G. D. Forney, Jr. 1973. "The Viterbi Algorithm." *Proc. IEEE* **61**: 268–78.

I. M. Jacobs and E. R. Berlekamp. 1967. "A Lower Bound to the Distribution of Computation for Sequential Decoding." *IEEE Trans. Inform. Theory* **IT-13**: 167–74.

J. P. Odenwalder. *Error Control Handbook.* San Diego, CA: Linkabit, Inc.

A. J. Viterbi. 1971. "Convolutional Codes and Their Performance in Communication Systems." *IEEE Trans. Commun. Technol.* **COM-19**: 751–72.

W. W. Wu, D. Haccoun, R. E. Peile, and Y. Hirata. 1987. "Coding for Satellite Communication." *IEEE J. Sel. Areas Commun.* **SAC-5**.

Algebraic Coding

E. R. Berlekamp. 1984. *Algebraic Coding Theory. Revised.* Laguna Hills, CA: Aegean Park Press Books.

R. Blahut. 1983. *Theory and Practice of Error Control Codes.* Reading, MA: Addison-Wesley.

D. Chase. 1972. "A Class of Algorithms for Decoding Block Codes with Channel Measurement Information." *IEEE Trans. Inform. Theory* **IT-18:** 170–82.
G. C. Clark and J. B. Cain. 1981. *Error Correcting Codes for Digital Communications.* New York: Plenum Press.
G. D. Forney, Jr. 1966. "Generalized Minimum-Distance Decoding." *IEEE Trans. Inform. Theory* **IT-12:** 125–31.
V. D. Goppa. 1988. *Geometry and Codes.* Dortrecht, the Netherlands: Kluwer Academic Publishers.
R. Lidl and H. Niederreiter. 1983. *Finite Fields.* Reading, MA: Addison-Wesley.
J. H. van Lint. 1982. *Introduction to Coding Theory.* New York: Springer-Verlag.
J. H. van Lint and G. van der Geer. 1988. *Introduction to Coding Theory and Algebraic Geometry.* Berlin: Birkhauser.
J. H. van Lint and T. A. Springer. 1987. "Generalized Reed–Solomon Codes from Algebraic Geometry." *IEEE Trans. Inform. Theory* **IT-33:** 305–9.
F. J. MacWilliams and N. J. A. Sloane. 1978. *The Theory of Error-Correcting Codes.* Amsterdam: North-Holland.
T. R. N. Rao and E. Fujiwara, 1989. *Error-Control Coding for Computer Systems,* Englewood Cliffs, NJ: Prentice Hall.
W. W. Peterson and E. J. Weldon, Jr., 1972. *Error-Correcting Codes, 2nd ed.,* Cambridge, MA: MIT Press.
I. S. Reed and G. Solomon. 1960. "Polynomial Codes over Certain Finite Fields." *J. Soc. Ind., Appl. Math.* **8:** 300–4.

Concatenation

S. Dolinar. 1988. "A New Code for Galileo." TDA Progress Report 42-93. Pasadena, CA: Jet Propulsion Laboratory. 83–96.
G. D. Forney, Jr. 1966. *Concatenated Codes.* Cambridge, MA: MIT Press.
J. P. Odenwalder. *Error Control Handbook.* San Diego, CA: Linkabit, Inc.
J. Justesen. 1972. "A Class of Constructive Asymptotically Good Algebraic Codes." *IEEE Trans. Inform. Theory* **IT-18:** 652–56.
K. Y. Liu and J. J. Lee. 1984. "Recent Results on the Use of Concatenated Reed–Solomon / Viterbi Channel Coding and Data Compression for Space Communications." *IEEE Trans. Commun. Technol.* **COM-32:** 518–523.
R. A. Scholtz. 1982. "The Origins of Spread-Spectrum Communications." *IEEE Trans. Commun. Technol.* **Com-30:** pp 822–854.
M. K. Simon, J. K. Omura, R. A. Scholtz, B. K. Levitt. 1985. *Spread-Spectrum Communications. Vols 1–3,* Rockville, MD: Computer Science Press.
J. H. Yuen and Q. D. Vo. 1985. "In Search of a 2-dB Coding Gain." TDA Progress Report 42-83. Pasadena, CA: Jet Propulsion Laboratory. 26–33.

Interleaving

W. W. Wu, D. Haccoun, R. E. Peile, and Y. Hirata. 1987. "Coding for Satellite Communication." *IEEE J. Sel. Areas Commun.* **SAC-5,** vol. 4, pp. 724-748.

Ring Theory

A. A. Albert, ed. 1963. *Studies in Modern Algebra.* The Mathematical Association of America. Englewood Cliffs, NJ: Prentice–Hall.

B. Hartley and T. O. Hawkes. 1970. *Rings, Modules, and Linear Algebra.* London: Chapman and Hall.

K. Ireland and M. Rosen. 1982. *A Classical Introduction to Modern Number Theory.* New York: Springer-Verlag.

I. N. Stewart and D. O. Tall. 1987. *Algebraic Number Theory.* 2d ed. London: Chapman and Hall.

Lattices

A. R. Calderbank and N. J. A. Sloane. 1987. "New Trellis Codes Based upon Lattices and Cosets." *IEEE Trans. Inform. Theory* **IT-33**: 177–195.

J. H. Conway and N. J. A. Sloane. 1988. *Sphere Packings, Lattices, and Groups.* New York: Springer-Verlag.

N. D. Elkies, A. M. Odlyzko, J. A. Rush. 1991. "On the Packing Densities of Superballs and Other Bodies." *Invent. Math.* **105**: 613–639.

G. D. Forney, Jr. 1988. "Coset Codes—Part I: Introduction and Geometrical Classification." *IEEE Trans. Inform. Theory* **34**: 1123–1151.

G. D. Forney, Jr., R. G. Gallager, G. R. Lang, F. M. Longstaff, and S. U. Qureshi. 1984. "Efficient Modulation for Band-Limited Channels." *IEEE J. Sel. Areas Commun.* **SAC-2**: 632–647.

I. Peterson. 1990. "Curves for a Tighter Fit." *Science News* **137**: 316–317.

C. E. Shannon. 1948. "A Mathematical Theory of Communication." *Bell Syst. Tech. J.* **27**: 379–423; 623–656.

N. J. A. Sloane. 1977. "Binary Codes, Lattices, and Sphere Packings." In *Combinatorial Surveys: Proceedings of the Sixth British Combinatorial Conference,* edited by P. J. Cameron. London: Academic Press.

T. M. Thompson. 1983. *From Error-Correcting Codes through Sphere Packings to Simple Groups.* Washington, D.C.: Carus Mathematical Monographs, Mathematical Association of America.

G. Ungerboeck. 1982. "Channel Coding with Multilevel/Phase Signals." *IEEE Trans. Inform. Theory* **IT-28**: 55–67.

FEC, ARQ, and Networks

D. Bertekas and R. Gallagher. 1987. *Data Networks.* Englewood Cliffs, NJ: Prentice-Hall.

D. Chase. 1985. "Code Combining—A Maximum-Likelihood Decoding Approach for Combining an Arbitrary Number of Noisy Packets." *IEEE Trans. Commun. Technol.* **COM-33**: 385–393.

S. Lin and D. J. Costello, Jr. 1983. *Error Control Coding: Fundamentals and Applications.* Englewood Cliffs, NJ: Prentice-Hall.

R. E. Peile and R. A. Scholtz. 1990. "Adaptive Channel/Code Matching Using Hidden Markov Chains." *Twenty-Fourth Asilomar Conference on Signals, Systems, and Computers.* Conference Proceedings.

A. S. Tanenbaum. 1988. *Computer Networks.* 2d ed. Englewood Cliffs, NJ: Prentice-Hall.

Author Index

N. M. Abramson, 61
R. L. Adler, 61
A. V. Aho, 60, 61
A. A. Albert, 417
I. Anderson, 417
T. M. Apostol, 110, 129
S. Arimoto, 368
R. B. Ash, 60, 61

E. S. Barnes, 413, 415
L. A. Bassalygo, 306, 307
T. Bayes, 248
E. R. Berlekamp, 193, 204, 240, 306, 307, 380, 416
D. Bertekas, 418
P. Billingsley, 60, 61, 241
G. D. Birkhoff, 189, 194, 204
R. E. Blahut, 61, 416
R. C. Bose, 375
L. Breiman, 60
J. Brillhart, 189, 204
N. G. de Bruijn, 60
G. Burton, 26, 60
H. O. Burton, 307
H. Bustamante, 416

J. B. Cain, 416
A. R. Calderbank, 404, 417
P. J. Cameron, 418
A. Cayley, 413
D. Chase, 375, 395, 397, 416, 418
P. L. Chebychev, 52, 257
H. Chernoff, 323–336
N. Chomsky, 60, 61

G. C. Clark, 416
J. H. Conway, 307, 389, 413–417
D. Coppersmith, 61
D. J. Costello, Jr., 416, 418
T. Cover, 61
H. Cramer, 241
I. Csiszar, 60, 61
L. J. Cummings, 204

L. D. Davisson, 241
M. Delbruck, 205
P. Delsarte, 295, 307
K. Dessouky, 205
J. S. Devitt, 199, 205
L. E. Dickson, 413
P. G. L. Dirichlet, 110, 111, 113, 129
S. Dolinar, 417

W. L. Eastman, 199, 205
F. G. Eisenstein, 398, 402, 404
P. Elias, ix, 124, 126, 223, 224, 226–231, 236, 241, 269–273, 277, 295, 306
N. D. Elkies, 415, 417
L. Euler, 191, 203
S. Even, 128, 129, 130, 205

R. M. Fano, 60, 61, 128, 129, 317, 346, 368
A. Feinstein, 60, 61, 368
W. Feller, 61, 241, 242, 257, 307
A. Fodor, 60, 61
G. D. Forney, Jr., 307, 368, 375, 383, 415, 416, 417
J-B-J Fourier, 286, 292
E. Fujiwara, 416

Galileo, 417
R. G. Gallager, 60, 61, 242, 357, 368, 417, 418
F. M. Gardner, 205
G. van der Geer, 379, 416
F. Ghazvinian, 205
E. N. Gilbert, 129, 199, 205, 273–275, 301, 306, 307
M. J. E. Golay, 276, 306, 307, 375, 414
S. W. Golomb, 60, 61, 101, 149, 173, 180, 181, 182, 199, 205, 241, 242, 306, 307
V. D. Goppa, 416
B. Gordon, 149, 173, 180–182, 199, 205
T. Gosset, 412
R. L. Graham, 205
S. Guiasu, 61

D. Haccoun, 416, 417
J. Hadamard, 280
W. C. Hagmann, 205
R. W. Hamming, 251–307, 345, 374, 378, 389, 412
G. H. Hardy, 129
B. Hartley, 399, 405, 406, 417
R. V. L. Hartley, 59, 60
M. Hasner, 61
F. Hausdorff, 241
T. O. Hawkes, 399, 405, 406, 417
Y. Hirata, 416, 417
A. Hocquengheim, 375
D. A. Huffman, 95, 98, 101, 102, 116, 129, 219, 242
W. J. Hurd, 26, 60, 61
A. Hurwitz, 412

K. Ireland, 417

D. M. Jackson, 205
I. M. Jacobs, 416
E. T. Jaynes, 128–130
F. Jelinek, 61, 129
B. H. Jiggs, 150, 199, 205
S. M. Johnson, 307
J. Justesen, 374, 417

G. A. Kabatiansky, 295, 307
M. Kac, 123, 126, 129, 130
I. Kang, 416
S. Karlin, 61
R. M. Karp, 129, 130

J. Katz, 60, 61
J. G. Kemeny, 61
W. B. Kendall, 199, 205
A. L. Khintchin, 60, 61
L. Kleinrock, 61
A. N. Kolmogorov, 129, 130
T. Korner, 60, 61
L. G. Kraft, 85–88, 90, 92, 96, 128, 130, 181, 231
M. Krawtchouk, 295
H. W. Kuhn, 368
S. Kullback, 60, 61

J-L. Lagrange, 215, 218, 352
G. R. Lang, 417
P. S. Laplace, 111
H. J. Larson, 61
J. J. Lee, 417
P. M. Lee, 60, 61, 254–256, 264, 265, 269, 275, 277, 278, 279, 306
Y. W. Lee, 417
J. Leech, 413–415
D. Lehmer, 204
A. Lempel, 116–129, 130, 207, 230, 236
V. I. Levenshtein, 135, 142, 156, 170, 177, 199, 205, 295, 307
B. K. Levitt, 417
J. C. R. Licklider, 26, 60, 61
R. Lidl, 416
S. Lin, 416, 418
W. C. Lindsey, 205
J. van Lint, 150, 205, 306, 307, 379, 416
K. Y. Liu, 417
F. M. Longstaff, 417
L. H. Loomis, 241, 242

F. J. MacWilliams, 239, 242, 262, 289, 291, 291, 306–308, 374, 389, 404, 413–418
H. B. Mann, 307
M. Mansuripur, 62
R. S. Marcus, 128, 130
R. H. Margenau, 368
P. Martin-Löf, 129, 130
R. J. McEliece, 62, 293–295, 308
B. McMillan, 46, 47, 60, 61, 88, 92, 93, 96, 128, 130, 181, 207
D. E. Muller, 384, 413

I. Newton, 239
C. Nguyen, 416

Author Index

H. Niederreiter, 416
Y. Niho, 119, 205
E. Norwood, 129, 130

J. P. Odenwalder, 371, 416, 417
A. M. Odlyzko, 417
J. K. Omura, 417
L. H. Ozarow, 129, 130

A. Papoulis, 62
M. A. Parseval, 286
E. Parzen, 62
G. W. Patterson, 68–70, 74, 83, 128, 130, 132, 136, 140, 143
R. E. Peile, 416–418
I. Peterson, 417
W. W. Peterson, 416
M. S. Pinsker, 60, 62
M. Plancherel, 286
V. Pless, 293–295, 306, 308
M. Plotkin, 265–269, 271, 286, 306, 308, 309
S. D. Poisson, 219, 238
N. U. Prabhu, 62
V. R. Pratt, 130

S. U. Qureshi, 417

T. R. N. Rao, 416
D. K. Ray-Chaudhuri, 375
Lord Rayleigh, 372
I. S. Reed, 199, 205, 240, 301, 370, 376–387, 413, 416
B. Reiffen, 368
H. Reisel, 189, 205
A. Renyi, 60, 62
F. M. Reza, 62
G. H. B. Riemann, 238, 379
O. Rice, 372
M. Riesz, 129
G. Roch, 379
M. Rodeh, 129, 130
E. R. Rodemich, 293–295, 306
M. Rosen, 417
H. Rumsey, Jr., 293–295, 306, 308
J. A. Rush, 417

A. A. Sardinas, 68–70, 74, 83, 128, 130, 132, 135, 140, 143
K. S. Schneider, 129, 130
M. P. Schützenberger, 128, 130

R. A. Scholtz, 177, 181, 189, 193–195, 199, 205, 387, 417, 418
J. Selfridge, 204
C. E. Shannon, 26, 46–47, 59–60, 62, 93, 100, 128, 130, 207, 223, 224, 226–230, 241, 335, 342, 344, 345, 367, 368, 377, 383, 385, 386, 388, 389, 418
B. O. Shubert, 61
M. K. Simon, 417
D. Slepian, 368
N. J. A. Sloane, 239, 242, 262, 300, 307, 308, 374, 389, 404, 413–418
J. L. Snell, 61
G. Solomon, 240, 301, 370, 376–385, 416
T. A. Springer, 416
I. N. Stewart, 417
J. J. Stiffler, 205, 307, 308
J. Stirling, 331
R. M. Storwick, 199, 205, 262
G. Strang, 62
D. D. Sullivan, 307
P. L. M. Sylow, 194

D. O. Tall, 417
A. S. Tanenbaum, 418
B. Tang, 150, 205
H. M. Taylor, 61
T. M. Thompson, 418
A. Tietäväinen, 276, 308
A. W. Tucker, 368
B. Tuckermann, 204, 306
H. Tverberg, 60, 62

J. D. Ullman, 60, 61
G. Ungerboeck, 388, 418

H. S. Vandiver, 189, 194, 204
R. R. Varshamov, 306, 308
T. Verhoeff, 306, 308
A. J. Viterbi, 307, 370–374, 384, 416
Q. D. Vo, 41
D. C. van Vorhis, 242

S. Wagstaff, Jr., 204
G. E. Wall, 413, 415
V. K. W. Wei, 205
L. R. Welch, 149, 189, 193–195, 199, 205, 293–295, 306–308
E. J. Weldon, Jr., 416
T. Welch, 129, 130, 199

M. S. Wesley, 417
N. Wiener, 60, 62, 124
J. Wolfowitz, 60, 62, 368
W. W. Wu, 416, 417
A. D. Wyner, 123, 126, 129, 130, 233, 236

J. H. Yuen, 417

N. Zierler, 306
A. Zipf, 27
J. Ziv, 116–130, 207, 230, 231

Subject Index

a priori, 3–5, 14, 16, 94, 250
a posteriori, 14, 15, 16, 248, 249, 332
Abelian group, 399
Abscissa of convergence, 110, 112
Ack: *see* Positive acknowledgment system
Acquisition delay, 141, 151, 160–162
Adaptive routing, 396
Additive white Gaussian noise (AWGN), 370, 372, 384, 388–389
Algebraic
  curve, 379, 415
  geometry, 274, 379, 415
Algorithm
  Berlekamp, 240
  capacity computation, 363
  construction of equidistant codes, 281, 283
  decoding, 374
    algebraic, 372, 377
    bounded distance, 375–378, 381
    Chase, 375
    generalized minimum distance, 375
    hard decision, 375
    maximum-likelihood decoding: *see* Likelihood
    minimum distance, 354
    random code, 337
    sequential decoding, 373, 374
    soft decision, 375
  Euclid, 240
  Huffman, 98, 101, 102
  lexicographic code, 155
  Lempel–Ziv (LZW), 110–128, 207, 230, 236

Algorithm (*Cont.*)
  Levenshtein, 135–140, 142, 156, 170, 177
  Sardinas–Patterson, 68, 69, 74, 83, 128, 135, 140
  Viterbi, 299, 307, 370–374, 384–388
Alphabet
  input, 64
  output, 64
Alternative division ring, 413
Ambiguity, 134, 170
Analysis
  real, 224
  statistical, 243
Analytic number theory, 238
ARQ: *see* Error detection; Retransmission schemes
Association schemes, 295, 307
Asymptotic
  code rates, 263
  Elias bound, 310
  equipartition property and theorem, 46, 47, 50, 76, 128
  transmission rate, 160, 161, 259, 266, 268, 272, 274, 341
Asymptotically optimal codes, 230–239, 241
Automata theory, 60
Automatic code generation, 119
Automatic power control, 382
Automation of bounded synchronization delay codes, 189–196
Average recurrence time, 123
Auxiliary equations, 359
AWGN: *see* Additive white Gaussian noise

Subject Index

Axiom, 5, 7, 9, 13
   grouping, 7, 12, 23
Axiomatic approach, 4, 13, 60

Bayes' law, 18, 248
BER: *see* Bit error rate
Berlekamp's algorithm, 240
Binary demarcation point, 227
Binary symmetric channel: *see* Channel, binary symmetric
Binomial distribution: *see* Distribution, binomial
Bit error rate, 372, 376, 379, 380
Bits, 9
Boolean function, 167
Bound
   Chernoff, 323–330
   Elias, 263, 269–273, 277, 306, 310
   Fano, 317, 318
   Gilbert, 263, 273–275, 301, 302, 306
   Hamming, 260–265, 269, 275, 306
   on information rate, 315–322
   JPL: *see* WMR
   Plotkin, 263, 265–269, 280, 306, 309
   Singleton, 378
   sphere-packing: *see* Hamming
   union, 330, 331, 339, 343
   WMR, 263, 293–295, 306
Bounded distance: *see* Algorithm, decoding
de Brujin graph, 60
BSC: *see* Channel, binary symmetric
Buffer overflow, 129
Burst errors, 299, 301, 373, 374, 381, 394

Cartesian product, 240
Capacity, 94, 368
   attaining codes, 223–226
   band-limited Gaussian channel, 388
   binary erasure channel, 365
   computational algorithm, 363
   concatenated codes and, 383
   effective, 160, 161
   fixed-rate sources, 91
   input-uniform channel, 358
   input–output uniform channel, 358
   memoryless channel, 321, 366
      computation, 358–367
   noiseless channel, 78, 81, 100
   noisy channel: *see* Chapter 6
   nonsynchronized channel, 160

Capacity (*Cont.*)
   symbol, 341
   symmetric, 365
   unbalanced ternary channel
   variable symbol duration channel, 106, 111, 129
   zero error, 368
Cayley–Dickson integers, 413
CCSDS: *see* Consultative Committee on Space Development Services
Channel
   additive error, 248
   binary, 366
   binary erasure, 247, 260, 322
   binary symmetric, 246, 256, 258, 259, 293, 310, 321, 345, 376, 391
   fading, 299
   continuous, 368
   deterministic, 245
   discrete communication, 64
   fading, 298, 299, 372, 393
      correlated, 372
      Rayleigh, 372
      Rician, 372
   feedback, 305
   finite state, 295–301
   input uniform, 246, 247, 358
   Markov: *see* Markov
   matrix, 245, 246, 255, 345, 351, 357, 359, 361
   with memory, 320
   memoryless, 320, 357, 367
   multipath, 372
   noiseless, 64, 78
   output uniform, 246, 247
   pure intersymbol interference, 297–299
   stationary, memoryless, synchronized, 245, 303, 322, 333, 346, 347
   symmetric, 365
   unbalanced ternary, 366
   variable symbol duration, 105, 129
Chase code combining, 395
Chebychev's inequality, 52, 257
Chernoff bound: *see* Bound, Chernoff
Chord, 348
Clock recovery, 372
Cluster, 33, 34
Code, 65
   adaptive, 377
   algebraic geometry, 274, 379

Subject Index

Code (*Cont.*)
  asymptotically optimal, 230–239, 232, 234, 241
  binary, 370, 374
  block, 65, 100, 315, 370, 374–377
  Bose–Chaudhuri–Hocquengheim (BCH), 375
  codesigned coding and modulation, 370, 388–389, 402
  comma, 132, 141
  comma-free: *see* Comma-free codes
  compact, 95, 98, 102
  concatenated, 370, 377, 383–387, 393–394
  constructable, 261
  convolutional, 370–374, 384–386, 394
  coset, 404, 414
  distinct, 90
  Elias and Shannon, 223–229
  equidistant, 260, 280–284, 305
  Golay, 276, 306, 375, 413, 414
  Hamming, 275, 276, 292, 306
  Huffman, 95, 102
  inner, 383–387
  Justesen, 374
  Linear, 407
  maximal distance separable (MDS), 378, 379, 396
  nested, 405–407, 412
  nonsingular, 65
  optimal, 230–237, 241
  outer, 383–387
  perfect, 260, 263, 264, 275–280, 306, 309
    in the Lee metric, 278, 279, 306
  prefix, 102, 232
  prefix-free, 83, 84, 88
  random, 337, 338, 344, 345, 367
  Reed–Muller (RM), 384, 413
  Reed–Solomon (RS), 240, 301, 370, 377–386, 392, 394
  repetition, 275
  symbol, 374, 377
  synchronization, 131–206
  systematic, 380, 394, 396
  tree, 83, 85, 88, 94, 95, 97, 100, 128, 129
  trellis-coded modulation (TCM), 370, 388
  U_f, 66, 72, 83, 89, 90, 95, 134
  U_i, 66, 72, 132, 134, 135, 143, 144, 147, 171
    fixed word length, 147–158, 160–162
    variable word length, 164–169

Code (*Cont.*)
  U_s, 66, 72, 83, 88, 89, 90, 134
  uniquely decodable, 65, 89, 90, 92, 128
  universal, 223–239, 241
Code-word
  generators, 337, 341, 347
  length, 92, 100, 114, 232
Coding theorems: *see* Theorem
Comma, 132, 141
Comma-free codes
  fixed word length, 147–158
  lexicographic, 155–157, 196, 199
  maximal, 148–151, 182–188
  path-invariant, 153–155, 199
  prefixed, 151–152, 199
  variable word length, 164–169
Communication, reliable, 310, 319, 320, 322
Complexity of a sequence, 117, 127, 129
Concatenation of segments, 134
Constellation, 388
Constraint length, 371
Construction
  Scholtz, 178–181
  shortest add-length suffix (SOLS), 182–188
Continuous function, 9
Convex U, 325
Coordinate array of lattice: *see* Lattice
Coset, 401–404
Cyclic equivalence class, 145
  occupancy, 174
  representations, 189–196
Cyclically
  distinct, 171
    nonperiodic code-word sequences, 172
  equivalent as word sequences, 171

Decipherability, 210, 231
Decoding
  erasure, 243
  optimal algorithms, 243
  table, 252
Deinterleaving: *see* Interleaving
Delay
  acquisition, 141, 151, 160–162
  decoding, 142
Dictionary
  maximum size, 262
  run length, 220, 221
  U_f, 66, 68, 72, 83, 88–90, 95, 134

Dictionary (*Cont.*)
   U_i, 66, 72, 132, 135, 143,144, 147, 171
     existence of, 170–174
     fixed word length, 147–158, 160–162
     lexicographic, 155–157, 196
     maximal comma-free, 148–151
       binary, 150
       of odd word length, 182–188
       ternary, 150
     path-invariant comma-free, 153–155
     prefixed comma-free, 151–152
   U_s, 66, 72, 83, 88–90, 134
Dirichlet series, 110, 111, 129
Distance
   Euclidean, 255, 388–389, 398–415
   Hamming, 251, 252, 255, 256, 264, 265, 270, 271, 273, 275, 287, 303, 306, 345, 374, 378
     enumeration, 284–293
   Lee, 255, 256, 264, 265, 269, 275, 278
   minimum, 253, 255, 378
     fractional, 263, 266, 268, 273
Distribution
   binomial, 238, 256, 257, 329
   continuous, 208
   countably infinite, 207–242
   finite, 208
   Gaussian, 218
   geometric, 211–218, 219, 230, 231
   infinite discrete, 207–242
   m^{th} order, 123
   marginal, 340
   of moments, 226
   Poisson, 219, 238
   tilted, 325
   uniform, 344
   of waiting times, 226
Divergence inequality: *see* Inequality, divergence

Efficiency, 100
Eigenvalue, 32, 33
Eigenvector, 32
Eisenstein integers, 398, 402, 404
Elias, 231
Elias and Shannon coding, 223–229, 241
Elias bound: *see* Bound, Elias
e-linked: *see* Sequences, linked
Elliptic curves, 415

Entropy, 5, 10, 64, 314, 320
   conditional, 31, 311, 349
   finite, 210, 212, 213
   function, 3, 4, 12, 13, 350
   infinite, 210, 212, 213
   maximal, 250
   of English, 22, 26
     of homogeneous Markov source, 30, 40, 47
   of a geometric distribution, 209
   of a language, 46
   list length and, 46, 47
   physical, 82
   of Poisson distribution, 219
   of sequence, 23
     normalized, 25
   of stationary source, 26, 47, 80, 93, 119
   statistical, 82
   vs mean, 211–218
Equalization, 299, 372
Equidistant code: *see* Code, equidistant
Equivalence class, 283; *see also* Nonperiodic equivalence class (NPCE)
Equivocation, 15, 17, 25, 318, 320, 349
Erasure
   channel, 247
   decoding, 243
Ergodic, 50, 123, 126, 237
   information, 333–335, 337, 339, 340, 341, 346, 347
Error
   detection, 299, 302–306
   generator, 314
   indicator, 315
Estimation
   error, 315
   problems, 310–315
Euclid's algorithm, 240
Euclidean
   distance: *see* Distance, Euclidean
   domain: *see* Ring theory
   space: *see* Space, Euclidean
Euler totient function, 191, 203

Factorization of $2^n - 1$, 189
FEC: *see* Forward Error Correction
Fields, 370, 374, 389
Finite state machine, 372
Finitely generated: *see* Ring theory
Fourier transform, 286

Fractional
    expansion, 225
    minimum distance, 263
Frequency-hopping radios, 385-387
Function
    convex, 348-357
    real-valued, 286

Gain of a lattice, 415
Games, coin-tossing, 210
Gaussian
    integers, 398, 402-403, 413
    noise: *see* Average white Gaussian noise
GCD: *see* Greatest common divisor
Generating function
    information, 237-241
    moment, 110, 110, 128, 241, 324-326, 329
    semi-invariant, 324
Generator matrix for lattice: *see* Lattice
Genus, 379
Geometry: *see* Algebraic geometry
GF: *see* Fields
Gilbert bound: *see* Bound, Gilbert
Gosset lattice: *see* Lattice, Gosset
Greatest common divisor, 38, 283
Group
    abelian, 399
    cosets, 194
    cyclic, 189, 190, 194
    direct product of, 189
    factor group, 194
    orbits, 190
    Sylow subgroup, 194
    of units, 189-196, 286

$H[\ \ ]$, 3, 4, 52
Hadamard matrix, 280
Half-Leech lattice: *see* Lattice, half-Leech
Hamming
    bound: *see* Bound, Hamming
    distance: *see* Distance, Hamming
    metric: *see* Metric, Hamming
    sphere: *see* Sphere, Hamming
    weight: *see* Distance
Hard decision, 370, 372
Hausdorff dimension theory, 241
HDLC, 396
Hexagonal lattice: *see* Lattice

Huffman
    algorithm, 98, 101
    codes, 95, 101, 102, 219
Hurwitz quaternion integers, 412
Hybrid error control, 389
Hyperplanes, 348

Ideal: *see* Ring theory
Identity
    MacWilliams, 289, 291, 292, 306
    Pless power moment, 293-295, 306
Imprimitive elements, 191
Independent, 4
Inequality
    Chebychev, 52, 257
    divergence inequality, 93, 366
    Fano, 317, 318, 346, 368
    Golomb and Gordon, 173, 180, 181
    Kraft-McMillan, 85-88, 92, 93, 128, 181, 231
    triangle, 252
    Wyner, 233, 236
Infinite
    countably, 207
    fractional expansion, 225
Information
    gain, 14
    lexicographic, 55
    rate, 80
        of a noiseless channel, 82
        of a noisy channel, 315-322
    semantic, 55
    Stationary Markov source information, 347
    syntactic, 55
Inner code: *see* Code, inner
Instantaneous decoding, 84, 88, 223
Integer compression, 119
Integral domain: *see* Ring theory
Interleaving, 299, 300, 302, 372, 382-384
Intersymbol interference, 297, 298, 307, 372
Interval: *see* Theorem, nested interval
Irreducible element: *see* Ring theory
Isomorphic, 407, 408
Iterative construction of synchronisation
        codes, 164, 168, 185, 199
    sufficiency of, 174, 199

Jet Propulsion Laboratory (JPL), 384-386
    bound: *see* Bound, WMR

Joint Tactical Information Distribution System (JTIDS), 387

Kac's lemma, 123, 126, 129
Kolmorogov complexity, 129
Kraft–McMillan inequality: *see* Inequality, Kraft–McMillan
Krawtchouk polynomials, 295
Kuhn–Tucker conditions, 368

Lagrange multiplier, 215, 218, 352
Laplace transform, 111
Lattice, 398–415
  algebraic geometry
  Barnes–Wall, 413
  block, 388
  coordinate array for, 410–411
  decomposable, 413
  definition, 404
  depth-j, 405, 411
  dihedral, D_n, 412, 413
  gain of a lattice, 415
  generator matrix for, 410–411
  Gosset, 412, 413
  half-Leech, 413, 414
  indecomposable, 413
  Leech, 414, 415
  primary decomposition of: *see* Ring theory; *see also* Sphere packings
Law of large numbers, 326, 328, 330
Lee metric: *see* Distance, Lee
Leech lattice: *see* Lattice, Leech
Lempel–Ziv algorithm, 110–128, 207, 230, 236
Leningrad paradox, 208–211, 241
Lexicographic information, 55
Likelihood
  conditional, 250
  incremental, 332, 333
  maximum-likelihood
    decision rule, 250, 370
    decoding algorithm, 370, 372, 384
    sequence estimation, 372
Linear transformation, 348
Linked sequences: *see* Sequences, linked
List length, 46
LZW: *see* Algorithm, Lempel–Ziv

MacWilliams identities: *see* Identities, MacWilliams

Markov
  channel-state, 298, 299
  cluster diagram, 34, 35
  clusters of states, 33
  encoding, 101
  entropy of, 22, 30, 40, 320, 346
  ergodic, 40, 50, 237
  homogeneous, 22, 30, 35, 38
    channel, 297
    source, 27, 35
  information rate, 347
  irreducible, 22, 35, 38
  list length, 46, 47
  models for English, 54
  n^{th} order, 22, 25, 29, 345
    upper bound on, 24
  period of, 38
  phase sets, 38, 39
  recurrence time, 38
  steady-state probability vector, 35
  stationary probability vector, 35
  transition probabilities, 28, 299, 361
  trellis diagram, 28, 73, 137
Mars, 384
Maximum-likelihood decision rule: *see* Likelihood
Maximum distance separable (MDS) code: *see* Code, MDS
Mean
  finite, 210, 212, 213
  infinite, 210, 212, 213
  sample, 326, 328
Measure, 239
  preserving map, 224, 237, 239
  of a set, 224
  space, 241
  theory, 239, 241
Metric
  Hamming, 251, 252, 275, 280
  Lee, 254, 269, 278, 279
Minimal encoding, 231
Minimum distance, 253
Mobius inversion, 147
  function, 147
Modulation: *see* Code, codesigned coding and modulation
Modules: *see* Ring theory, modules
Moment
  noncentral, 324
  r^{th}, 219
  second, 218

Subject Index

Multinomial, 379
Multipath: *see* Channel, multipath
Multiplier Lagrange, 215, 218
Mutual information, 15, 17, 19, 77, 80, 94, 310, 311, 316, 317, 321, 341, 351, 353, 357, 360

NASA, 371, 384
Nats, 9
NCPE: *see* Nonperiodic cyclic equivalence class
Necklace, 203
Nested interval theorem, 224
Networks, 395–397
Newton's identities, 239
Node
　initial, 84, 90, 98
　interior, 84, 86, 90, 97, 98
　terminal, 84, 86, 97, 98
Nonperiodic cyclic equivalence class (NPCE), 145, 146, 149, 152, 155–157, 171, 172, 177, 179, 182, 192–193, 195–198
Norm, 239

Observation boundary, 142
Optimal decoding algorithms, 244
Orbit: *see* Group, orbit
Order ideal: *see* Ring theory
Orthogonal
　columns, 154
　polynomials, 295

Packet coding, 395–397
Packetized data, 382
　missing packets, 395–397
　optimal length, 391–392
Palindrome, 56, 57, 120
Paradox
　Leningrad, 208–211, 241
　St. Petersburg, 208, 211, 241
Partition function, 128
Pattern matching, 124
Perfect code: *see* Code, perfect
Period
　of a n-tuple, 144
　of a sequence of code words, 134
　of a source, 38
　of a state, 38
Periodic n-tuples, 144
Permutation matrix, 39

Phase sets, 38, 39
Phase shift keying modulation, 372, 384
Pless power moment identities, 293–295
Plotkin bound: *see* Bound, Plotkin
Poisson: *see* Distribution, Poisson
Polynomial representation, 168, 169
Power budget, 370
Power-limited, 388–389
Prefix, 68, 94, 102, 134
　construction, 166
Prime (in ring): *see* Ring theory
Prime decomposition: *see* Ring theory
Primitive
　elements, 191
　NPCE class, 189
　NPCE classes, 192
Principal ideal domain: *see* Ring theory
Probability of correct code-word transmission, 303–304, 315, 317
　distribution: *see* Distribution
　of error, 315, 317, 318
　of an incorrect code-word transmission
　　detectable, 303–304, 315
　　nondetectable, 303–304, 315
Production, 55
Projections, 192
Pseudolinear averaging, 240
PSK: *see* Phase shift keying
Punctuation, 162

Quantum mechanics, 14
Quaternion: *see* Hurwitz

Random process (or stochastic process), 25
Random coding: *see* Code, random
Rayleigh fading channel, 372
Real analysis, 224
Recurrence times, 38, 123, 124
Recursion, shortest possible, 240
Repetitive parity: *see* Retransmission schemes, Type 2 hybrid
Reproducible extension, 127
Retransmission schemes (ARQ), 389–397
　chase code combining, 395, 397
　go-back N, 390–391
　negative acknowledgment, 390–391
　positive acknowledgment, 390–391
　selective repeat, 390–391
　sliding window, 391
　stop and wait, 390–391

Retransmission schemes (ARQ) (*Cont.*)
   type 1 hybrid, 391–393
   type 2 hybrid, 393–395
Rician fading channel, 372
Riemann zeta function, 238
Ring theory, 370, 389, 397–415
   alternative division ring, 413
   commutative ring, 399, 405
   Euclidean domain, 399–401, 406
   ideal, 400
   integral domain, 401
   irreducible, 400
   module, 389, 397, 399, 405
      cyclic, 399, 406, 408
      finitely generated, 400, 406, 408
      order of, 406, 408
      torsion, 407
      torsion-free, 407
   order ideal, 400, 406
   prime, 400
   prime decomposition, 399, 400, 404
   principal ideal domain (PID), 389, 397, 400, 401, 405, 412
   unique factorization domain (UFD), 399–401, 408
   unit, 400
RM: *see* Code, Reed–Muller
Root of unity, 286
RS: *see* Code, Reed–Solomon
Run-length coding, 219–223, 241

Sardinas–Patterson algorithm, 68, 69, 83, 128
Satellite industry, 370, 371
Seg, 68, 70, 71, 72
Semiordered partition, 171, 173
Sequence
   complexity of, 117, 127
   linked, 331–336, 337, 339, 340, 343, 345, 346
   periodic ambiguous, 135
   typical, 47, 50, 52, 76, 77, 331, 336, 343
Sequential decoding algorithm, 373
Shift register, 281
Soft-decision, 370, 372, 373, 384, 394
Shortest odd-length suffix construction, (SOLS), 182–188
Source
   controllable rate, 92, 114, 128
   entropy, 47, 91, 93, 317, 320
   ergodic, 40, 119, 237

Source (*Cont.*)
   fixed rate, 75
   homogeneous n^{th} order Markov, 27, 101, 152, 320, 346
   infinite discrete, 207
   information, 22, 119
      memoryless, 24
      rate, 80, 91
   stationary, 26, 119
   variable duration symbol, 105, 108, 109, 129
Space
   convex, 348–359
   Euclidean, 348, 349
Sphere
   Hamming, 261, 270, 274, 275, 279
   Lee, 264, 279
Sphere-packing, 306, 307, 370, 389, 397, 412; *see also* Lattice
   bound: *see* Bound, Hamming
   bounds, 295
   concepts, 270
St. Petersburg paradox, 208, 211, 241
State
   probability vector, 29
   sequence, 372
Stationary
   random process, 25, 126
   stationary state probability vector, 33, 39
   steady-state probability vector, 35, 39
Statistic, sufficient, 313
Statistical
   analysis, 243
   theory, 60
Steady-state probability vector, 35
Stirling's approximation, 262, 331
Stochastic process, 25, 126
   time continuous, 60
Subinterval, 225
Sufficient statistic, 313
Suffix, 68
   construction, 166
Supersymbol, 301, 302
Symbol
   erasure, 378, 379
   error, 378
Synchronizable
   fixed word length codes, 147–164
   hierarchy of codes, 140

Subject Index

Synchronizable (*Cont.*)
    statistically, 140–142
    variable word length codes, 164–169
Synchronization
    acquisition delay, 141, 142, 151, 158, 160–162
    automated bounded, 189–196
    cost, 141
    information, 66
    loss, 142
    points of word, 66, 140
Synchronizer, 164, 166, 167
Syndromes, 239
Systematic code or decoder: *see* Code, systematic

TCM: *see* Code, trellis-coded modulation
Theorem
    asymptotic equipartition, 46, 47, 50, 76, 128
    central limit, 376
    Chinese remainder, 192
    coding (general), 368
    convolution, 286, 288
    data-processing, 310–315
    Elias (1975), 124
    finite mean implies finite entropy, 213
    Golomb and Gordon, 173, 180, 181, 199
    Golomb, Gordon, and Welch, 149
    inner-product, 286, 288, 292
    inversion, 286, 292
    maximization of convex function, 354
    McMillan, 88
    nested interval, 224
    Parseval or Plancherel, 286
    Riemann–Roch, 379
    Scholtz, 177, 199
    Shannon, 93
    Shannon–McMillan, 46, 47, 207
    Shannon's coding theorem for noisy channels, 335–346, 376
    SOLS construction, 183
    Tietavainen, 276
    Wyner and Ziv, 124
Thermodynamics, 82
Time series analysis, 60
Torsion module: *see* Ring theory
Transcendental nature, 359

Transform
    Fourier, 286
    relation, 286
Transform-based decoder, 380
Transformational grammars, 60
Transition
    probabilities: *see* Markov
    probability matrix: *see* Markov
Transmission rate, 160, 161, 259, 266, 268, 272, 274, 341
Trellis-coded modulation: *see* code, trellis-coded modulation
Trellis diagram, 28, 137
Typical sequences, 47, 50, 57, 76, 77, 323

UFD: *see* Ring theory
U_m: *see* Group, units
Uncertainty function, 5; *see* Entropy
Uncertainty, 4
    a posteriori, 16,17
    a priori, 16, 17
    average reduction, 15, 17; *see also* Mutual information
    maximum, 11
    reduction, 16
    residual, 318–320
Union bound: *see* Bound, union
Unique decodability, 67, 89, 92, 128, 132
Unique factorization domain,: *see* Ring theory
Unit: *see* Ring theory
Units of uncertainty, 2, 3
Universal code, 223–239, 241

VA: *see* Algorithm, Viterbi
Voyager, 384, 385

Waiting times of Elias–Shannon coding, 226–230
Weight
    enumeration function, 288
    Hamming: *see* Distance, Hamming
WMR bound: *see* Bound, WMR

Z (the integers), 398–415
Z_m, 190
Zero divisor: *see* Ring theory
Zeta function,: *see* Riemann zeta function
Zipf law, 27